Writing in the Life Sciences

Writing in the Life Sciences

A Critical Thinking Approach

Laurence Greene

New York   Oxford
OXFORD UNIVERSITY PRESS
2010

Oxford University Press, Inc., publishes works that further
Oxford University's objective of excellence
in research, scholarship, and education.

Oxford New York
Auckland Cape Town Dar es Salaam Hong Kong Karachi
Kuala Lumpur Madrid Melbourne Mexico City Nairobi
New Delhi Shanghai Taipei Toronto

With offices in
Argentina Austria Brazil Chile Czech Republic France Greece
Guatemala Hungary Italy Japan Poland Portugal Singapore
South Korea Switzerland Thailand Turkey Ukraine Vietnam

Published by Oxford University Press, Inc.
198 Madison Avenue, New York, New York 10016
www.oup.com

Oxford is a registered trademark of Oxford University Press.

Library of Congress Cataloging-in-Publication Data
Greene, Laurence S., 1960-
Writing in the life sciences: a critical thinking approach / Laurence Greene.
    p. cm.
Includes bibliographical references and index.
ISBN 978-0-19-517046-7
1. Communication in medicine.   2. Communication in biology.
3. Communication in biotechnology.   4. Communication of
technical information.   I. Title.
R118.G74 2010
610.69'6—dc22
2009021102

9 8 7 6 5 4 3 2 1
Printed in the United States of America
on acid-free paper

Contents

Chapter 7: Revising Sentences • 315

Chapter 8: Rhetorical Goals for Scientific Papers • 379

Preface

Anyone who has reflected on what determines excellence in scientific communication will appreciate the view that success is the product of extensive content and discourse knowledge, advanced critical thinking and problem-solving skills, and mastery of the many complex procedures involved in planning, drafting, and revising presentations and papers. It follows that instructional books on scientific writing should be designed to provide essential knowledge and, *especially*, to support students in carrying out the cognitive processes that engender superior documents. The reality, however, is that most books on the subject take a descriptive and prescriptive approach that emphasizes products over processes. The mainstream pedagogical method entails describing the elements of exemplary scientific documents in their completed forms. This product-based approach to teaching and learning scientific writing is certainly of value to students who are new to the genre. They expressly want to know what scientific writing is, how it's structured, and what objective features characterize its format and style. But one might argue that, for student writers, the truly essential question is, "How do you actually go about producing excellent scientific papers?" *Writing in the Life Sciences: A Critical Thinking Approach* is designed to answer this question in its multifaceted aspects.

OVERVIEW

After the upcoming introduction, which elaborates on the book's unique pedagogical methods, the first seven chapters are organized by interdependent stages of a systematic approach to scientific writing. At each stage, I describe and demonstrate constituent activities, highlighting their underlying critical thinking and problem-solving demands. Demonstrations of processes and products are based on writing projects in the life sciences—the research disciplines that focus on the structure, function, and behavior of living organisms.

All of the chapters that involve planning, drafting, and revising scientific content share a common theme: Activities for working with content are ideally guided by appropriate genre-specific rhetorical goals. These are the writer's statements of intention for what to do and say, as well as for how audiences should respond to the content. As an example, consider the following rhetorical goal for the introduction section to a grant application:

To convince readers that the proposed study is based on an unanswered research question or an unsolved problem in society.

Here's another sample rhetorical goal, this one for the discussion section of a standard research paper:

To demonstrate how the reported study's findings and conclusions relate to those from previous studies on the issue.

An extensive catalog of conventional rhetorical goals for the primary types of scientific papers is presented in Chapter 8.

The book ends with two appendices: The first appendix presents guidelines for preparing and delivering scientific presentations (talks and poster presentations); the second appendix presents a glossary of basic grammar terms.

Highlights of the book's chapters are summarized as follows.

Chapter 1: Defining Your Writing Project

This chapter is intended to support students in developing comprehensive representations of the rhetorical situations that define their writing projects. Accordingly, the chapter features activities for understanding assignments and instructions to authors, selecting paper topics and refining research issues, analyzing audiences, searching for scientific literature, and reading the literature to gain topic and research knowledge. The products of these early planning activities apply directly to carrying out ensuing activities for planning, drafting, and revising scientific papers.

Chapter 2: Developing a Goal-Based Plan

Our goal-directed approach is based on the principle that student writers benefit most from learning how to develop rhetorical goals *independently* in order to meet specific individual needs. Chapter 2 serves this principle by guiding students through activities for developing effective rhetorical goals and integrating them into sound plans for papers. The activities include adopting and adapting established discourse conventions, using exemplary published articles to derive goals, applying task and audience analyses, and exercising a special type of goal-focused cognition called *helicopter thinking*.

Chapter 3: Generating Content

Of course, the essential challenge of writing in any genre is generating content—that is, figuring out what to say in a document. This process fundamentally involves developing information, ideas, and arguments that successfully contribute to achieving the writer's rhetorical goals. Chapter 3 presents activities for generating content through goal-directed brainstorming and various critical reading and thinking skills. Included are activities for synthesizing research outcomes, evaluating the methods of published studies, and assessing strengths and weaknesses of scientists' arguments in the published literature. Applying this chapter's activities, students will learn how to generate content in the form of rough notes that are integrated into goal-based plans.

Chapter 4: Organizing Content and Writing a Draft

As explained in more detail in the book's upcoming introduction, a primary reason for taking a process-based approach to scientific writing is to avoid becoming overwhelmed by juggling too many planning, drafting, and revising activities simultaneously. Toward helping writers reduce this all-too-common and troublesome cognitive burden, our approach separates activities for generating versus organizing content. In Chapter 4, students will learn how to apply key principles for ordering and integrating the elements of goal-based plans. The objective is to produce an *organizing plan* that outlines the rhetorical goals of major sections and subsections and prepares writers for drafting papers with strong global unity and coherence.

Somewhat like the process of organizing content, drafting involves translating rough ideas and mental images into structured forms. In the section of Chapter 4 that focuses on writing first drafts, guidelines are presented for drafting titles, abstracts, section headings, paragraphs, sentences, and graphics (tables and figures). In addition, the chapter addresses how to cite references and avoid unintentional plagiarism.

Chapter 5: Revising Document Design, Global Structure, and Content

Across Chapters 5, 6, and 7, the book takes a *divide-and-conquer* approach to revision, decomposing the entire process into the following five levels of discourse: document design, global structure, content, paragraphs, and sentences. The three chapters on revision are organized by common problems that arise in drafts of scientific papers. For each problem, I describe and demonstrate procedures for active detection, diagnosis (determining underlying causes), and resolution. After introducing the book's divide-and-conquer and problem-solving approaches to revision, Chapter 5 focuses first on revising for matters of document design, defined as manuscript preparation details such as font type and size, line spacing, margin width, page numbering, presentation of graphics, and citation format. The chapter then addresses revision of global structure, or the overall unity and coherence of a document's major sections and subsections. Next, the bulk of the chapter presents procedures for revising content-level problems including ambiguous, underdeveloped, and inaccurate information and ideas; content that falls short of accomplishing key rhetorical goals and that fails to adequately address concerns of audience; and logical fallacies that weaken scientific arguments. Chapter 5 ends by presenting guidelines for collegial peer review.

Chapter 6: Revising Paragraphs

This chapter presents procedures for detecting, diagnosing, and solving problems that compromise paragraph unity; coherence, defined in terms of the flow of meaning and logic across a paragraph's successive sentences; cohesion, defined by surface links and transitions across sentences; and variety of sentence structure, phrasing, and tone. Many of the diagnostic routines engage writers in deep analyses of sentence-to-sentence relationships in meaning, logic, and structure.

Chapter 7: Revising Sentences

In this chapter, procedures for revising problems at the sentence level are organized by the following categories: logic and clarity, style and structure, basic grammar, word choice, punctuation and mechanics, and uses of language for describing individuals by their ethnic and racial groups, ages, and sex. In keeping with the theme of revision as problem solving, Chapter 7 highlights many sentence-level flaws that cannot be corrected simply by applying hard-and-fast rules for grammar.

Chapter 8: Rhetorical Goals for Scientific Papers

As introduced in this preface, rhetorical goals are statement of intention for content and its influences on readers. Chapter 8 presents sets of conventional rhetorical goals for the major sections of the primary types of scientific papers, including the various forms for reporting original research (research papers, lab reports, theses, and dissertations), review papers, and research proposals and grant applications. For each general goal, the chapter presents a set of associated strategies, which are more detailed and specific ways to accomplish a rhetorical goal. The applications of selected goals and strategies are demonstrated in model excerpts from published journal articles as well as student papers.

HOW TO USE THE BOOK

For courses that are devoted exclusively to scientific writing, students will benefit most from reading the book's chapters and learning about their associated activities in linear order. This does *not* mean that, once learned, all of the activities must be undertaken as the book presents them. Although scientific writing is indeed a complex step-by-step process, the best steps and the ideal order for taking them depend on individual circumstances, specifically on the details of different rhetorical situations. This theme is reinforced throughout the book.

For students who are taking writing-intensive science courses or working independently on thesis or dissertation projects, selected chapters can be used as separate reference guides. For example, Chapters 2 and 3 will independently serve students who specifically need guidance in developing rhetorical goals and applying them to generate draft content. Most of the material in the three chapters on revision (Chapters 5, 6, and 7) can be used in stand-alone mode. In addition, Chapter 8 will serve as a comprehensive reference for students seeking directly applicable rhetorical goals for their papers.

COMPANION WEB SITE

A companion Web site for the book is located at http://www.oup.com/us/greene. The site provides links to many helpful online resources for scientific communication, including university-based online writing centers, databases of scientific literature, science magazines and blogs, and the Web sites of professional organizations and publishers in the life sciences. Among other resources for instructors, the site includes guidelines for assigning and grading student papers, suggestions for designing and administering classroom experiences based on the book's pedagogical approach, and PowerPoint slides that reproduce the book's key lessons.

ACKNOWLEDGMENTS

The main inspiration for this book comes from my rewarding experiences in teaching scientific writing to undergraduate and graduate students at The University of Colorado at Boulder from 1993 to 2008. I am truly grateful to my former students for everything they taught me about the value of a university education, about scientific communication, about the life sciences, and about life in general. Thanks also to the faculty of the CU-Boulder Department of Integrative Physiology for supporting a progressive curriculum in scientific writing. Special thanks to Marie Boyko, Bill Byrnes, Dale Mood, and Russ Moore.

For their many insightful comments and useful suggestions for revision, I thank the book's peer reviewers: John C. Abbott, The University of Texas at Austin; Dale Casamatta, University of North Florida; Lydia B. Daniels, University of Pittsburgh; Mark Eberle, Fort Hays State University; Clayann Gilliam Panetta, Christian Brothers University; Bruce Jaffee, University of California—Davis; Sarah-Hope Parmeter, University of California—Santa Cruz; Dorothy Raffel, George Mason University; Christopher Sawyer-Lauçanno, Massachusetts Institute of Technology; Rachel V. Smydra, Oakland University; and Karina Stokes, University of Houston—Downtown.

I am delighted that this book has been published by Oxford University Press. Thank you to OUP's production staff, to Production Editor Jaimee Biggins, and to Assistant Editor Cory Schneider. I am especially grateful to Executive Editor Jan Beatty for her inspiring encouragement and valued advice through every challenging phase of this project.

To my family, I cannot thank you enough for your love and support.

An Introduction to Writing in the Life Sciences

Given this book's primary audience—undergraduate and beginning graduate students who lack extensive experience in scientific communication—it is fitting that we begin by raising the following question: *With all of the interesting subjects to study and the many important skills to gain in higher education, what is the value of learning to write successfully in science?* From the viewpoint of career scientists, the answers are obvious. All practicing scientists know very well that the quality of their livelihood is strongly connected to the quality of their writing. Consider the passions and pursuits of professionals in the life sciences—researchers who study the structures, functions, and behaviors of living organisms. They aim to explain the mysteries of nature and to enhance the conditions of life for plants, animals, and humans. Of course, successful outcomes demand advanced topic knowledge and consummate skill in the technical aspects of designing and conducting research. To achieve their highest goals, however, life scientists must ultimately *communicate* their discoveries and innovations. The predominant form of scientific communication is writing.

The reasons and rewards for writing skillfully are highly motivating for career scientists. In universities and private research institutions, most scientists are expected to obtain the funding that is necessary to support their research programs. Nowadays, the costs of maintaining a first-rate life science laboratory can run easily in the millions of dollars per year. The most brilliant plans for designing and carrying out cutting-edge research will never come to fruition unless they are communicated clearly, coherently, and convincingly in written grant applications. Moreover, the most revolutionary research breakthroughs will be practically meaningless if they are not effectively communicated, through various types of scientific papers, to targeted audiences that can learn, apply, and benefit from them. By necessity and as a consequence of their motivated efforts to communicate successfully, elite scientists are elite professional writers.

If you're a student headed toward a career in research, an appreciation of the preceding points may aptly inspire your efforts to learn to write successfully in science. Then again, if your professional interests lie elsewhere, consider some other meaningful benefits of the endeavor at hand. There are, of course, the immediate rewards of good grades on the scientific papers you're writing to fulfill course requirements. Taking a longer view, you may come to appreciate that the skills to be gained are absolutely essential for continued success in school and in the working world. I say this because scientific writing strongly emphasizes the following skills:

- Critically evaluating the accuracy, validity, and reliability of new information;
- Organizing content in highly logical and coherent patterns;
- Shaping prose to meet the needs, expectations, and values of intended audiences;
- Expressing ideas with impeccable clarity and precision; and
- Constructing original arguments with the most objective and convincing evidence and reasoning.

These critical thinking and communication skills are indeed fundamental to success in all academic disciplines. Moreover, in the Information Age, they top the list of what employers are looking for in all college graduates. As you might imagine, these skills that underlie scientific thinking and writing are especially valuable in science-based career areas, such as the health professions, science education, and environmental management.

It's not even a stretch to say that scientific writing skills can also enhance the quality of our personal lives. Consider, for example, everyday decisions that we face about matters of health for ourselves and for those who are close to us. How many hours of sleep do we need every night to prevent illness and to function at our best? Are claims for the effectiveness and safety of the latest fad diets legitimate? Is it really necessary for children to be immunized? For diseases that have not been cured with conventional medicine, are alternative treatments—such as acupuncture, herbal supplements, and homeopathy—worthwhile? In the popular media, as well as in the scientific literature, the answers to these questions (and many others like them) are often downright contradictory. Take the question of whether low-carbohydrate diets are effective and safe for promoting weight loss and optimal cardiovascular health. The results and conclusions of some published studies support significant advantages of these popular diets. The outcomes of other studies indicate no effects at all. Still others back the claim that people who go on low-carbohydrate diets are at significant risk for weight gain and cardiovascular disease. The published study outcomes, as well as popular reports about them, can be completely baffling—*unless*, that is, you know how to think critically about them. This entails knowing how to interpret study data independently, how to evaluate strengths and weaknesses of the research methods used to obtain the data, and how to critique the arguments that support debatable claims.

These skills, guiding good decisions about our own well-being and that of family and friends, are central to writing effective scientific papers.

So, even if you don't intend to pursue a career in science, there is much to be gained through mastering the craft of scientific writing.

INTENDED AUDIENCES

This book is intended primarily for undergraduate students, specifically those who are writing papers for science courses, for scientific writing courses offered through English departments or college-wide writing programs, and for independent study and thesis projects. A secondary audience is graduate students who are undertaking thesis and dissertation projects, as well as writing for publication and grants, without having had formal training in scientific communication. Practicing scientists may also find the book of value—if not for help with their own papers, then for learning the best approaches to supporting their trainees in developing advanced writing skills. Focusing on writing in the life sciences, the book is especially tailored to students and scientists working in various fields of psychology and the specialized subdisciplines of biology, which include biochemistry, botany, genetics, ecology, medicine, physiology, and zoology.

THE CULTURE OF SCIENCE AND SCIENTIFIC COMMUNICATION

To develop advanced communication skills in any specialized discipline, you must understand its culture. The culture of science is defined by what career researchers do in their daily work, by what they value and aspire to, by how they think, and by how they interact with each other and society. These elements underlie the conventions and qualities of successful scientific writing. Of course, the day-to-day activities of scientists involve planning, conducting, and communicating research. This work is essentially inspired by the great value that the scientific community attaches to education and discovery. Most practicing scientists began college in their late teens and, in a sense, never left. Long after many years of undergraduate, graduate, and postdoctoral education, veteran researchers spend much of their time doing what students do: reading, writing, acquiring knowledge, and sharpening their intellectual and practical skills. If you've had the good fortune to observe elite scientists at work, you know very well that they love to learn. This passion naturally inspires a strong value of communication. Most scientists are eager to talk about their work, and they do so regularly in e-mail correspondences with colleagues, through publications and oral presentations targeted to their scientific communities, and via the public media of Web sites, radio, and television.

By and large, scientists are effective communicators and especially conscientious writers. This may come as a surprise if your view has been influenced by the long-standing stereotype of scientists as universally terrible writers. I would argue that the stereotype is misinformed and unfair. One of its sources is the general writing expert who, from a position far afield of science, complains about excessive jargon in scientific publications that are explicitly intended for scientific audiences. Jargon is primarily defined as the language and vocabulary used in specialized trades, professions, and groups. Jargon thus reflects the technical but quite normal and accepted mode of communication in distinct discourse communities. Much of the criticism of "jargon abuse" in scientific writing is unfounded because the critics are in no position to judge whether the technical language and vocabulary are indeed necessary or not for the intended audience. Interestingly enough, other sources that stereotype scientists as bad writers are the introduction sections to books that teach scientific writing. You will find no such sweeping generalizations here.

It's really no surprise that scientists value communication and tend to be highly proficient at it. Many of the intellectual skills that apply to conducting first-rate laboratory and field studies are fundamental to producing superlative scientific papers. The design and execution of research and writing projects depend on highly creative, organized, analytical, precise, and reflective thinking. This is not to say that all scientists deserve prizes for their literary achievements. Instead, my point is that the culture of science strongly values effective communication and, through its literature, demonstrates a commitment to developing the requisite skills.

When leading scientists talk about their most important discoveries, they often reveal that their original inspirations were creative intuitions and pure gut feelings. But while subjective insights may supply the seeds for growing scientific knowledge, its cultivation ultimately depends on objective and rational thinking. This is certainly reflected by the content and language of effectively written scientific papers. For example, their claims and conclusions are always closely accompanied by well-developed supporting knowledge, evidence, and reasoning from established research. Strong scientific arguments are devoid of logical fallacies such as irrelevant evidence, appeals to emotion, and personal attacks on proponents of opposing views. To avoid bias, skilled scientific authors acknowledge viable arguments that oppose their own, weighing the counterarguments' strengths and weaknesses in the course of refuting them. Even the choices of individual words are influenced by the values of objectivity and rationality. For instance, when scientists present their conclusions from studies on evolving and debatable research issues, they don't use the word *prove*, as in the following sentence.

> This study's data definitely prove that broccoli consumption prevents colon cancer.

Instead, scientists use words that reflect the rational boundaries of their studies' methods and outcomes within the larger context of the research field:

> For this study's targeted population, the data indicate that broccoli consumption significantly reduces the risk of developing colon cancer.

The objective and rational approaches to thinking in science naturally complement the high standards that the scientific community sets for professional integrity. Scientific writing and publication processes are based on many obligatory ethical practices, including the following.

- Giving appropriate credit to primary sources of knowledge—in other words, avoiding plagiarism.
- Fairly recognizing the contributions of all coauthors.
- Acknowledging or, if necessary, completely avoiding conflicts of interest.
- Reporting study outcomes truthfully and without any distortion—that is, avoiding the fabrication and falsification of data.
- Disclosing all procedures and analyses of published studies to enable peer scientists to replicate them.
- Basing reviews of peers' papers on objective evaluations of their scientific merit and writing quality rather than on personal biases.

That the scientific community takes its codes of ethics seriously is evidenced by the numerous mechanisms that have been developed to maintain and enforce them. Every established research institution has an independent office devoted to monitoring the professional conduct of its faculty members. In addition, federal grant agencies take on this charge. At the United States National Institutes of Health, the Office of Research Integrity (http://ori.dhhs.gov/) oversees investigations of scientific misconduct. The counterpart at the National Science Foundation is the Office of Inspector General (http://www.nsf.gov/oig/). The policing efforts of these agencies, which are described at length on their Web sites, are intended to prevent the negative consequences of unethical practices in scientific research and communication. As one of many examples, consider that fabricated or falsified data in published reports of pharmaceutical studies can have life-and-death implications in society. Scientists who violate codes of ethics face severe penalties, which may include bans on conducting future research, termination of their employment, and even legal action taken against them.

Another hallmark of the culture of science is the progressive mix of collaboration and competition that defines how scientists interact with each other. Collaboration is essential in research because major team efforts are required to design and execute high-quality studies. These days, a leading team of life scientists includes content specialists in various research disciplines, laboratory technicians, computer scientists, statisticians, and even communication experts. Beyond the interactions of individual team members, scientific collaboration commonly extends to competing laboratories. Around the world, teams of researchers working to solve the same problems regularly share data, instrumentation, and ideas. Of course, given the competitive nature of science, each team vigilantly guards its most promising information and breakthroughs.

Collaboration and competition also characterize the scientific writing and publication processes. Unlike many authors in purely creative genres, scientists generally don't hole up in solitude and secrecy to work on their writing projects. Most published scientific articles are coauthored, so the writing process involves frequent exchanges of plans, draft material, and suggestions for revising documents. When scientists submit articles for publication or grant applications for funding, journal editors or officers of grant agencies recruit other scientists who are topic experts to serve as referees. These *peer reviewers* are instructed to critically evaluate papers for the merit of their scientific content, their potential for advancing important knowledge and resolving meaningful issues, and their qualities of written communication. Peer reviewers grade scientists' writing, just as professors do for students' writing. (If you think that your professors grade strictly, consider that in the highest-quality scientific journals only 5%–10% of submitted articles are accepted for publication.) In the end, the quality of coauthored and peer-reviewed papers greatly surpasses what individual scientists can produce on their own. Throughout the book, I suggest ways for students to apply the professional practices of collaborative writing and peer review.

OUR APPROACHES TO SUCCESSFUL SCIENTIFIC WRITING

The main feature that distinguishes this book from others on the subject is its emphasis on writing as *thinking*. In one way or another, all of the book's lessons are meant to support students in carrying out the advanced cognitive processes, or thinking routines, that underlie the production of superior scientific papers. The highlights of our unique approaches are introduced as follows.

A Critical Thinking Approach

As reflected by the book's subtitle, critical thinking is central to our methods for mastering the craft at hand. Critical thinking is evaluative thinking, interpretive thinking, judgmental thinking, analytical thinking, diagnostic thinking, and independent thinking. These modes of cognition underlie successful outcomes in the most challenging writing activities. Consider just a few examples:

- Excellent choices of scientific paper topics depend on *evaluating* whether candidate topics address important knowledge gaps and meaningful research issues.
- Going beyond simply parroting the scientific literature demands skill at *independently interpreting* research evidence (data) and *judging* the quality of research methods.
- Effective outcomes in organizing papers rely on deeply *analyzing* the logical and structural relationships among their elements of content.
- Revising drafts successfully entails *diagnosing* many features of content, structure, grammar, and style.

From the outset, perhaps the most important message about critical thinking is that it is *learned* thinking rather than an innate ability. Accordingly, my overall goal in this book is to help students learn about, practice, and ultimately master the critical thinking skills that are so essential to successful scientific writing.

A Process-Based Approach

Most books about scientific writing are basically designed to answer the following question: *What is it?* Their main instructional method thus involves describing the features, both exemplary and problematic features, of scientific documents. This product-based approach can indeed be useful, so we'll apply it to some extent here. Along the way, we'll examine excerpts from papers authored by scientists as well as students, identifying strengths to emulate and weaknesses to resolve. At its core, however, this book aims to answer a very different question about scientific writing: *How does one do it skillfully?* Accordingly, our main instructional method concentrates on describing and demonstrating the essential processes that bring about exemplary documents. A process-based approach is especially powerful because scientific writing is an exceptionally complex skill. It cannot be acquired solely by examining finished products. Instead, mastery requires decomposing the overall process into its constituent activities and gaining proficiency in the thinking routines that underlie them.

The overall scientific writing process is composed of numerous planning, drafting, and revising activities that demand specialized modes of thinking. In planning, for example, we develop some of our ideas through exploratory and unstructured brainstorming. Other ideas are generated through focused and systematic critical reading and thinking activities. Then, to organize our ideas in logical patterns, we must think in highly structured ways. And at every stage of the process, we are challenged to perform the mental gymnastics of thinking from the perspective of our readers. Whereas the many activities in scientific writing are distinct in their cognitive demands, their products must be woven together seamlessly. This skillful integration depends on what psychologists call *metacognition*, which in writing entails constantly monitoring one's progress, applying outcomes of completed activities to carry out ensuing ones, and deciding on the next best steps to take. In our step-by-step approach to scientific writing, we'll map out the overall process by 10 stages, which are illustrated in Figure I.1 (page xxvi). The book's first seven chapters are organized by these stages and the activities they encompass.

The lines connecting the stages represent the interdependence of activities in the writing process. As one example of these complex interrelationships, consider the stage labeled *Organize Content*, which we'll cover in Chapter 4. The key product of this stage is a linear outline. Take a moment to reflect on *how* you would actually go about developing a well-structured and functional outline for a scientific paper. What materials will you need, and what must you know, think about, and do in the outlining process?

- Your assignment may include explicit details and hints about how to structure your paper, so you'll need to have analyzed the instructions carefully.

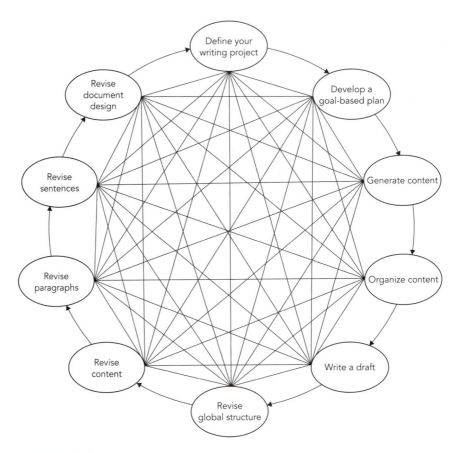

FIGURE I.1 The overall map of our process-based approach to scientific writing.

- There may be established conventions for organizing content in the type of scientific paper you're writing; if you haven't already learned the conventional guidelines, you'll need to do so before outlining.
- The best ways to order the elements of your outline will certainly depend on your readers' needs, expectations, and values; thus, a prerequisite for outlining is analyzing, or characterizing, your audience.

These three scenarios and their related activities are central to the stage labeled *Define Your Writing Project* in our process map. We'll cover this stage in Chapter 1.

In addition to materials for building the general structural framework of your hypothetical outline, you'll need rough content to flesh it out. Appropriate content for scientific papers is developed through various critical reading and thinking activities. We'll walk through these activities in Chapter 3, which is devoted to the stage labeled *Generate Content* in our overall process map. Of course, it doesn't make sense to develop an outline unless you're going to put it to good use. The applications come in the stages labeled *Write a Draft* (Chapter 4), *Revise Global Structure* (Chapter 5), and *Revise Content* (Chapter 5). The take-home message here is that any given activity in the writing process may (a) rely on input from numerous preceding activities and then (b) yield products that are necessary to carry out successive ones.

One of the best reasons for taking a process-based, step-by-step approach to scientific writing is to avoid the frustration and disappointing outcomes that result when we attempt to do too many planning, drafting, and revising activities all at once. This is a defining characteristic of novice writers, who have not learned how to decompose the overall process and are therefore easily overwhelmed by its cognitive demands. In contrast, experts apply their extensive procedural knowledge to concentrate on individual activities or manageable sets of them, flexibly monitoring their progress and integrating products along the way. Through learning such a systematic method and adapting it for your purposes, you'll come to realize something of a paradox about scientific writing: Whereas it's certainly challenging, *it ain't necessarily rocket science!*

This book is by no means intended to prescribe its process-based approach as the only effective one. It would be a great mistake to think that excellent scientific papers depend on carrying out all of the book's activities exactly as described, demonstrated, and ordered. Skillful writing demands choosing and applying appropriate procedures to meet unique situational demands. So I encourage you to experiment with the book's activities, to adopt those that work well for you *as is*, and to adapt others to meet your specific needs. Along the way, I'll offer suggestions for tailoring our process-based approach accordingly.

A Problem-Solving Approach

Thinking back to your earliest lessons in writing, perhaps you recall grade school English teachers, red marking pens in hand, commanding countless rules. A few examples might jog your memory.

- Verbs must agree with their subjects in person and number.
- Infinitives must not be split.
- Sentences should never end with prepositions.
- Don't ever begin sentences with the word *because*.
- Always avoid the use of first-person in academic writing.
- Use active voice, rather than passive voice, in all cases.

Many rules for writing are essential to learn and follow, especially those that are well-established conventions for grammar and punctuation in Standard English. However, writing instruction that only emphasizes *rule-following* can be counterproductive. Some of the so-called rules that we learned in grade school are not actually accepted conventions that expert writers regularly apply. Take the "rule" to avoid beginning sentences with the word *because*. Some grade school teachers enforce this directive to steer students away from sentence fragments:

I am not going to school. Because today is Saturday.

But if a subordinate clause beginning with *because* is joined to an independent clause, there is no good reason for avoiding sentences such as the following:

> Because today is Saturday, I am not going to school.

Another pitfall to strict rule-following approaches is that many rules are revised or discarded as language naturally evolves to serve more functional purposes. Take the rules to avoid splitting infinitives and ending sentences with prepositions. Most contemporary grammarians agree that these rules are outdated and that, when applied mechanically, the rules can undermine effective communication. Here's a case in point:

> Regarding the causes of honeybee population declines, many researchers view the immunodeficiency theory as one for which to argue strongly.

The infinitive (*to argue*) is not split and the sentence does not end with a preposition. But it's an awkward sentence to read and, as a result, the main message is a bit obscure. Here's a revision:

> Regarding the causes of honeybee population declines, many researchers view the immunodeficiency theory as one to strongly argue for.

The revised sentence contains a split infinitive (*to strongly argue*), and it does end with a preposition (*for*). But it is noticeably more readable and clear in meaning.

Most activities in the scientific writing process are far too complex to be guided by hard-and-fast rules. This is because scientific writing is essentially a problem-solving process. The best answers to the most difficult questions are never simple statements of rules. Instead, they are well-developed justifications and reasons that account for many situational factors. So this book's approach is to pose challenging problems in the writing process and to work through the advanced procedures for identifying, diagnosing, and solving them.

A Goal-Directed Approach

One of the most distinguishing features of well-written scientific papers is that they clearly reveal their authors' communication goals. As readers, our responses to goal-directed documents are extremely favorable. At every turn, we clearly understand how their information and ideas serve their overall purposes. And we immediately grasp what their authors want us to know, believe, think, and do. It's no coincidence when readers effortlessly recognize writers' goals for scientific papers. Instead, it's a product of a goal-directed approach to the writing process. Our method for fostering this approach emphasizes a powerful type of communication goal, called a *rhetorical goal*. For now, just note that rhetorical goals are statements of purpose that guide writers in generating and shaping effective content. Chapter 2 defines rhetorical goals in great detail and explains how to develop them and incorporate them into useful plans for scientific papers. In Chapters 3, 4, and 5, you'll learn how to use rhetorical goals to guide key activities for generating, organizing, and revising content. Then, Chapter 8 presents the rhetorical goals for scientific writing and demonstrates their applications in model papers.

An Audience-Centered Approach

All good books on writing hail an audience-centered approach. By analyzing the needs, expectations, and values of their intended readers, writers can solve all sorts of difficult problems. In scientific writing instruction, however, the skills of audience analysis are sometimes relatively underemphasized. This is true for school projects in which the targeted audience comprises only one member: the science professor, who already knows everything about the students' paper topics and who is reading mainly to evaluate content for its completeness and accuracy. This artificial situation fails to engage student writers in developing and shaping the content, structure, and style of their papers to address the varied interests and purposes of readers in real-world settings.

Another undermining influence is the misconception that scientists write only to highly specialized audiences comprising their peers in narrowly defined research fields. The reality is that to excel in their writing projects, career scientists must skillfully apply audience analysis, accounting for the concerns of diverse groups of readers. For instance, the instructions for writing many federal and private grant applications explicitly call for authors to speak to varied audiences. For the introductions to grant applications, where scientists must summarize their proposed research projects, the instructions are often to address lay audiences, which may include government officials and even the public. Sections devoted to reviewing previous studies and discussing the significance of proposed projects may need to be tailored to a general scientific readership. The core sections that describe planned research procedures and analyses should target peers who are technical experts.

Especially in the life sciences, an audience-centered approach to writing is becoming increasingly essential. A key reason is that today's most progressive life science studies emphasize interdisciplinary issues and methods; thus, research and writing projects demand interdisciplinary communication. For example, to learn more about how genes affect behavior, geneticists are talking with psychologists. To develop effective drug treatments, molecular physiologists are working side-by-side with clinicians. To prevent extinctions of animal species endangered by newly formed viruses and bacteria, ecologists are interacting with immunologists. In addition, given the increasing availability of scientific information on the Internet, more and more life scientists are taking on the rewarding challenges of communicating with the public. To give an audience-centered approach its due, we'll work through many writing problems that are ideally solved through audience analysis.

A Discipline-Specific and Content-Rich Approach

Under ideal circumstances, we hone our writing skills by working with discipline-specific content, conventions, and practices. The book's approach is predominantly based on the life science disciplines of ecology, physiology, nutrition, movement and exercise science, gerontology (the study of aging), and psychology. Each chapter begins with a synopsis of a research topic in one of these disciplines. For example, Chapter 1 begins by introducing the topic *Endangered and Extinct Animal Species*. To demonstrate the writing processes and products that are central to each chapter, I draw examples from its research topic. For instance, a key activity in Chapter 1 involves searching for scientific literature. Accordingly, the chapter demonstrates how to search for articles and books about animal conservation. I chose research topics that are comprehensible to lay audiences, generally interesting, and important to society. It's quite possible, however, that you are writing papers in life science disciplines that are not represented by the chapter topics and examples in this book. If so, I encourage you to view the book's science content primarily as a vehicle for learning the book's most important lessons, which naturally focus on the skills of scientific writing. Regardless of how they are demonstrated, the essential skills that are presented as follows can be handily adapted and applied to writing in all life science fields.

CHAPTER 1

Defining Your Writing Project

A Natural Disaster: Endangered and Extinct Animal Species

In this century's first decade, a leading ecological conservation agency reported that nearly 25% of all mammal species worldwide were extinct or threatened with extinction (http://www.iucnredlist.org/). For birds, amphibians, and fishes the reports ranged between 14% and 31%. Whereas die-offs of animal species occur naturally over the course of millions of years, the extinction rate in recent history has been estimated to be 100 to 1,000 times greater than normal. The sad fact is that many animal species have recently become extinct or are currently endangered due to human intervention—that is, as a result of the environmental pollution we cause, our exploitation of natural resources and habitats, and our introduction of nonnative predators into established animal communities.

For many reasons, research on endangered animal species is vital to all living organisms, including human beings. We depend on animals as direct sources of food and medicine. In addition, animals indirectly maintain food webs that include the plants that we eat and use to treat disease. Animals protect our environment and keep our soil fertile. Of course, beyond serving our basic survival needs, animals have unsurpassed aesthetic and educational value. They thrill and amaze us, especially when we are young or at least young at heart. In the life science fields of ecology and biological conservation, researchers seek to better understand the factors that threaten the survival of animal species. The ultimate goal of this research is to devise methods for reversing current population declines and for preventing future extinctions.

Research on animal species conservation sets the backdrop to this chapter, which covers the essential first steps in planning scientific writing projects for successful outcomes.

INTRODUCTION

Leading ecologists and conservation biologists take an all-inclusive approach to studying the determinants of survival versus extinction in animal species. The approach accounts for factors such as the following.

- The genetic makeup of the endangered species under investigation.
- How the species responds to potentially harmful environmental conditions, such as pollution and harsh changes in climate.
- Whether the species can mount a successful defense against life-threatening pathogens, including bacteria, viruses, and fungi.
- The species' ability to evade dangers posed by native and nonnative predators.

When ecologists and conservation biologists turn their attention to *writing* about their research, like experienced authors in all scientific disciplines, they also take a comprehensive approach. It's one that fully accounts for the following determinants of success in written scientific communication.

- Whether paper topics are based on important and novel issues.
- The extent to which authors understand the research, theories, and arguments that define their paper topics.
- How strictly documents follow discipline-specific conventions for matters of content, structure, format, language, and style.
- Whether authors attend to specific instructions, guidelines, and evaluation criteria for manuscript preparation.
- How effectively authors address the concerns of intended readers, who may be peer scientists, professors, government officials, or the lay public.

Taken together, these determining factors compose what writing instructors call the *rhetorical situation*. Depending on the context, the word *rhetoric* has specialized meanings, some of which happen to directly contradict one another. For example, in politics and advertising, rhetoric refers to communication that may be deceitful, pretentious, and lacking in true substance. In sharp contrast, the words *rhetoric* and *rhetorical* usually have extremely positive connotations in scholarly communication. When writing instructors describe an aspect of a document as rhetorical, they mean that it successfully follows discipline-specific conventions or that it effectively addresses concerns of its intended audience. In this chapter, you will learn how to analyze and account for the factors that define your rhetorical situations in scientific writing. In other words, the chapter covers the process of *defining your writing project*.

ABOUT THE PROCESS

Our approach to defining scientific writing projects covers six activities, which are represented by the labeled rectangles in Figure 1.1 and previewed as follows.

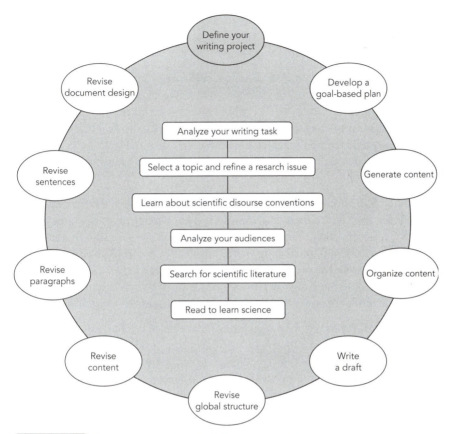

FIGURE 1.1 Process map for *defining your writing project.*

Activity 1: Analyze your writing task. A logical first step in the writing process is to take stock of the instructions, guidelines, and evaluation criteria that come with your assignment. This initial task analysis guides many important decisions in later stages of planning, drafting, and revising.

Activity 2: Select a topic and refine a research issue. Fitting and engaging topics can make a world of difference in the quality of scientific papers. But the process of choosing interesting topics and narrowing their focus appropriately for an assignment is by no means trivial. In this activity, you will learn strategies for selecting and refining excellent paper topics.

Activity 3: Learn about scientific discourse conventions. If you lack experience in scientific communication, part of defining your writing project will involve learning about the genre's discourse conventions. These are the established guidelines that scientists follow in generating content, structuring and formatting documents, and using language. Of course, this entire book is partly intended to teach these conventions. In this chapter, we begin with an introduction to the structure and goals of the three most common types of scientific documents: research papers, review papers, and research proposals.

Activity 4: Analyze your audiences. Who are your intended readers? What do they know and need to know about your paper's topic? What preconceptions and values do they hold about your ideas and arguments? These are a few of the questions that guide an activity that writing instructors call audience analysis. It's an absolutely essential activity because many decisions in the writing process depend on considering readers' needs, expectations, and values.

Activity 5: Search for scientific literature. Nearly all scientific papers build on well-established foundations of knowledge that is stored in books, journal articles, and Web sites. Searching for scientific literature is thus fundamental to defining writing projects. As you will learn in this activity, successful searches demand specialized knowledge and skills that apply to judging the credibility of various forms of scientific literature and to using computer resources such as online databases.

Activity 6: Read to learn science. Excellence in scientific writing obviously depends on strong basic communication skills. At least equally important, however, is extensive science content knowledge. If you are not already an expert on your paper's topic, you will face the challenges of learning about the concepts, arguments, and research studies that define it. This activity is for meeting these challenges by applying advanced strategies for reading and taking notes on published sources of scientific knowledge.

It makes good sense to begin the process of defining a writing project by analyzing your task, and of course, you must select your paper's topic before you can start learning about it. Otherwise, the order for carrying out this chapter's activities is fairly flexible. For example, while you are learning about your topic through reading the scientific literature, you might also spend time on audience analysis and on learning the discourse conventions for the type of paper you are planning to write.

As you are working on these project-defining activities, don't lose sight of their role in the overall writing process. The products of a sound project definition—such as notes that you take about your topic, about important instructions for your assignment, or about characteristics of your intended readers—will guide later stages of planning, drafting, and revising. Many examples of these applications are presented in upcoming chapters.

ANALYZING YOUR WRITING TASK

All scientific writing projects begin with assignments and directions, which are usually given by a professor, an editor, or a grant agency. Taking the first steps to successful final papers, experienced authors carefully study their assignments and directions. That is, they analyze their writing tasks. This activity entails (a) attending to instructions, guidelines, and evaluation criteria and (b) seeking clarification when assignments and directions are vaguely or incompletely defined.

Attending to Instructions, Guidelines, and Evaluation Criteria

When writing for publication or grants, scientists prepare their manuscripts according to detailed and often extensive directions, which are included in documents that are commonly titled *Instructions for Authors*, *Author Information*, or *Author Guidelines*. These documents are usually posted on the Web sites of scientific journals and grant agencies. (A scientific journal is a regularly published collection of literature in a defined research field.) The directions address all matters of manuscript preparation, including how to set margins, space lines, use acronyms and abbreviations, number pages, cite references, design figures and tables, and much more. For career scientists, the implications of failing to account for instructions, guidelines, and evaluation criteria can be severe. For instance, federal grant agencies do not accept research proposals that exceed prescribed page limits. Imagine spending many months working on a grant application in which you have requested millions of dollars to support a vital research project and then having your manuscript returned because it exceeds the maximum length specification by a quarter of a page. It happens in the real world of scientific writing.

Professors often use author-instruction documents from scientific journals and grant agencies as models for class writing assignments. Like career scientists, students must read instructions carefully to account for all details. A helpful strategy is to extract from your assignments only those instructions and guidelines that you do not normally follow in preparing academic papers. If, for example, your assignment calls for double-spacing the text and you always do that automatically, don't sweat the detail. But if you usually single-space your papers and your computer's default is set accordingly, make sure to note the different requirement. As another example, suppose that in English classes you are accustomed to citing references according to Modern Language Association (MLA) format. If you are planning to write a paper for a psychology class, your instructions will likely be to use the citation format of the American Psychological Association (APA). To avoid point deductions for incorrect citation format, note the requirements for your project ahead of time.

In addition to analyzing instructions for manuscript preparation, experienced scientific authors carefully attend to the evaluation criteria that are included in their assignments. These are the high standards by which papers will be graded, deemed appropriate for publication, or awarded with research funding.

Task analysis is a prime example of a writing activity that ain't exactly rocket science but is still quite cognitively demanding. Some assignments come with endless lists of *I*'s to dot and *T*'s to cross, which can be an overwhelming amount of information to absorb and to act on. An organized and efficient approach to managing the process is to record your task analysis in checklists and note forms like the samples shown in Figure 1.2 (page 6).

Instructions for General Manuscript Preparation

- Number of pages or words: _____minimum _____maximum
- Information to include on title page: _____
  _____
  _____

- Margin width:_____
- Line spacing:_____
- Page numbering
  - Placement: ☐ Top of page ☐ Bottom of page
  - Alignment: ☐ Right ☐ Left ☐ Center
  - Number first page? ☐ Yes ☐ No
- Font typeface: _____
- Font size: _____

Instructions for Preparing Tables and Figures

- Tables
  - Placement: ☐ Integrated into the text ☐ On separate pages at the end of the text
  - Placement of table title: ☐ Above table ☐ Below table
  - Use gridlines for table cells: ☐ Yes ☐ No
  - Other instructions for tables: _____
    _____
    _____

- Figures
  - Placement: ☐ Integrated into the text ☐ On separate pages at the end of the text
  - Placement of figure captions: ☐ Above figure ☐ Below figure
  - Use of color for lines and columns of graphs: ☐ Yes ☐ No
  - Suggested aspect ratio of graphs (height/width of axis lines): _____
  - Other instructions for figures: _____
    _____
    _____

Instructions for Documenting Sources

- Number of required citations: _____
- Required citation format: ☐ American Psychological Association
  ☐ American Physiological Society ☐ Council of Science Editors
  ☐ Other: _____

FIGURE 1.2 Sample task analysis checklists and note forms

Instructions for Submitting Manuscripts

- Binding: ☐ Unbound ☐ Stapled ☐ Paper clipped
- Number of copies to submit:_____
- Submit electronic copy to: _____ (e-mail address)
- Deadlines
  - Draft 1: _____ (date) at _____ (time)
  - Draft 2: _____ (date) at _____ (time)
  - Final Paper: _____ (date) at _____ (time)

Evaluation Criteria

1. _____
2. _____
3. _____
4. _____
5. _____
6. _____

FIGURE 1.2 *continued.*

Seeking Clarification of Problematic Assignments and Directions

As writers we always appreciate assignments and directions that are crystal clear and easy to follow. But we should be prepared for ones that are ambiguous and insufficiently detailed. Take the following brief instruction for a paper in an environmental health science class.

> Illustrate relationships between a common environmental pollutant and species outcomes.

On the receiving end of this instruction, would you confidently understand what the professor is expecting? What precisely does he mean by *relationships*? Which sorts of relationships would you be writing about? How exactly are you supposed to *illustrate* the relationships? Moreover, what types of species *outcomes* are appropriate to discuss? Without satisfactory answers to these questions, you might overlook important requirements and standards for excellence. Right from the start, your potential for writing a strong paper would be diminished.

Skillful writers routinely raise questions to clarify confusing and incomplete instructions. For students, this means taking the perspective that academic writing should not be a guessing game of trying to figure out what professors are looking for. The professional approach is to ask for clarification directly. At the start of their writing projects, experienced scientists do not hesitate to e-mail and phone journal editors and grant agency officers to double-check their understanding of vague instructions to authors. Indeed, scientific journals and grant agencies strongly encourage this dialogue because they genuinely want the authors to succeed. The same goes for college professors, whose job is to help students write outstanding papers. You can bet that your professors do not want to read poorly written ones, especially when flaws are actually attributable to misunderstandings about assignment instructions. So when your assignments are not crystal clear, a visit to your professor's office hours, or a quick e-mail message like the one that follows, can get your project started on the right track.

> From: Justine.Pierce@somecollege.edu
> To: Amby.Guous@somecollege.edu
> Subject: Clarification of our writing assignment
>
> Dear Professor Amby Guous,
> I'm planning to write my term paper on the effects of lead-contaminated river sediment on nervous system diseases in Canadian geese. Reading over the assignment that you gave us, I'm not exactly sure if I clearly understand the instruction to "Illustrate relationships between a common environmental pollutant and species outcomes." Given that this is an environmental health science class, I'm interpreting the instruction to mean that we should write about how the environmental pollutant influences certain health outcomes. So I'm planning to discuss how lead exposure can cause oxidative stress in the brain, which destroys brain cells and leads to several nervous system disorders and diseases. Would you please let me know if my topic and plan are on the right track for the assignment?
> Thank you,
> Justine Pierce

SELECTING A TOPIC AND REFINING A RESEARCH ISSUE

Some assignments for scientific papers come with preset topics, eliminating a major aspect of defining a writing project. However, when our assignments do not prescribe topics, the ones we choose can significantly influence the quality of our writing experiences and their outcomes. Meaningful and interesting topics motivate us to take on advanced challenges, and they naturally enhance the quality of what we write. In contrast, inconsequential and dull topics sap our motivation and result in predictably uninteresting and dry papers. You might be surprised by all of the advanced knowledge and skill that underlie selecting a knockout topic for a scientific paper. Our approach involves identifying topics and then transforming them into focused *research issues*. Whereas topics are simply the general subjects on which papers are based, research issues are the unanswered questions and unsolved problems that motivate scientific studies and writing projects.

To better understand the distinction between scientific topics and research issues, consider a writing project on the effects of water pollution on the endocrine (hormonal) system in fish. Recent studies have revealed that chemical pollutants, specifically ones found in waterways near sewage treatment plants and industrial factories, can disrupt normal hormonal function in various fish species. The endocrine-disrupting chemicals include polychlorinated biphenyls (PCBs) and dioxins, which are used to make pesticides, plastics, and industrial products. Some of these chemicals affect the production of estrogen, the hormone that influences the development of sex-specific characteristics and reproduction in females. Studies have revealed evidence of abnormal reproductive function in female fishes living in polluted waterways. A paper on this research might be based on the following topics.

- Water pollution and hormone function in fish.
- Hormone-disrupting chemicals in waterways.
- Reproductive function in fish exposed to environmental pollutants.

In a more refined form, the paper might be inspired by one of the following research issues.

- We know that endocrine-disrupting chemicals, such as PCBs and dioxins, cause reproductive abnormalities in female fish; however, we do not know the specific biological mechanisms, or the underlying causes, by which these chemicals exert their effects.
- In the early development of rainbow trout, at which critical stages do endocrine-disrupting pollutants affect sexual development?
- What are the critical concentrations at which endocrine-disrupting pollutants cause reproductive dysfunction in female fishes?

Notice the qualities that distinguish the topics from the research issues. The topics are broadly defined, whereas the research issues are narrowly focused. The topics are conveyed by noun phrases that generally tell what the paper is about, whereas the research issues are expressed as engaging problem statements and intriguing questions.

The best scientific papers are intended to contribute resolutions to focused research issues rather than to simply talk about broadly defined topics. A problem with purely topic-based papers is that they do not afford opportunities to go beyond summarizing and paraphrasing established knowledge. In contrast, issue-based papers naturally inspire writers to engage in interesting debates, to make original arguments, and to advance scientific knowledge. It's no simple task to transform a general topic of interest into a powerful and practical research issue. For extra help in this process, follow the guidelines presented next.

Brainstorm Topics That Inspire Your Interest and Enthusiasm

For engaging and productive writing projects, take the necessary time to brainstorm topics that you are genuinely interested in and passionate about. In your science classes, which subjects have especially captured your attention and spurred your desire to learn more? In everyday life, what experiences might lend themselves to appropriate topics for your assignment? In careers that you are considering, are there any relevant topics that you are keen to write about? Be sure to note every potentially interesting topic that comes to mind, whether it is an absolute favorite or not. As you work through the process of refining a research issue, you may find that topics at the top of your list are not actually suitable for your assignment. In this case, you will need to turn to backup topics.

Ask Knowledgeable Experts for Advice on Hot Research Issues

Some research issues stand out for their importance and the enthusiasm they generate in science and society. These are called *hot* research issues. Examples in the life sciences are unresolved problems and unanswered questions involving the Human Genome Project, the potential uses of stem cells in medicine, new methods for treating and preventing cancer, and the influences of climate change and chemical pollution on animal life and the environment. If you are new to writing about scientific research, how do you find out which issues are hot and which are not? A good way to start is to ask people who may know, beginning with your science professors and graduate teaching assistants. They will likely be glad to share their knowledge and advice about hot issues in their research fields.

Learn about Hot Research Issues from the Scientific Literature

In selected sections of published scientific articles, authors commonly discuss the major knowledge gaps, unanswered questions, and unsolved problems that currently define their research. In doing so, authors point directly to the hottest issues in their fields. If you read the published literature with the specific aim of gleaning hot research issues to write about, you will certainly find them. The best places to look are in the discussion sections of research papers and the conclusion sections of review papers.

Learn about Hot Research Issues on the Internet

Without question, the Internet has become a hotbed for information about hot issues in science. On Web sites featuring topics that interest you, look for discussions of current controversies, debates, and areas that are getting a lot of attention in the scientific community and society. Your search need not be limited to scientific Web sites. You will find information about hot research issues on the Web sites of educational institutions, bloggers, businesses, nonprofit organizations, and various news media.

Use Your Task Analysis to Refine Your Research Issue

For an ecology class, a student named Phil has an assignment to write a five-page paper on any topic concerning animal conservation. Phil's professor has given the class a list of her evaluation criteria. She is reserving the highest grades for papers that

(1) address the roles that humans have played in species endangerment and extinction;
(2) deeply explain the primary causes of current threats to species' survival; and
(3) use previous research to develop an argument for how to prevent future population declines.

To refine an excellent research issue, Phil must solve a few immediate problems. How many animal species should he write about? Which human-caused contributors to species declines should his paper address? Phil considers that the assignment is relatively short and that his professor is looking for *deep* explanations of the threats to survival. Accordingly, he wisely decides to focus his research issue narrowly. It's better to write about only a few animal species, Phil reasons, in order to reach his professor's high standard. If he broadens the issue to encompass too many species, Phil's explanations will be superficial. Reflecting on the criterion to explain the *primary* causes of current threats to species' survival, Phil narrows his focus on just a few human-caused contributors. Finally, he isolates specific causes of species declines that have been confirmed, through published research, to be preventable or reversible. This refinement of the issue will enable Phil to meet his professor's third criterion in the list. The point of this story is that skilled writers refine their research issues by applying task analysis.

Add a Novel Twist to Your Selected Research Issue

Every scientific writing project presents an opportunity to pose a novel research issue and to contribute to its resolution in unique ways. This sort of originality is a hallmark of excellence in science and scientific communication. After all, no one wants to read the same old papers about well-worn topics over and again. One way to add a novel twist is to focus on relatively underdeveloped areas of research fields. For example, many articles have been published on the effects of endocrine-disrupting chemicals on reproductive function in fish; in comparison, few have focused on these effects in other animals. A novel issue for a new paper would thus focus on the effects of endocrine-disrupting chemicals on reproduction in amphibians, birds, or mammals.

The novelty of a research issue depends partly on the audiences to which it is targeted. Suppose, for example, that most conservation biologists know all about the effects of industrial pollution on reproductive function and toxicity in fishes. But maybe the research has not yet been synthesized for public health officials, who are concerned with the well-being of people who may be eating contaminated fish. This situation prompts a novel research issue simply by adapting established scientific knowledge for a new audience.

Check Ahead for the Availability of Scientific Literature on Your Selected Research Issue

By following this practical bit of advice, you will avoid the counterproductive and frustrating experience of choosing an excellent research issue but then having trouble obtaining published literature on it. It's possible that your campus library does not subscribe to all the scientific journals that publish the articles you need. You may be able to order the essential articles through interlibrary loan, but they might not arrive for weeks. Of course, this is a major problem if your paper's deadline is fast approaching. The take-home message here is to make sure that the scientific literature you need is easy to obtain before making a final decision about a research issue. Strategies for searching for and obtaining scientific literature are presented later in this chapter.

Make Sure That You Have Sufficient Time and Resources to Learn the Science on Your Research Issue

Here's another obstacle worth avoiding: If you chose an advanced and complex research issue and you lack background knowledge to understand it, you might need to spend an inordinate amount of time just learning the science. This will reduce the time you can spend on all of the other essential activities for planning, drafting, and revising your paper. Before deciding on a research issue, read a representative set of articles on it. Make sure that you understand the basics and that you will have sufficient time and good resources to learn the more advanced science, if necessary.

LEARNING ABOUT SCIENTIFIC DISCOURSE CONVENTIONS

If you are new to writing in science, you will need to define your writing projects by learning how professional researchers define theirs. In other words, you will benefit from learning about scientific discourse conventions, which are the established guidelines that scientists follow to establish the objectives, structure, format, and style of their writing. A good way to begin learning these conventions, in addition to reading the rest of this book, is to examine published scientific texts. Suppose that you have an assignment to write a review paper, which is a standard type of journal article. If you have never written one before, consider the advantages of analyzing published review papers. In general, how long are they? How are their major sections structured? What communication goals do their authors seek to accomplish in them? How formal is their tone of language? As summarized in Table 1.1, there are different forms of written scientific communication, which vary in their length, structure, and objectives. Most of the primary types of scientific papers are published in journals that are accessible through the Internet. Table 1.2 (page 14), which describes a few leading journals in the life sciences, includes instructions for how to locate and download their published articles.

TABLE 1.1 Common Forms of Written Communication in the Life Sciences

Communication form (approximate number of manuscript pages)	Main objectives and overall structure
Research papers (15–30)	Report on original research, either a single study or a series of several related studies on a focused issue. Major sections are organized by the IMRAD format.
Review papers (30–60)	Summarize, synthesize, and critically evaluate existing knowledge on a focused topic or research issue. Major sections may be devoted to discussing subtopics of an overall topic or to presenting evidence and reasoning in support of arguments.
Research proposals and grant applications (up to 50)	Request approval and funding for research projects. Major sections are devoted to presenting background information, proposing research methods, detailing and justifying budget plans, and describing potential beneficial outcomes of the research.
Meta-analyses (10–25)	Synthesize the outcomes of numerous studies on a focused research issue by applying a special form of statistical analysis to their results. Major sections are organized by the IMRAD format.
Case studies (5–10)	Report on observations or experiments involving a single individual who is suffering from a disease or disorder.
Technical notes and methods papers (5–10)	Describe new instruments, procedures, and analyses for conducting research.
Editorials and commentaries (2–4)	Discuss current issues in research fields. Journal editors may invite peers to contribute these pieces.
Letters to the editor (1–2)	Authored by readers who comment on, review, and critique articles in recent issues of a journal. When an article is negatively criticized in a letter to the editor, the authors of the article are usually given the opportunity to write letters in response.
Books (hundreds of pages)	Present a broadly defined scientific topic in a full-length treatise.
Book reviews (1–2)	Summarize and critique recently published scientific books.

Abbreviation: IMRAD, Introduction, Methods, Results, and Discussion.

The following sections introduce the general structure, objectives, and contents of the three major types of papers that life scientists write: research papers, review papers, and research proposals. Most student writing assignments are based directly on these paper types or on closely related versions of them.

TABLE 1.2 Examples of Leading Journals in Various Life Science Disciplines

Journal title	Types of papers published and topics covered
Discipline: Biochemistry	
Annual Review of Biochemistry	Review papers on all topics in biochemistry.
Biochemistry	Research papers and review papers focusing on the biochemical structure, function, and regulation of molecules and genes.
FEBS Journal (Federation of European Biochemical Societies)	Research papers and review papers on all topics involving molecular life science. Special focus on topics including bioinformatics, genomics and proteomics, and the molecular biology of disease.
Discipline: Conservation Biology	
Animal Conservation	Various types of papers on topics including population biology, evolutionary ecology, population genetics, and biodiversity.
Conservation Biology	Various types of papers involving the study and preservation of species and habitats.
Journal of Applied Ecology	Various types of papers on topics including conservation biology, global change, pollution biology, wildlife and habitat management, and the implications of genetic modifications of living organisms.
Discipline: Medicine	
Journal of the American Medical Association (JAMA)	Various types of papers on a wide range of issues in biomedical science and public health.
The Lancet	Various types of papers covering the biomedical sciences. Includes sets of specialty journals on infectious diseases, oncology, and neurology.
The New England Journal of Medicine	All of the major types of scientific papers, in addition to specialty papers on matters of health policy and legal issues in medicine.
Discipline: Physiology	
Annual Review of Physiology	Review papers on all topics in physiology.
Journal of Physiology (London)	Research papers and review papers on physiological studies at the molecular, cellular, tissue, organ, and organ systems levels. Emphasizes human and mammalian physiology.
American Journal of Physiology	Research papers and review papers published in seven specialty journals, covering cell physiology, gastrointestinal physiology, cardiovascular physiology, endocrinology, and more.

TABLE 1.2 *continued*

Journal title	Types of papers published and topics covered
	Multidisciplinary Journals
Nature	Various types of papers in all fields of science and technology.
Proceedings of the National Academy of Sciences of the United States of America (PNAS)	Various types of papers in the biological, physical, and social sciences.

Instructions for downloading sample articles from these journals:
1. Enter a journal title into an Internet search program such as Google.
2. Navigate to the journal's Web site.
3. On the home page for the journal's Web site, look for links to past issues. If your campus library subscribes to the journal you are searching, you will likely have access to articles in most, if not all, of its past issues. However, you may need to access the journal through your library's Web site or your campus's computer network.
4. On the Web pages for the journal's past issues, look for the links to "Full Text" and "PDF" versions of articles. The full text versions typically open as Web pages. The PDF versions, which are easier to read, open in Adobe's Acrobat Reader software, which is free to download from http://www.adobe.com.

Research Papers

In the mid-1990s, large populations of amphibians—a class of animals that includes frogs, toads, and salamanders—began to die in mountainous regions of Australia, Central America, and the United States. In Central America, within a few months after the first deaths, more than 50% of the species became extinct. The remaining population declined to 20% of normal. The causes of death were not immediately evident, although researchers speculated on several viable culprits, including pollution, climate change, and infectious diseases. Circumstantial evidence led to the hypothesis that a disease called *chytridiomycosis* might be responsible. Chytridiomycosis is caused by a parasitic fungus called *Batrachochytrium dendrobatidis*. In 1998, an international team of researchers led by Karen Lips from Southern Illinois University began a 7-year study to determine whether the spread of *B. dendrobatidis* was indeed causing the amphibian population declines (Lips et al., 2006). Lips and coworkers set up their field laboratory in El Copé, Panama, where the amphibian community had been thriving and *B. dendrobatidis* had not yet spread. In the fall of 2004, hundreds of frogs, toads, and salamanders were found dead in El Copé. Nearly all had been infected by *B. dendrobatidis* and had signs of chytridiomycosis. Given the coordinated timing of the mass die-offs, observations of the parasitic fungus, and evidence for chytridiomycosis, the researchers concluded that the events were causally related rather than just coincidental.

To communicate their breakthrough findings, Lips and coworkers wrote a type of document called a research paper (or a research report). Scientific research papers have next to nothing in common with the library research papers, or book reports, that we all wrote in grade school. Research papers report original studies; thus, they are the primary and most precise sources of scientific knowledge.

Consider the study conducted by Karen Lips and her coworkers. To get any closer to the investigation than reading the research paper that reports it, you must have directly witnessed the international team of scientists as they collected and analyzed their data in Panama and their labs around the world.

Lips et al.'s paper was published in a journal called *Proceedings of the National Academy of Sciences of the United States of America*, or *PNAS* for short. You can obtain the entire paper from the journal's Web site (http://www.pnas.org/). On the home page, look for a link to past issues. (When I accessed the site in the spring of 2009, the link was titled "Archives.") Lips et al.'s research paper is in the February 28, 2006 issue, pages 3165–3170. With this information you can navigate to the article and download a PDF version of it.

Most research papers follow a structure called the *IMRAD format*. IMRAD is the acronym for the paper's major sections: Introduction, Methods, Results, and Discussion. The sections are usually ordered as just listed, although sometimes the methods section follows the results or discussion section. The general contents of the IMRAD sections are as follows.

- Introduction sections present the research issues, hypotheses, and purposes that motivated reported studies, and they address the studies' importance to society and the scientific community.
- Methods sections describe and justify the procedures for collecting and analyzing study data.
- Results sections present the study data and statistical analyses.
- Discussion sections interpret the results to support conclusions that resolve the research issues and hypotheses of studies.

Several types of student papers are typically modeled on the structure, objectives, and contents of standard research papers. For example, if you have conducted experiments in laboratory classes, your assignments for writing lab reports may have been based on the IMRAD format. In addition, senior theses and graduate-level dissertations are usually written in the form of research papers.

Research papers are the most common type of article published in most life science journals. In a typical edition of a journal, at least 75% of the articles are research papers. Journals typically restrict the maximum length of research papers to as few as 3,000–4,000 words, or approximately 10–15 double-spaced manuscript pages. At the high end, the limit is roughly 7,000–8,000 words. The journal PNAS restricts research papers to 47,000 characters, which is in the neighborhood of 7,500 words, or approximately 25 double-spaced manuscript pages. Consider that Lips et al.'s research paper, which reported a very complex 7-year study, was approximately 5,000 words long, or only about 15 double-spaced manuscript pages. Because they are relatively short and they tend to be extremely dense with technical information, research papers can be challenging to read. If you lack experience with this form of scientific literature, you will benefit from learning a strategic reading approach. We will work through a step-by-step activity for reading and comprehending research papers later in the chapter.

Review Papers

Whereas research papers are the primary sources of scientific knowledge, the outcomes reported in any single research paper usually do not completely resolve its motivating issue. A case in point is the 2006 research paper authored by Lips and coworkers. Considering the study's impressive results, one might confidently pinpoint chytridiomycosis, the disease caused by *B. dendrobatidis*, as the cause of amphibian deaths in Central America. However, by 2006, more than 60 research papers had been published on chytridiomycosis in amphibian communities, and all of their conclusions did not unanimously support those of Lips et al. In addition, hundreds of papers have been published on *other* suspected causes of declining amphibian populations around the world. The culprits include habitat destruction, global warming, acid rain and other forms of environmental pollution, and the introduction of nonnative predators. Theories have also been based on interactions between causal factors. For example, researchers have speculated that changes in weather patterns have enabled the *B. dendrobatidis* fungus to emerge and thrive, and then to infect and kill amphibians.

The sort of indeterminate state of knowledge that describes the causes of amphibian die-offs is common, as well as commonly expected, in many life science fields. It makes sense to expect that the answers to life's complex questions are neither straightforward nor obvious. Few scientific puzzles can be solved immediately by the outcomes of a single study as reported in a single research paper. This situation demands a form of communication that summarizes and synthesizes the findings and conclusions from the numerous research papers that have been published on a given issue. This type of document is called a review paper or a review of literature. A thorough review paper on an extensively researched topic might include information from hundreds of studies, which are all cited in the paper's text and reference list. In most life science journals, review papers are outnumbered by research papers. However, some specialized journals contain only review papers (see examples in Table 1.2). Published review papers typically range between 10,000 and 25,000 words, or approximately 30–80 double-spaced manuscript pages.

There are several subtypes of review papers, called *informative reviews*, *conceptual reviews*, *critical reviews*, and *position papers*. The subtypes generally have the same major sections: an introduction, a body, and a conclusion. As in research papers, the introduction sections to review papers present their motivating issues and purposes, and they address the importance of the issues to society and science. The conclusion sections of review papers summarize the key contents of their body sections. The conclusion is also the place for suggesting ideas for future studies to further resolve the issues at hand. As explained next, the main differences between the subtypes of review papers come in the body section.

Informative Review Papers

Informative review papers update readers about the state of knowledge on a focused topic or research issue. The body of informative reviews is thus often structured by subtopics of the overall topic or by various aspects of the central research issue. Take the example of an informative review paper on infectious diseases in declining amphibian populations. Each subsection of the paper's body might be devoted to a specific pathogen—that is, a disease-causing bacterium, virus, fungus, or parasite. Within a given subsection, the author might present research findings on the pathogen's life cycle, its means of transmission into amphibian communities, and its mechanisms for infecting the animal's body and causing death. This structure is consistent with the basic objectives of an informative review paper, which are to present scientific knowledge and research findings in a straightforward way.

For more complex research issues, a higher goal of informative reviews is to *synthesize* research. In science, synthesis means putting together the pieces of a puzzle, where the pieces are the methods, data, and conclusions of various studies on a topic or issue. Synthesizing research entails determining which puzzle pieces fit together seamlessly and which do not. When the outcomes of studies on an issue disagree—that is, when the puzzle pieces do not easily match—the challenge of synthesis is to explain *why* not. Chapter 3 demonstrates how to synthesize research in the process of generating content for scientific papers.

Conceptual Review Papers

Conceptual review papers go beyond informing readers about research by raising and supporting original concepts, which might be novel ideas, hypotheses, or theories. Take the concept that amphibian population declines might be caused by many interacting factors, rather than by a single factor. In the body of a conceptual review paper on this issue, the author might organize successive subsections by various hypotheses for the interacting causal factors. One subsection, for example, might focus on hypothesized interactions between global warming and the virulence of harmful viruses and bacteria. Another subsection might concern the mutual influences of habitat destruction and the influx of invasive predatory species. Each subsection would present supporting evidence for the hypothesis on which it is based.

Critical Review Papers

In critical review papers, authors evaluate the quality of studies that have been conducted on given issues. An overall goal for this paper type is to make arguments about whether the study conclusions are valid. Authors of critical reviews thus identify strengths and weaknesses in the methods of published studies. In addition, they critique the arguments that researchers have advanced for their conclusions. For example, a critical review might be written in response to research papers in which scientists have concluded that chytridiomycosis has *directly caused* the extinction of amphibian species. Several concerns have been raised about possible flaws in the methods of studies on this issue. For example, some of the studies may not have adequately controlled for other factors, besides chytridiomycosis, that might cause amphibian deaths. The author of a critical review paper on this issue might devote separate subsections to each potentially problematic method, explaining how it may have led to flawed results and conclusions. Other subsections might be devoted to evaluating strengths in the methods and conclusions of selected studies on the issue.

Position Papers

Position papers have the overall objective of convincing readers to accept claims on debatable research issues. This type of review paper takes the critical analysis of research to its highest level. In the introduction section, the author summarizes the contrasting arguments that have been advanced on an unresolved issue. The author then states his or her position as a formal claim. The body of the paper is structured by lines of evidence and reasoning that are intended to support the author's central claim and refute opposing claims. The content is generated by applying the skills of constructing scientific arguments—these skills are central to several activities demonstrated in upcoming chapters.

Research Proposals

In their studies on the tragic disappearance of amphibian species worldwide, conservation biologists have discovered numerous viable explanations. Focusing on the likely culprits—including infectious diseases, climate change, and pollution—these scientists will take the next steps toward their ultimate goals of reversing population declines and preventing future species extinction. From amphibian communities around the world, healthy and sick animals will be gathered for experiments. Treatments will be administered, data will be collected and analyzed, and eventually the study results will be published in research papers and review papers. There is much work to be done indeed. However, before this important research can begin, scientists must clear two tall hurdles. The first is getting approval to conduct their planned studies. All reputable public and private research centers have committees, called institutional review boards (IRBs), that critically evaluate proposed studies for their merit and to ensure that experimental subjects will be treated humanely. The second hurdle is obtaining financial support for the research. Consider the costs, for example, of a study to determine the effects of antimicrobial drugs on diseased frogs in Australia, Central America, and the United States. The scientists will need money to travel to and from the sites of amphibian communities, for housing and food at the field sites, to transport the animals back to their laboratories, to house and feed the animals, to purchase instruments for data collection, to pay research assistants, and for numerous other costs of conducting the research.

The form of written communication that scientists use to request approval and funding for research is generally called a research proposal. Documents intended specifically to apply for funding are called grant proposals or grant applications. Students who are planning thesis or dissertation projects may be required to write research proposals to seek approval from their advisors and universities' IRBs. Students may also have opportunities to write grant applications to campus programs and extramural organizations that fund student research. Research proposals are common writing assignments in undergraduate and graduate science classes. Professors like to assign proposals because they challenge students to develop original hypotheses, to design effective studies, and to convincingly argue that new research truly needs to be conducted.

Of the three major types of scientific papers, research proposals are the most variable in their structure and length. Guidelines for manuscript preparation are set by the institutions and agencies that review the proposals. A campus-based program that sponsors undergraduate student research, for example, might set a five-page limit on grant applications, while a federal agency might allow 30 pages or more. While they vary in structure and length, most research proposals in the life sciences should include the following core sections.

- An introduction that presents the research issue that motivates the proposed study, discusses the importance of the research, and states hypotheses and specific purposes.
- A section that presents background information about previous studies on the central research issue.
- A section devoted to describing and justifying the planned research methods.
- A section in which the author discusses anticipated beneficial outcomes of the proposed research.
- For grant applications, a section that presents the proposed budget with justifications of planned expenses.

Unlike research papers and review papers, research proposals are not published in journals. They are usually read only by a small audience who may be IRB staff, professors, peer reviewers, and officers of grant agencies. For many life scientists in U.S. universities and research institutes, the most important grant agencies are the National Institutes of Health (NIH) and the National Science Foundation (NSF). The NIH funds research that may immediately or ultimately lead to methods for preventing disease and improving health in humans. The NSF has a broader mission of promoting scientific progress in a wide variety of fields, including the life sciences as well as astronomy, atmospheric science, and physics. The Web sites for NIH (http://www.nih.gov/) and NSF (http://nsf.gov/) contain extensive resources for grant writers, including sample grant applications.

ANALYZING YOUR AUDIENCES

By definition, scientific communication is not a means of introspective self-expression. Indeed, scientific papers serve no practical purpose unless they meaningfully influence the knowledge, understanding, beliefs, and behaviors of their intended readers. Skillful writers take this point to heart by thoughtfully considering matters of audience to guide decisions about the content, structure, and style of their papers. This expert approach applies the products of a preliminary activity that writing instructors call audience analysis. It essentially involves characterizing intended readers, or forming mental representations of their needs, expectations, and values. Audience analysis engages our memory and imagination. That is, it's a process of recollecting and visualizing who your readers are, what they know about your paper's subject matter, what preconceptions they hold about your arguments, and so on. When you are writing to new groups of readers whom you do not know well, the challenge is learn about them. One approach is to gain insights by talking with people who are very familiar with your targeted audience. An even better tactic is to speak directly with your intended readers or with people who closely represent them. This direct approach is quite common in scientific writing. As I mentioned earlier, experienced scientific authors are accustomed to phoning and e-mailing peers, journal editors, and grant agency officials who represent their audiences. The authors ask pointed questions to gain information that will help them serve their readers best. Excellent student writers take the same approach in their interactions with professors and other audiences.

Key Questions for Audience Analysis

Described as follows, audience analysis is ideally guided by raising questions that focus our attention on key characteristics of intended readers.

Who Are Your Primary and Secondary Readers?

The first step to characterizing your readers is simply to note who they are. This is an especially important step for planning papers that are intended for several audiences with distinct characteristics. Career scientists often face the challenge of tailoring individual documents to address the concerns of varied groups of readers. A common example is a grant application submitted to a federal funding agency such as NIH or NSF. The readership includes peer researchers, who review grant applications for their scientific merit; officials of the grant agency, who make the final decisions about funding; and perhaps even representatives of Congress who want to ensure that taxpayer dollars are funding research that will benefit society. These three groups may differ vastly in their knowledge about the grant application's topic and in their perceptions about its meaningfulness in science and society.

Student authors also face the real-world challenges of writing individual papers to varied audiences. For a lab report in a biology class, for example, you might be instructed to address your paper to members of the scientific community. Obviously, however, you should consider another audience, which is your professor. In addition to reading student papers from the perspective of the scientific community, your professor will read as an educator of science and scientific writing. So, in a sense, you would have two audiences to address. When you are writing to more than one audience, it's a good idea to designate each as primary or secondary. Primary readers are those who will have the greatest use for your paper's information and ideas. They are also the readers who will make the final decisions about your paper's fate, such as its grade or whether it will be accepted for publication. Designations of primary versus secondary readers will help you make good decisions to settle diverging approaches to writing your paper. Whereas you should not completely overlook secondary audiences, you may need to serve the needs of your primary readers first and foremost.

How Are Your Readers Likely to Relate to You?

How we define our personal relationships strongly influences the content, structure, and language of our everyday communication. Scientific writing poses a similar situation. Will your readers view you as a familiar member of their research community or as an outsider? If the latter applies, a more formal tone than normal may be appropriate. When you are making arguments on debatable issues, will your readers view you as friend or foe? If they are likely to strongly oppose your position initially, you will do well to acknowledge and validate their views before presenting evidence and reasoning to refute them.

What Do Your Readers Know and Need to Know about Your Topic?

By considering what your readers already know about your paper's subject matter, you will make good decisions about appropriate content to include and develop. In addition, considerations of your readers' knowledge needs will reveal content to *exclude*, saving you valuable space and avoiding the negative reaction that readers have to being told what they already know.

Why Is Your Audience Reading Your Paper?

Scientific papers are read for different reasons and practical applications. Of course, science professors read to evaluate whether students have successfully acquired important knowledge and convincingly communicated original arguments. Journal editors read to decide whether submitted articles should be published. Scientists may read peers' articles to obtain data for supporting their arguments or to extract details about study methods in order to replicate them in upcoming experiments. For what purposes will your audience read your paper? The answers to this question will help you generate effective content and organize it appropriately. As one of many examples, suppose that you are writing an IMRAD-structured paper to a primary audience that is mostly concerned with practical applications of research. To address this audience's reading purposes, you might devote a considerable amount of content in your discussion section to presenting practical guidelines that follow from your study's results.

What Values, Preconceptions, and Biases Might Readers Hold about Your Research Issue?

This question is especially important for making arguments about controversial research issues. If your readers agree completely with your views, there is no need to emphasize content that preaches to the choir. Consider, however, the challenge of writing to audiences who are initially biased against your views. To identify the best evidence and reasoning to convince these readers to accept your arguments, you must understand and address your readers' values, preconceptions, and biases.

What Qualities of Written Communication
Do Your Readers Value Most?

Experienced scientific readers have strong views about the defining features of successful writing. This is especially true of professors, who greatly value student writing that demonstrates precise and succinct language, highly organized patterns of ideas, and well-reasoned arguments. It's definitely worth analyzing your audiences for what *specifically* they value in written communication. How strictly do they adhere to traditional rules of grammar? Do they appreciate a more or less formal tone of language? How do they feel about the use of graphics to present data? What is their response to arguments based on speculative, out-of-the-box thinking?

How Will Your Audiences Go about Reading Your Paper?

Student writers might very well expect professors to read papers from start to finish, progressing from one section to the next in linear order. But this is not necessarily how scientific audiences read published journal articles and research proposals. Studies on the reading habits of scientists reveal that they often do not progress through papers linearly and completely. For instance, after perusing the title and abstract of a published research paper, experienced readers may skip directly over the introduction and methods sections in order to read the results section. There they examine the tables and figures to glean the study's key outcomes. If the interpretations are not crystal clear, readers head for the discussion section. In this common pattern, readers may save the introduction and methods sections for last, or they might skip these sections completely.

To account for these nonlinear reading habits, scientific authors structure their papers' content strategically. Take the example of an introduction section to a research paper in which an author has presented an especially noteworthy rationale for her hypothesis. She is concerned, however, that experienced readers might skim the introduction or skip it completely. In the paper's discussion section, the author wants to ensure that readers understand the rationale for the hypothesis because it is an essential foundation to her conclusion. Of course, she cannot force the audience to read the introduction section where the rationale initially appears. A solution, derived through analyzing how scientific audiences go about reading journal articles, is to summarize the rationale early in the discussion section and to refer readers back to the introduction for elaboration.

Taking Notes on Your Audience Analysis

While audience analysis is extremely practical and powerful, its applications in the writing process can be cognitively demanding. Consider the challenges of applying audience analysis to guide the revision process. To revise your drafts successfully, you must consider whether their elements appropriately address concerns of audience. The first cognitive task, which is not at all trivial in the grand scheme of revision, is to *remember* to apply your audience analysis. Then, the real challenge begins as you evaluate your draft more deeply. Have you addressed the needs, expectations, and values of both primary and secondary audiences? Does the draft's tone fittingly reflect your relationship with readers? Have you fully accounted for readers' background knowledge? Does your draft acknowledge the preconceptions and biases that readers might hold about your research issue? To answer these questions and revise your draft accordingly, you must reflect back on the characteristics of your audience that you brainstormed in early stages planning. For example, to determine how well your draft's content accounts for readers' background knowledge, you must recall your earlier thoughts about what readers know and need to know about your topic. It's every bit of mental gymnastics with a high degree of difficulty.

To reduce the cognitive burden of applying audience analysis throughout the writing process, experienced writers take notes on their initial characterizations of readers. A sample form for taking audience-analysis notes is presented in Table 1.3. At the top of the form are spaces for identifying primary and secondary audiences. Then, the form's main rows are organized by our audience-analysis questions. For each question, the form includes spaces for noting (a) key characteristics of primary and secondary audiences and (b) strategies for tailoring a paper's content, structure, and style to account for the characteristics.

For an Honors biology course, a student named Anna is planning a research proposal. Anna will use the form in Table 1.3 to record her audience analysis.

TABLE 1.3 A Format for Taking Audience-Analysis Notes

Primary readers:

Secondary readers:

	Characteristics of primary audiences	Characteristics of secondary audiences	Strategies for applying audience analysis
How are my readers likely to relate to me?			
What do my readers know and need to know about my topic?			
Why is my audience reading my paper?			
What values, preconceptions, and biases might readers hold about my research issue?			
What qualities of written communication do my readers value most?			
How will my audiences actually go about reading my paper?			

Anna's research issue concerns the alarming worldwide declines in honeybee populations that have occurred in recent decades. In North America, for example, the number of honeybee colonies has decreased by more than 40% over the last 50 years. The inexplicable disappearance of honeybees has been termed Colony Collapse Disorder (CCD). Its implications for humans are drastic because honeybees pollinate crops that make up a significant portion of our diet. Among other suspected causes of honeybee deaths are pollution, pesticides, climate change, and pathogens such as viruses and parasites. Anna is proposing a study to test the hypothesis that these factors create physiological stresses that overwhelm the honeybee's immune system and thereby weaken its defenses against fatal pathogens.

Anna's professor has instructed students to prepare their proposals as if they would be submitted to NSF. Accordingly, Anna figures that her primary audience will be biologists whom NSF officials recognize as leaders in species conservation research. Anna reasons that her secondary audiences will be NSF officials and perhaps the staff of government agencies that oversee agricultural policies and practices. To focus her attention on these readers, Anna will quickly note who they are in the corresponding spaces of the form in Table 1.3. In part of her audience analysis, Anna will raise the question, "What do my readers know and need to know about my topic?" She reasons that her primary audience, the experts in conservation biology, will surely know a lot about CCD and its suspected environmental causes. However, they might lack specialized knowledge about the physiology of honeybees, specifically about how their immune systems respond to environmental stress. Reflecting on her secondary readers, Anna surmises that most will lack the specialized physiological knowledge. So in our note-taking form—specifically in the row for recording what readers know about or need to know about a paper's topic—Anna will note these characterizations. Then, in the last column of the same row, she will note focused strategies for applying her audience analysis. One of the strategies, for example, is to present the physiological concepts in fairly basic terms, with well-developed explanations that Anna's scientific and nonscientific readers will understand. This reminder will be a useful guide when Anna drafts and revises her paper. But first she will complete her notes for other relevant questions listed in the first column of Table 1.3.

Whether you need to take elaborate audience-analysis notes depends on the complexity of your writing project and on how well you already know your readers. Even if formal written notes would be overkill, it's still important to organize your thoughts by the format in Table 1.3. In other words, in your mind, you should characterize your intended readers by answering key audience-analysis questions; then, devise strategies for applying your audience analysis to writing your paper.

SEARCHING FOR SCIENTIFIC LITERATURE

Many activities in scientific writing involve working with published journal articles, books, and Web site materials. Among other applications, you will use published literature to refine your research issues and to acquire the essential knowledge that defines them, to obtain evidence and reasoning to support your arguments, and to relate your original study outcomes to those from previous studies on your selected issue. As we consider in this section of the chapter, it takes a fair amount of specialized procedural knowledge, along with detective-like skills and computer savvy, to locate and obtain useful scientific literature to support writing projects.

Evaluating the Credibility of Published Scientific Literature

The quality of any scientific paper obviously depends on the accuracy and dependability of its content. These features are directly determined by the credibility of the published literature from which authors derive knowledge, evidence, and reasoning. How can you be certain that the literature you are using is trustworthy, accurate, and representative of consensus views in the scientific community? Use the questions below to guide your critical evaluation.

1. How far removed is the literature from primary sources of scientific knowledge? As I explained earlier, the primary sources of scientific knowledge are published research papers, the documents that report outcomes of original studies. Next in line come published review papers, which are called secondary sources, because they summarize and synthesize information from research papers. Tertiary sources, scientific books and textbooks, glean information from primary and secondary sources. As scientific knowledge is removed farther and farther from primary sources, it's worth raising more critical questions about matters of credibility. Sometimes, for example, the authors of published review papers do not accurately interpret original study findings. This is not to say that all review papers, textbooks, and other nonprimary sources are automatically less credible than research papers. Instead, the take-home message is that unless you are absolutely certain about the credibility of information in secondary and tertiary sources, you should verify it. Do so by seeking the original research papers from which the information was obtained. In addition, to support your arguments, make sure to use evidence from primary sources rather than from second-hand summaries.

2. Has the literature been peer reviewed? As described in the book's introduction, peer review is the main quality-control system in scientific publication. Although the system is not perfect, it does filter out most articles that contain inaccurate information, lack significant new knowledge, are based on flawed methods and reasoning, and are poorly written. Accordingly, experienced writers limit their literature searches to peer-reviewed sources. All reputable scientific journals use the peer review system to select articles for publication. You can learn about the peer review policies of scientific journals on their Web sites (look for links to author information). If you are uncertain about whether literature you have obtained has been peer reviewed, ask your professor or teaching assistant.

3. How credible are the journals in which research papers and review papers are published? Some scientific journals have well-established reputations for publishing the highest-quality research. These are the journals with the strictest peer-review standards. Their articles address the most novel and meaningful research issues, they report experiments that apply revolutionary technologies, and they demonstrate the most effective features of written communication. If you rely on articles published in leading scientific journals, the credibility of your paper's ideas and arguments will be automatically enhanced. How do you know which journals publish the highest quality articles in your field? You can learn through your reading experience, through word-of-mouth recommendations, and by asking people who know the field. In addition, you can obtain published rankings of scientific journals. The most widely accepted rankings are based on measures called impact factors. An impact factor is a number that indicates how frequently, on average, published articles in a specific journal are cited by authors in all other journals. Impact factors thus reflect the influence that journals have in their scientific communities. For a given year, a journal's impact factor is calculated as follows.

Step 1: Count the total number of articles published in the journal over the previous 2 years.

Step 2: Count the number of times that the journal's articles were cited by authors in all other journals over the same 2-year period.

Step 3: Divide the sum from step 2 by the sum from step 1 to obtain the impact factor.

If a journal has an impact factor of 2.0, one of its average articles has been cited by authors in two recent articles published in other journals. A journal with an impact factor above 2 or 3 is generally considered to have a good reputation. Journals with scores ranging from 5 to 10 and beyond are highly respected for their influence and credibility. To learn more about impact factors, including limits to their interpretations and applications, search for *Journal Citation Reports* on the Web site of Thomson Scientific, (http://www.thomsonscientific.com/). If your campus library subscribes to this resource, you can use it to check the impact factors for all reputable life science journals. If your library does not offer free access to *Journal Citation Reports*, go to the Web sites of the journals that you want to check. On Web pages devoted to providing information for prospective authors, many journals display their impact factors.

4. How reputable are the authors? In all research fields, scientists recognize elite peers who are on the leading edge, conducting the most essential and impressive studies, and writing the most engaging articles, books, and Web site materials. Although an author's reputation should not be the sole means for evaluating the credibility of scientific literature, it's a start. In your literature searches, look for the names of scientists who are receiving lots of credit and praise for their contributions. It's also worth asking your professors and teaching assistants to point out the leading scientists in the fields you are writing about.

5. *How current is the source?* For some life science issues, especially those in dynamically evolving fields, "current" knowledge can become completely outdated within a matter of months. But for many other issues, knowledge that dates back many decades is actually relevant to new writing projects. So you should not automatically dismiss the credibility of older publications on your paper's issue. To judge the pertinence of relatively old articles, books, and Web site materials, pay attention to how current authors write about them. If, in discussing the present state of an unresolved research issue, authors consistently include information and ideas from older articles, the articles may indeed be relevant to your project.

Searching for Peer-Reviewed Journal Articles: Research Papers and Review Papers

As we have discussed, the most credible sources of scientific knowledge are, by and large, peer-reviewed research papers and review papers published in reputable journals. Advice on searching for these forms of scientific literature is presented as follows.

Ask Experts on Your Research Issue to Point the Way to Useful Journal Articles

This advice is to take a productive shortcut to pertinent published literature. If your science professors have written or collected journal articles on your paper's issue, they may be happy to share them with you. Or perhaps your campus is home to renowned scientists who specialize in research on the issue. In response to your well-presented requests, these experts might suggest articles for you to seek, hand you paper copies of articles from their files, or send you articles attached to e-mail messages.

Search the Specialty Journals in Your Field

Let's say that you are writing a review paper on the reproductive success of gray wolves that have been reintroduced to native habitats in the Rocky Mountains. To find articles on this issue, you can go directly to the Web sites of journals that specialize in publishing research on animal conservation. For example, articles on gray wolves are frequently published in a journal called *Conservation Biology* and in *The Journal of Wildlife Management*. To learn which journals specialize in research on your paper's issue, ask your professor. Or, into an Internet search program, enter "leading scientific journals in (your paper's general topic)." On the Web sites of specialty journals, you can use search programs and tables of content to seek relevant articles.

Use Comprehensive Databases of Journal Articles

Especially at the start of a writing project, you might not want to limit your literature search to only a few specialty journals. For a more inclusive approach, you will need to use comprehensive databases of scientific literature—specifically, databases that store information about journal articles along with digital copies of them. Your campus library likely has a computer-based catalog that enables access to such databases, which are available through the Internet. The defining information about journal articles comes in their citations. Every journal article in a database has a citation, which typically includes the following elements.

- The article's title and the name(s) of its author(s).
- The title of the journal in which the article was published.
- The year that the article was published.
- The volume, issue, and page numbers of the journal in which the article appears.
- A digital object identifier (DOI), which is a number that uniquely identifies an electronic version of the article (for accessing it over the Internet).

To demonstrate the basics of searching for journal articles by using comprehensive databases, I will take on a hypothetical assignment to write an informative review paper on the health risks that have been linked to eating mercury-contaminated fish. I searched for articles on this topic using an online database called *PubMed* (http://www.ncbi.nlm.nih.gov/sites/entrez). Maintained by the U.S. National Library of Medicine, PubMed is the largest and most inclusive database of journal articles in the life sciences. In 2009, it contained citations for over 18 million articles that have been published in more than 5,000 journals since 1948. Like other databases of scientific literature, PubMed displays lists of citations when users enter search terms. To begin looking for articles on my paper's topic, I used PubMed's Web interface to enter a phrase containing the following three search terms: *eating fish* AND *mercury* AND *health risks*. The connector word AND cues the database program to search for articles that have been indexed by all of the search terms that users enter. PubMed recognizes two other connector words: OR and NOT. If, for example, I had entered "eating fish AND health risks NOT mercury," the search program would have overlooked articles concerning mercury contamination. The instructions for using PubMed indicate that connector words should be capitalized. (Complete instructions, along with helpful tutorials, are accessible on the PubMed Web site.) Figure 1.3A (page 30) shows the first 6 of 28 citations returned from my initial query. The elements of a PubMed citation, which are similar to those of citations in other comprehensive databases, are highlighted in Figure 1.3B (page 30).

After clicking on the authors' names in a list of PubMed citations, users are directed to a page that displays detailed information about the corresponding article, including its abstract, or summary. An example detailed record for a PubMed article is shown in Figure 1.3C (page 31). Toward the middle of the figure, on the right-hand side, notice the icons labeled "FREE." These indicate that, in addition to storing citation information about the article, the database contains a full-text digital version, usually a PDF, that users can download at no cost. However, of the 18 million-plus journal articles in PubMed, not all are available in full-text digital form. And of those that have been digitized, not all are available for free. Soon I will explain options for obtaining journal articles that cannot be downloaded immediately and at no cost from online databases.

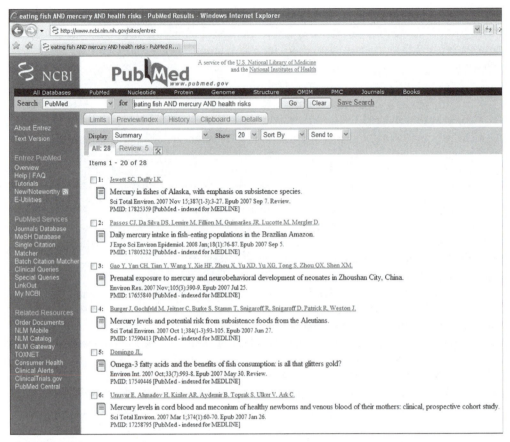

FIGURE 1.3A Citations returned from a literature search using the database PubMed.

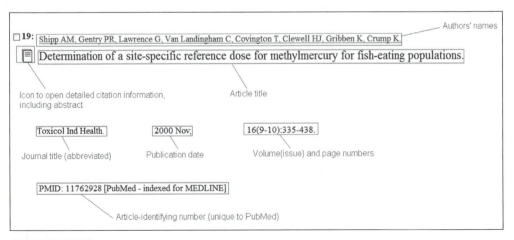

FIGURE 1.3B Elements of a journal article citation.

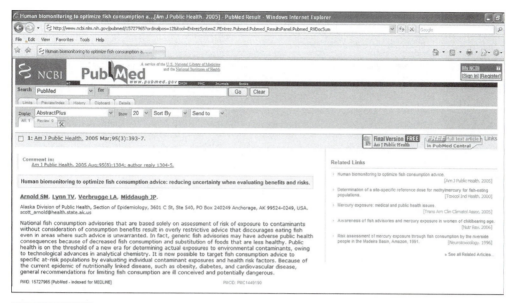

FIGURE 1.3C Detailed information about a journal article citation.

Several of the most commonly used databases of life science literature are described in Table 1.4 (page 32). All can be accessed through their Web sites, which have fairly similar user-interfaces for entering search terms, viewing citation records, and retrieving full-text articles. Some of the databases provide free public access, whereas others require paid subscriptions, which your institution's library might hold. As indicated in Table 1.4, databases of published literature vary by the academic disciplines that they cover. For example, Biological Abstracts and PsychInfo concentrate on topics in biology and psychology, respectively. Google Scholar covers all academic disciplines, including the life sciences. PubMed specializes in the biomedical and health sciences, but it also branches out to all other life science fields including biology, biochemistry, exercise science, nutrition, physiology, and psychology. Some databases, typically the smaller ones, provide access to full-text articles for every citation that they contain.

By entering the same search terms into different databases, you will find many overlaps in the article citations returned. But you may also find notable differences in the search results, because each database uses its own specialized search program. So, for a given literature search, you might use several databases to gain the advantages of each.

Every online database has its unique features and services, which are explained in instructional documents and tutorials on its Web site. However, the general process of searching for journal articles is similar across all comprehensive databases. Figure 1.4 (page 33) presents an overview of the essential steps, which are detailed as follows.

TABLE 1.4 **Databases of Life Science Journal Articles**

Database	Description of contents	Web site address
Biological Abstracts	Selected full-text articles from journals in all subdisciplines of biology, including biochemistry, biomedicine, biotechnology, genetics, botany, ecology, microbiology, pharmacology, zoology, and more.	http://scientific.thomson.com/products/ba
Biomed Central (BMC)	Links to full-text articles from more than 180 life science journals. Maintained by BMC, an independent publisher.	http://www.biomedcentral.com
BioOne	Links to full-text articles from more than 150 journals specializing in the biological sciences.	http://www.bioone.org
Google Scholar	Selected full-text articles and citation information from journals in all academic fields.	http://scholar.google.com
HighWire Press	Links to approximately 5 million full-text articles from over 1,100 journals. The largest database of free full-text life science articles in the world. Developed by Stanford University Libraries.	http://highwire.stanford.edu
PubMed	Approximately 18 million citations and selected full-text articles from more than 5,000 journals in all life science disciplines. Contains the MEDLINE database. Maintained by the U.S. National Library of Medicine.	http://pubmed.gov
PsycINFO	Links to selected full-text articles from journals in the behavioral sciences (including psychology and social science) and mental health.	http://www.apa.org/psycinfo
Scirus	Over 450 million science-specific Web sites and selected full-text journal articles.	http://www.scirus.com/srsapp

Note. The table reports numbers of full-text articles and journals based on database status in 2009.

Step 1: Brainstorm all relevant search terms describing your topic. The organizations that maintain databases regularly enter citation information about newly published journal articles. Although the process varies across organizations, it generally involves indexing articles by search terms, which are also called keywords. On the title pages of manuscripts that scientists submit for publication, they are commonly instructed to list four to six keywords that describe their topics. After a paper is published, the journal sends it to various databases for

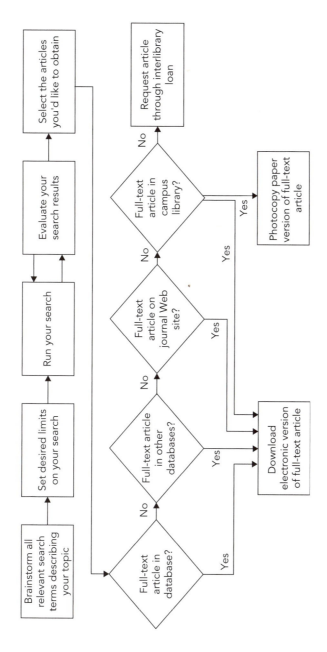

FIGURE 1.4 Steps to searching for and acquiring published journal articles.

indexing. The database staff may add a set of keywords that are more consistent with their organization's vocabulary for certain topics. Whether a database program finds relevant articles on your topic depends partly on whether your search terms match the indexed terms. So it's important to brainstorm at least a few synonyms for the keywords that describe your topic. For an organized approach to this step in the process, you might construct a matrix of search terms, such as the one illustrated in Table 1.5. To construct this matrix, I brainstormed synonymous search terms in three categories that reflect key aspects of my topic: fish eating, mercury, and health. In successive queries of PubMed, to get the most comprehensive lists of citations on my topic, I will enter different combinations of the synonomous search terms. Some database programs include a thesaurus that suggests keywords to match the vocabulary used to index articles. For example, PubMed has a thesaurus called MeSH, which stands for medical subject headings. When I typed mercury into the MeSH thesaurus, which is available through a link on PubMed's home page, I was informed that an appropriate keyword is *mercury poisoning*. So I included this keyword in my search matrix in Table 1.5.

TABLE 1.5 **Sample Search Term Matrix**

Category 1	Category 2	Category 3
Eating fish	Mercury	Health risks
Fish consumption	Mercury contamination	Health outcomes
Fish intake	Mercury poisoning	Disease

Step 2: Set desired limits on your search. Most database programs enable users to limit their searches for journal articles by language, paper type (e.g., review papers, research papers, and editorials), publication dates, and other features. At the start of a writing project, to gain a comprehensive and current overview of your topic, you might limit your search to review papers published within the last 5 or 10 years. Later, when you are looking for specific study data to support your arguments, you might limit your search to recent research papers.

Step 3: Run your search. Before the Internet, searching for literature meant going to the library. As you may have heard from older folks, people had to walk many miles to reach these buildings, barefoot and in the snow. Of course, the journey was uphill—both ways! Now, as we sit comfortably at our computers, online databases display citation information and full-text articles within seconds after we enter search terms. Successful online searches, however, still demand a fair amount of time and effort. This is because database programs usually are not perfect in finding all relevant articles, especially in response to just one set of search terms. For example, when I entered "eating fish AND mercury AND health risks," PubMed returned 28 citations. But when I changed the phrase simply by removing the word "risks," the program returned 68 citations. Many of the additional 40 articles were pertinent. To help a database find all relevant articles on your topic, you may need to run several searches using different combinations of keywords. Many databases have services that enable users to review their search history and to save the results of separate searches.

Step 4: Evaluate your search results. In evaluating search results, first note the number of citations returned. When I searched PubMed using the terms "fish AND mercury," 2,235 citations were returned. A quick scan of their titles indicated that most of the articles did not focus on my specific topic—human health risks associated with eating mercury-contaminated fish. So I refined my search terms, adding phrases such as *health* and *fish consumption*. In response, PubMed returned a more manageable number of citations, in the neighborhood of 25–75 for different keyword combinations. If your searches return hundreds or thousands of citations, either your paper topic is too broad or you are using search terms that are too general. So this step may involve revising your search terms as necessary.

To determine whether the articles returned by a database program are relevant to your writing project, begin by examining their titles for whether they match your research issue. If you know the reputations of leading scientists and journals in the research field, note the authors and journal titles to identify high-quality articles. In addition, check the publication dates to find the most current literature. To make the final decision about whether an article is relevant to your writing project, carefully read its abstract.

Step 5: Select the articles you would like to obtain. Online database programs enable users to mark citations of interest and then save them in computer files, print them out, or send them (to yourself and others) by e-mail. These features are useful for compiling references lists and obtaining full-text articles that are not stored in the database.

Step 6: Check for full-text articles in the database you are searching. In the best-case scenario, the database you are searching will provide links for downloading free copies of all the articles that are relevant to your writing project. But when a database does not contain full-text articles for every citation of interest, you will need to complete your search elsewhere. Steps 7–10 are options for doing so.

Step 7: Check for downloadable full-text articles in other databases. Suppose that in a search of the database PsycINFO, you find a citation and abstract for an article that fits your needs perfectly. But for some odd reason, PsycINFO does not provide a full-text copy of the article itself. In this scenario, you might try searching for the article in other databases, such as PubMed or Google Scholar. Rather than beginning your search from scratch by entering keywords, just enter the citation information into the search programs of the alternate databases.

Step 8: Check for full-text articles on journal Web sites. The Web sites of some scientific journals provide free access to articles that are not immediately available in comprehensive databases. From the citation information of the articles you need, identify the journal titles and then enter them into an Internet search program. After navigating to the home page of a journal's Web site, look for links to archives of articles or use the journal's search program to track down the article you want.

Step 9: Check for full-text articles in your campus library. Your campus library might subscribe to the journals containing articles that cannot be downloaded for free from databases. If the library subscribes to electronic versions of the journals you need, its Web site or computer-based catalog will provide links to them. If the library subscribes only to printed versions of the journals, its Web site or computer-based catalog will provide information, such as call numbers, to indicate where they are stored in the physical building. If you are not accustomed to searching for printed copies of journal articles in a library's stacks, ask a librarian for help.

Step 10: Request articles through interlibrary loan. This is the last resort to acquiring free full-text articles. Your campus library likely offers a helpful interlibrary loan service for ordering books and journal articles that it does not hold. Don't delay making your requests, because the delivery for some materials may take up to several weeks.

Use the Reference Lists of Journal Articles on Your Research Issue

When queried with the appropriate search terms, database programs return complete lists of relevant journal articles. But, as mentioned earlier, the programs are not always perfect. An effective complementary search strategy is to use the reference lists of recently published articles. Let's say that you have found a recently published review paper that precisely targets your research issue. If the paper was written by a leading scientist who knows the research field thoroughly, its reference list will include citations for most, if not all, of the pertinent journal articles that have been published on the issue. Perusing the reference list, you might identify citations that your database searches did not uncover. You can then seek full-text copies of the associated articles by entering their citation information into the search programs of databases, journal Web sites, or your campus library's catalog.

Searching for Scientific Books

Most scientific books fall into one of the following four categories.

1. Popular books, which cover relatively broad topics of general interest to society; are authored by scientists, science journalists, or practitioners in fields that rely on science; and are primarily intended for the lay public.

2. Edited volumes of research papers and review papers, which typically focus on narrowly defined research issues, are authored by scientists and edited by leading researchers in their fields, and are intended primarily for scientific audiences. Some edited volumes are exactly like journals because they are published periodically, often annually, in numbered series. Other edited volumes are published one time only.

3. Monographs, which cover focused topics, are authored by scientists, and are primarily intended for scientific audiences. They are usually organized by chapters that cover subtopics of an overall topic or research issue.

4. Textbooks, which cover entire scientific disciplines, are authored by leading researchers or science educators, and are intended primarily for students.

As sources of knowledge and research data, scientific books have strengths as well as weaknesses. They can be excellent resources for acquiring basic content knowledge and comprehensive insights into research fields. However, because they tend to cover broadly defined subjects, scientific books can be limited in depth and detail. For example, in describing seminal research studies, textbooks typically gloss over key details about the procedures, results, and statistical analyses. These details might be necessary to develop convincing arguments in your own papers. This is another situation in which it's best to use secondary and tertiary sources to learn the basic science that defines a topic and then to seek primary sources for verification and details.

The credibility of scientific books depends partly on the qualifications of their authors. Popular books, for example, may be written by scientists, journalists with a background in science, or practitioners in fields that rely on science. Many books authored by science journalists and practitioners accurately convey complex knowledge in understandable and practical ways. But the credibility of books authored by nonscientists must be judged by how accurately they communicate original research and consensus knowledge in the scientific community. If, for example, the authors of a popular book challenge views that are widely accepted in science, their arguments must still be judged by scientific standards.

The best places for student writers to begin searching for scientific books are the computer-based catalogs of campus libraries. In addition to its catalog of holdings, your library might also subscribe to databases of books stored in libraries around the world. The largest of these databases is called *Worldcat* (http://www.worldcat.org/). Beyond its many millions of book citations, Worldcat stores information about dissertations, videos, and CDs. Of course, another great place to search for scientific books is the World Wide Web. You might try Google's book-searching program (http://books.google.com). Another option is to browse online book stores.

Some scientific organizations publish high-quality, full-text books that can be accessed through the Internet for free. For example, the National Center for Biotechnology Information (NCBI) has an online "Bookshelf" program that is available through PubMed's Web site. In 2008, the NCBI Bookshelf contained

more than 80 books on life science topics including cell biology, health care, ecology, endocrinology, and genetics.

Searching for Scientific Literature on Web Sites

The World Wide Web has developed into a storehouse for truly vital scientific information. Maps of the human genome can be accessed through the Web site of The National Human Genome Research Institute (http://www.genome.gov/). The latest results from clinical research on human diseases can be downloaded from the Web site of the National Institutes of Health (http://www.nih.gov/). These organizations, and others like them, also provide high-quality and easily accessible educational resources on scientific topics. In additional, highly credible scientific information is contained on the Web sites of educational institutions, research centers, hospitals and other clinical settings, government agencies, professional and nonprofit scientific organizations, and science-based businesses.

More than any other medium for scientific literature, however, Web sites demand critical evaluation. The credibility and usefulness of scientific literature on the Web truly runs the gamut. In searching for Web sites on your topic, within a few mouse clicks you will come across content that is inaccurate and misleading (if not completely fabricated) and content that reflects the most progressive science in the world. You will not have to search long before coming across a site with just the information you need. But after searching a little more, you will find other sites that present the information differently. Statistics will not be exactly consistent. Explanations of concepts will vary. Which sites will you trust? A general guideline for searching Web sites for scientific information is to steer toward .edu, .gov, and .org sites and away from .com sites. But there are certainly exceptions. If you do come across apparently useful information on a commercial site, evaluate whether its details and perspectives serve anything more than the company's financial interests. Look for whether the .com site acknowledges alternative perspectives, especially ones held by established scientists. The same sort of critical analysis applies to .edu, .gov, and .org sites. An especially useful search strategy is to ask professors and teaching assistants to point you to Web sites of reputable scientific organizations, research centers, clinics, and government agencies.

READING TO LEARN SCIENCE

A fundamental challenge in scientific writing is identifying appropriate content and developing it in ways that advance knowledge, offer new insights, and make convincing arguments. Reaching these supreme goals requires a deep understanding of paper topics, which you may learn through science courses, special lectures and seminars on your campus, or slide and video presentations on educational Web sites. For students as well as career scientists, however, the primary activity for learning the science that defines a writing project is reading published

literature. This section of the chapter presents strategies for reading to learn science in the early stages of the writing process.

Solving Comprehension Problems

For several reasons, students may face challenges in reading literature that contains the most useful information for scientific writing projects. Published resources that teach the basics—science textbooks, for example—are often *too* basic. But sources of more advanced knowledge, such as peer-reviewed journal articles, are often written for practicing scientists who are already topic experts and have considerable experience in reading these specialized forms of communication. The following advice is for helping student writers solve problems that can limit comprehension of published literature that is intended for advanced scientific audiences.

1. You're just not sure what to focus on. Because many scientific articles are densely packed with technical details, all readers are challenged to figure out what *exactly* to focus on. A skilled approach requires thinking proactively about the information you need to extract from an article—that is, the information to build a foundation of topic knowledge and to begin developing ideas for your paper. The next section of this chapter presents strategies for focusing attention to guide this proactive reading.

2. Words and phrases are unfamiliar or confusing. Sometimes a complete understanding of a passage in a scientific article can hinge on a short phrase or even a single word. So it's always worth checking the definitions of unfamiliar and confusing terms. The Internet makes this a cinch. For definitions of nonscientific terms, refer to reputable online dictionaries. For scientific terms, consult a discipline-specific online dictionary, such as the examples listed in Table 1.6 (page 40).

3. Your grasp of a complex concept is shaky. An obligatory approach to understanding difficult passages in scientific articles is to peruse them repeatedly. Talk yourself through the complex material, and try explaining it to others, asking for feedback on anything they do not understand in your account. Another useful comprehension strategy is to draw out complex concepts and processes in flowcharts, diagrams, and maps.

4. You follow everything in the text, but the author has left out key information to complete your understanding. Due to length restrictions of published scientific articles, authors are not always able to fully develop their ideas, explanations, and arguments. In response to knowledge gaps in articles, experienced readers take detailed notes about what they still need to learn. Then, they seek out the appropriate resources for filling in the gaps.

5. Data presented in support of conclusions don't seem to add up. It's not uncommon for conclusions in scientific articles to be supported with complex data that, at least on the surface, are confusing. The mathematical and logical connections between

TABLE 1.6 **Examples of Life Science Dictionaries**

Dictionary name	Maintained by	Web site address
Biochemistry Dictionary	Dr. Jeff D. Cronk at Gonzaga University	http://guweb2.gonzaga.edu/faculty/cronk/biochem/dictionary.cfm?letter=front
Biology-Online	Biology-Online.org	http://www.biology-online.org/dictionary/Main_Page
Biotech Life Science Dictionary	Dr. Andrew Ellington at the University of Texas at Austin	http://biotech.icmb.utexas.edu/pages/dictionary.html
Dictionary of Cancer Terms	U.S. National Institutes of Health	http://www.cancer.gov/dictionary/
Diabetes Dictionary	U.S. National Institutes of Health	http://diabetes.niddk.nih.gov/dm/pubs/dictionary/index.htm
Dictionary of Genetic Terms	U.S. Department of Energy Office of Science	http://www.ornl.gov/sci/techresources/Human_Genome/publicat/primer2001/glossary.shtml
Medline Plus	U.S. National Library of Medicine	http://www.nlm.nih.gov/medlineplus/mplusdictionary.html
Mental Health Dictionary	U.S. Department of Health and Human Services	http://mentalhealth.samhsa.gov/resources/dictionary.aspx
Talking Glossary of Genetic Terms	U.S. National Institutes of Health	http://www.genome.gov/glossary.cfm

the data and conclusions are not immediately evident. In this situation, skilled readers do the math for themselves. They double-check the author's calculations, and they do their own calculations to evaluate whether the arguments are sound.

6. *The material is completely baffling.* If your reading experience elicits this response, count yourself among every conscientious reader who has, on occasion, struggled to understand scientific literature. Then, try to figure out *why* the material is completely baffling. Do you lack the necessary background knowledge? Or is the literature incomprehensible because it is poorly written? If you lack essential knowledge, seek appropriate resources to acquire it. If it's a matter of poor writing on the author's part, abandon the text and seek more intelligible literature on the topic.

Reading and Taking Notes on Published Research Papers

As I mentioned a moment ago, the comprehension challenges posed by published scientific documents are ideally met by a proactive approach to reading. Ahead of time, you must know what you are looking for, at least in a general sense. Skillful readers take a proactive approach by raising well-targeted questions, or what we will call *focusing questions*, to isolate essential information from the literature. Figure 1.5 presents focusing questions for reading published research papers, the documents that report original studies. We will concentrate this activity on research papers because they tend to be the most challenging type of scientific articles to comprehend, especially for students.

To demonstrate how to apply the focusing questions, I will refer to selected parts of a published research paper titled "Reversing Introduced Species Effects: Experimental Removal of Introduced Fish Leads to Rapid Recovery of a Declining Frog." Authored by biologist Vance Vredenburg of San Francisco State University, the article appeared in the May 18, 2004 issue of the journal *PNAS* (Vredenburg,

Introduction Section

1. What was the study's motivating research issue?
2. Why was the study worth conducting?
3. What was novel and unique about the study?
4. What hypotheses guided the study?
5. What were the specific purposes of the study?

Methods Section

6. What key characteristics describe the study's subjects and materials?
7. How was the study designed?
8. What were the study's main procedures and analyses?
9. What were the independent variables and how were they measured?
10. What were the dependent variables and how were they measured?

Results Section

11. What were the study's key data?
12. What were the outcomes of the statistical analyses that the researchers performed?

Discussion Section

13. What were the author's main conclusions and arguments for them?
14. How did the author relate the study's outcomes to those from previous studies on the issue?
15. What methodological limitations and strengths did the author address?
16. What suggestions for future research were offered?

FIGURE 1.5 Focusing questions for reading research papers.

2004). With the article's citation information, you can download a full-text copy from the PNAS Web site (http://www.pnas.org/). Vredenburg's study concerned the endangerment of an amphibian species called *Rana muscosa*, more commonly known as the mountain yellow-legged frog. In lakes of the Sierra Nevada Mountains in California, the population of *R. muscosa* has declined drastically over the last five decades. Vredenburg and his research team focused their study on the possible predatory influence of nonnative trout. For more than a century, lakes in the Sierra Nevada Mountains have been stocked with trout for recreational fishing. Applying painstaking research methods, Vredenburg's team removed all of the trout from five lakes in Kings Canyon National Park. Then, over the next 8 years, the scientists counted *R. muscosa* frogs and tadpoles in numerous stretches of the lakes' shorelines. They compared the numbers to those in neighboring lakes that still contained nonnative trout. Vredenburg reasoned that if the *R. muscosa* population increased in the lakes from which trout were removed, trout predation would be a verified cause of *R. muscosa* declines.

In the sections that follow, we will apply the focusing questions listed in Figure 1.5 to parts of Vredenburg's research paper. Given the activity at hand, which is reading to *learn* science, we will not be critiquing the study or trying to figure out exactly what to write about it. Instead, at this early stage in the writing process, our primary objective is to identify and note essential information for deeply understanding the topic's defining science. The activity naturally entails taking organized notes to enhance learning and to enable easy access to vital information later in the writing process. You might be accustomed to taking notes on scientific literature by marking important passages with a fluorescent highlighter or underling them with a pencil or pen. A potential problem with this approach is that the markings do not directly reflect the relevance of the content. Rereading articles that you have marked lavishly in bright fluorescent colors, perhaps you have experienced the frustration of forgetting *why* you thought that the highlighted passages were important in the first place. One solution is to make notes in the published paper's margins to indicate the significance. For example, next to content that reflects a study's hypotheses, you might simply note "hypotheses." But if you expect to write at length about a published article in your own paper, it's usually best to take fairly well-developed notes in your own words. Doing so will help you learn the material more deeply, get on the right track for drafting, and avoid inadvertently plagiarizing in your paper. (More detailed suggestions for how to take effective notes and avoid plagiarism are offered in Chapters 3 and 4.)

Focusing Questions for the Introduction Section

In a research paper's introduction section, the key information to focus on includes the motivating research issue, importance, novelty, hypotheses, and purposes of the study being reported.

1. What was the study's motivating research issue? Research issues, as defined earlier, are the questions and problems that inspire scientists to conduct studies and write about them. In introduction sections, authors commonly present their research issues in one of three forms: as direct questions that studies were intended to answer, as explanations of problems that studies were intended to solve, or as purposes that studies were intended to achieve. A study's motivating research issue serves as a cornerstone for understanding a research paper. In Vredenburg's introduction section, the second paragraph conveys the study's central issue. Take a few minutes to read the paragraph, which is presented in Figure 1.6. Your task is to note the specific questions and problems that inspired Vredenburg's study. For practice, try putting the study's motivating issue into your own words, which you can compare to my notes that follow.

Vredenburg explains several key problems that reflect his study's central research issue. The problems are directly revealed in the last three sentences

Predatory fishes are a major force structuring amphibian assemblages, particularly in permanent bodies of water because they can alter the distribution and abundance patterns of amphibians by extirpating local populations (21–23). Nonindigenous fishes have been extensively introduced into many naturally fishless areas on every continent except Antarctica (19), and permanent bodies of water in mountainous areas are often targets for introductions (20, 24). Although some salmonids such as Atlantic and Coho salmon (*Salmo salar* and *Oncorhynchus kisutch*) were introduced to establish commercial fisheries (25), others such as rainbow trout (*Oncorhynchus mykiss*), brown trout (*Salmo trutta*), and brook trout (*Salvelinus fontinalis*) were intended for recreational fishing (20, 24, 26). In the western U.S., an area experiencing severe amphibian declines, trout are reared in fish hatcheries and subsequently delivered by airplanes to remote regions, many designated as "wilderness" (19, 20). Thousands of lakes (> 7,000) are stocked with trout on a regular basis (20), and trout now occupy up to 95% of larger, deeper mountain lakes in the western U.S. (20). Salmonids, especially trout, are highly effective predators (27), readily establish self-sustaining populations, and successfully colonize new habitats (28). These predatory fish exert strong effects on aquatic food webs (26, 29, 30). Surveys have shown that, where introduced trout are present, amphibians are often absent, thus possibly implicating introduced trout in amphibian declines (12, 31, 32). Introduction of nonnative trout continues on a large geographic scale, yet few studies have attempted to directly assess the effect introduced trout predation may be having on threatened amphibian populations. In addition, with the exception of a single study (33), no evidence exists to suggest whether impacts of nonnative trout are reversible.

FIGURE 1.6 A paragraph that conveys a study's central research issue. From the introduction section to biologist Vance Vredenburg's (2004) research paper on the predatory effects of nonnative trout on *R. muscosa* frogs and tadpoles. Copyright National Academy of Sciences, U.S.A.

of the excerpted paragraph. In my notes, which follow, I have paraphrased the research issue in a set of questions. You may find that this method—that is, noting research issues as questions that scientists sought to answer in their studies—enhances your understanding and memory of this fundamental element of research papers.

RESEARCH ISSUE
1. *Has the introduction of nonnative trout into lakes in the Sierra Nevada Mountains directly caused the observed population declines of* R. muscosa *frogs?*
2. *If nonnative trout have indeed threatened* R. muscosa *populations through their predatory behaviors, is it possible to reverse the declines by removing the fish from the lakes into which they were stocked?*

These notes would be quite handy if I were actually planning to write a paper on population declines of amphibians such as *R. muscosa*. My paper, for example, might involve summarizing Vredenburg's study and comparing it to other related studies. So I would definitely need the noted information about Vredenburg's research issue.

2. Why was the study worth conducting? In introduction sections, scientists address the reasons for conducting their studies. In doing so, they seek to convince readers that the studies were truly important and therefore worth conducting. The motivating reasons may focus on the problems—in nature, society, and science—that studies were originally intended to solve.

By taking notes on the reasons that an author offers for conducting a study, you will gain a more enriched understanding of its underlying research issue. In addition, the notes may inspire you to develop original ideas about why your paper's topic is important.

3. What was novel and unique about the study? Of the many published research papers on your topic that you might read, each will be distinguished by its novel and unique features. Among other distinctive qualities, a given study might have focused on a subject population that was never investigated before. Or it might have implemented new technologies for collecting and analyzing data. Authors often summarize their studies' distinguishing features in the introduction sections to research papers. By noting what makes a published study novel and unique, you will better comprehend its place in the larger research field. Your notes will be especially useful if your paper involves comparing and contrasting previous studies.

4. What hypotheses guided the study? A scientific hypothesis is a statement that conveys

- specific outcomes, or results, that a researcher anticipated before planning and conducting a study;

- tentative explanations for observed outcomes of previous studies—explanations that a researcher sought to test through conducting his or her current study; or
- tentative explanations for a researcher's anticipated results in a current study.

In research papers that report hypothesis-driven studies, hypotheses are presented in the introduction section. Like research issues, hypotheses are noteworthy cornerstones for understanding much of the content in published research papers.

5. What were the study's specific purposes? Toward the end of introduction sections, authors commonly present purpose statements, or the detailed aims of their studies. Notes about a study's specific purposes will serve as a checklist of sorts, guiding your reading of the sections that follow the introduction. After the introduction section, for example, the methods section describes the procedures used to achieve a study's specific purposes. In addition, the discussion section communicates whether, and to what extent, the purposes were indeed achieved.

Focusing Questions for Methods Sections

Methods sections are usually the most dense and technically detailed parts of research papers. The following questions focus attention on the defining features of a study's methods, which include characteristics of the subjects studied, the main variables that were manipulated and measured, and the procedures for collecting and analyzing data.

6. What key characteristics describe the study's subjects and materials? Methods sections usually begin with descriptions of the subjects and materials studied. The characteristics worth noting are ones that relate to your paper's topic. Suppose, for example, that you are writing a review paper on trout predation and declining frog populations. In a published research paper on this topic, the results indicate that trout consume recently hatched tadpoles but not tadpole eggs. It would be worth noting the specific species of trout that were studied, because the predatory behaviors observed might be unique to that species. Perhaps other species of trout eat tadpole eggs rather than hatched tadpoles. Or maybe some trout species are not predators of tadpoles at all. By noting relevant characteristics of a study's subjects and materials, you will avoid making hasty and sweeping generalizations in your own paper. You will also gain a deeper understanding of the larger research field by comparing study outcomes according to key characteristics of the subjects and materials investigated.

7. How was the study designed? In scientific research, the term *design* refers to the overall organization and implementation of study methods. In general, studies have either an *experimental* design or an *observational* design. In an experimental design, scientists administer treatments to subjects or manipulate their environments and tasks. In an observational design, subjects' natural behaviors and conditions are observed and recorded without experimenters' direct interventions.

As presented in Figure 1.7, an excerpt from the methods section of Vredenburg's paper conveys information about the study's design features.

The first subsection of Vredenburg's methods section, headed "Premanipulation Frog and Fish Distribution," describes an observational component to the study. In 1996, the author and his assistants observed *R. muscosa* frogs and tadpoles in their existing conditions by counting their numbers in lakes that contained introduced trout and in lakes in the test region that had never been stocked. The second subsection in Figure 1.7, headed "Trout Removal Experiment," describes an experimental component of Vredenburg's study. Beginning in 1997, the research team removed nonnative trout from five lakes. This was the experimental manipulation. At regular intervals through 2003, *R. muscosa* frogs and tadpoles were counted on the shorelines of

(1) the five lakes from which nonnative trout were removed, called the *trout removal lakes*;

(2) eight lakes that still contained nonnative trout, called the *fish control lakes*; and

(3) eight lakes that had never been stocked with trout, called the *fishless control lakes*.

Methods

Premanipulation Frog and Fish Distribution. The Sixty Lake Basin (36.8186° north, 118.4251° west; 3,000- to 3,500-m elevation) Kings Canyon National Park, CA was selected as the study area because large populations of frogs were suspected to be in the area (unpublished data). In 1996, 1 yr before experiments began, 50 lakes and ponds in the study area (Fig. 1) were surveyed for *R. muscosa* and introduced trout. Standard visual encounter surveys (45) along shorelines were conducted for postmetamorphic *R. muscosa* (adults plus juveniles) and tadpoles (all size classes combined) during the warmest time of day, between 1000 and 1500 hours. The distribution of introduced trout was determined by using visual surveys for ephemeral water bodies and by using sinking monofilament gill nets in all permanent lakes (>1.5-m depth) (12). A single hand-deployed gill net was set for 8–24 h in each lake and was set perpendicular to shore. I compared the mean number of postmetamorphic frogs and tadpoles per 10 m of shoreline in lakes with introduced trout vs. in fishless lakes using a one-way ANOVA (46).

Trout Removal Experiment. To test whether introduced trout limit the size and distribution of R. muscosa populations, trout were removed from five lakes by using 35 hand-deployed gill nets (47). Removals began July 20, 1997, July 15, 1998, and August 15, 1999, in lakes 1, 2, and 3, respectively (Fig. 1). In August 2001, the National Park Service began removing trout from lakes 4 and 5 (Fig. 1).... To

FIGURE 1.7 Key information from Vredenburg's (2004) methods section. Copyright National Academy of Sciences, U.S.A.

assess the consequences of fish removal on frog populations, I conducted counts of *R. muscosa* in the trout removal lakes ($n = 5$) and in a subset of fish-containing lakes ($n = 8$; "fish controls") and fishless lakes ($n = 8$; "fishless controls") from 1997 to 2003 (Fig. 1). Counts were conducted in the 21 lakes approximately every 2 wk from 1997 to 2001, twice per summer in 2002, and three times in 2003 (see Fig. 3). For the statistical analyses, in fish removal lakes, multiple counts from each lake were averaged, and one value was used for each year for each lake whereas, in the fish control and fishless control lakes, an average of all years was used for each lake in the comparison.... I compared the mean density of postmetamorphic frogs and tadpoles in experimental trout-removal lakes ($n = 5$) vs. in fish controls lakes ($n = 8$) and fishless controls ($n = 8$) by using three statistical tests. First, to ensure that the five lakes I nonrandomly selected for removal were not different from fish control lakes in their premanipulation condition, I used a *t* test to compare the density of postmetamorphic frogs and tadpoles between these two lake types 1 yr before fish removal. Second, to determine whether fish removal led to an increase in frog populations, I compared the density of postmetamorphic frogs and tadpoles in the fish-removal lakes vs. in fish control lakes 1 yr after fish eradication began (*t* test). Finally, to determine whether frog populations in fish removal lakes reached the levels seen in fishless control lakes, I compared the density of postmetamorphic frogs and tadpoles in the fish-removal lakes vs. fishless control lakes (*t* test) 3 yr after eradication began....

FIGURE 1.7 *continued.*

This experimental design enabled Vredenburg to determine whether *R. muscosa* counts would increase in the trout removal lakes, which would implicate nonnative trout in causing the population declines that inspired the study.

A study's design can significantly influence the validity of its results and conclusions. So it's always worth noting design features. Such notes guide writers in evaluating strengths and weaknesses in published study methods, which is an essential activity for generating effective content for scientific papers. This activity is described and demonstrated in Chapter 3.

8. *What were the study's main procedures and analyses?* By noting the general design features of a study, you gain a bird's-eye view of how it was conducted. To fill in the details, you must narrow your focus on the study's more specific methods, including the following.

- The detailed procedures by which subjects were assigned to groups, conditions, or treatments.

- The instruments and techniques that were used to collect and analyze data.
- Key events in the timeline over which treatments were administered and data were collected and analyzed.
- The detailed statistical analyses that were performed.

Through a concerted effort to comprehend these details, you will more easily grasp the study's outcomes because you will understand how they were actually derived.

9. *What were the independent variables and how were they measured?* In research, measurable or classifiable quantities and characteristics are called *variables*. They fall into two general categories: *independent variables* and *dependent variables*. In studies with experimental designs, independent variables are the factors that scientists intentionally manipulate. For example, in biomedical experiments, common independent variables are doses of drug treatments. (Independent variables are also called *treatment* variables.) In studies with observational designs, independent variables are designated categories or conditions that may influence or be associated with the outcomes of interest.

One aim of Vredenburg's observational study was to determine whether there were fewer *R. muscosa* in lakes containing nonnative trout versus lakes that had never been stocked. So the independent variable was *trout presence*. As described in the first subsection of Vredenburg's methods section (Figure 1.7), the independent variable was measured "by using visual surveys for ephemeral water bodies and by using sinking monofilament gill nets in all permanent lakes (>1.5-m depth) (12)." Unless you are an experienced researcher in the field, you probably do not fully understand this method from its brief description. I raise this point to illustrate a common comprehension barrier in reading methods sections. As mentioned earlier, scientific journals restrict the length of published research papers. Often, authors simply do not have sufficient space to detail every single step in their procedures and analyses. The reader's solution is to seek additional resources for more thorough explanations. In the excerpted sentence above, Vredenburg guides his readers to this extra help by citing a published article that elaborates on the visual survey method for determining trout presence. If you need to know precise details about the method, you would obtain article #12 as cited in the paper's reference list.

In the experimental component of Vredenburg's study, the primary aim was to determine the effects of eliminating trout on *R. muscosa* numbers. As described in the second subsection of the methods section (Figure 1.7), the research team manipulated the independent variable, trout presence, by removing the fish from five experimental lakes.

10. *What were the dependent variables and how were they measured?* Dependent variables are the targeted outcomes of research, or the results that are applied to resolving research issues and testing hypotheses. The main dependent variable in Vredenburg's study was the density of the *R. muscosa* population in the different groups of lakes. The research team measured this variable by counting *R.*

muscosa frogs and tadpoles in numerous 10-m stretches of the study lakes' shore-lines. Then they calculated the mean values, or the arithmetic averages. Using statistical tests, Vredenburg determined whether *R. muscosa* density differed in the fish removal, fish control, and fishless control lakes.

The precision, validity, and reliability of study results and conclusions depend on the methods used to measure dependent variables. So it's especially important to note these methods before critically evaluating study outcomes.

Focusing Questions for Results Sections

The results sections of research papers present study data and the outcomes of statistical tests. The following questions focus attention on these elements.

11. What were the study's key data? Results sections are often packed with data, but not all of it is essential to note and remember. To avoid getting bogged down by inessential details, isolate the most relevant data—in other words, the results that contribute directly to resolving a study's research issue and testing its hypotheses. For an example, let's recall Vredenburg's research issue, which concerned whether (a) nonnative trout have directly caused the drastic decline of *R. muscosa* popu-lations and, if so, (b) whether the decline could be reversed by removing trout from the Sierra Nevada Mountain lakes under study. The most relevant results for resolving this issue were the counts of *R. muscosa* frogs and tadpoles in the differ-ent groups of lakes over the 8-year study. Vredenburg presented the correspond-ing data in two line graphs, which are reproduced in Figure 1.8 (page 50).

The top graph (Graph A) illustrates the presence of *R. muscosa* frogs in the different groups of lakes between 1996 and 2003. The plotted data indicate frog density, measured as the mean number of frogs per numerous 10-m stretches of shoreline. Toward the bottom of Graph A, notice the five numbered lines marked by filled triangles; these lines designate frog density in the five fish removal lakes. On the graph's horizontal axis below the yearly dates, the numbered lines indicate the periods over which trout were removed from each lake. The results demonstrate a noteworthy pattern: Soon after trout were removed from a given lake, the frog population increased, as indicated by the upward slope of the plot lines. Compare these lines (the ones marked by the filled triangles) with the line marked by shaded squares at the bottom of Graph A. This line designates the average density of frogs in the fish control lakes, the ones from which trout were not removed. In these lakes, the number of frogs remained close to zero over the entire study. Toward the top of the graph, the line marked by shaded circles designates the average density of frogs in the fishless control lakes, the ones that never contained trout. Their populations remained high throughout the study. The bottom graph in Figure 1.8 (Graph B) provides the same information for tadpoles, or larval *R. muscosa*.

From the apparent increases in frog density in the fish removal lakes, as rep-resented in Graph A of Figure 1.8, we might jump to the conclusion that non-native trout predation has indeed been responsible for population declines of *R. muscosa* in the Sierra Nevada. We might also conclude that *R. muscosa*

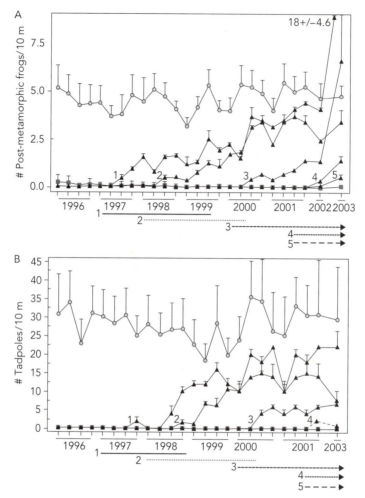

Fig. 3. Density (mean± SE) of postmetamorphic *R. muscosa (A)* and larval *R. muscosa (B)* in 21 lakes from 1996 to 2003. Filled triangles designate fish removal lakes (*n* = 5); numbers correspond to lake numbers in Fig. 1. Shaded circles are fishless control lakes (*n* = 8), and shaded squares are fish control lakes (*n* = 8). Horizontal lines at the bottom of each figure indicate the trout removal period for each of the removal lakes, and numbers correspond to individual lake numbers. No tadpole counts were conducted in 2002.

FIGURE 1.8 The results from (figure and caption) Vredenburg's (2004) trout removal experiment. Copyright National Academy of Sciences, U.S.A.

populations can recover completely if nonnative trout are eliminated. This argument would seem to be supported by the extent to which frog density increased. In Figure 1.8, notice that the plot lines for the five fish removal lakes approach, and in some cases extend beyond, the line indicating frog density in the fishless control lakes. Our conclusions, however, would be a bit hasty. As we consider in the next focusing question, final conclusions should not be based only on visible trends in graphs; in addition, they should not be hurriedly drawn solely from apparent differences in group means. The most valid conclusions depend on interpreting statistical analyses of study data.

12. What were the outcomes of the statistical analyses that the researchers performed? As mentioned earlier, Vredenburg's research team performed an observational "premanipulation" study before conducting their trout removal experiment. The baseline study involved counting *R. muscosa* in the five lakes that were scheduled for trout removal and the eight fish control lakes. The outcomes of this baseline analysis were reported in the paper's results section as follows.

> In the premanipulation comparison, the number of postmetamorphic frogs in lakes 1–5 did not differ from fish controls [0.048 ± 0.037 ($n = 5$) and 0.09 ± 0.03 ($n = 8$) (mean ± SE) postmetamorphic frogs per 10 m in removal lakes and fish control lakes, respectively; $p = 0.39$, t test].

The sentence is written in the language of statistics, the branch of mathematics that scientists apply in collecting, organizing, and analyzing data. Here and in Chapter 3, we will cover the basics for understanding research statistics in order to write about them effectively.

The take-home message from Vredenburg's analysis is that before the trout removal experiment there was no statistical difference between the number of frogs in the fish removal lakes (lakes 1 through 5, from which trout would be taken later) and the fish control lakes (in which trout would remain). This baseline analysis is quite important. To understand why, imagine a scenario in which the number of frogs differed in the scheduled fish removal lakes versus the fish control lakes before the experiment began. For some reason that had nothing to do with trout presence, suppose that many fewer frogs lived in the lakes scheduled for trout removal. Let's say that the researchers overlooked this difference because they failed to perform a baseline analysis of frog density. After the trout were removed, the researchers found no difference in the number of frogs in the fish removal lakes and the fish control lakes. This result would support the conclusion that nonnative trout have *not* been responsible for *R. muscosa* deaths. If the trout were to blame, we would expect the number of frogs to have been significantly *greater* in the trout-removed lakes. In other words, the frog density should have increased. Consider, however, that in this hypothetical scenario frog density actually *did* increase. Before the experiment began there were fewer frogs in the lakes scheduled for trout removal. So the conclusion that absolves nonnative trout would actually be flawed. As explained next, Vredenburg's results indicate that this problematic scenario did not occur in the study.

Let's look more closely at the excerpted text reporting Vredenburg's baseline analysis. Two key statistics to focus on are the values **0.048** and **0.09**. These are the mean numbers of frogs per 10-m stretches of shoreline from the fish removal and the fish control lakes, respectively, before trout removal. The numbers are obviously very small, reflecting the population declines that initially inspired the study. We can surmise that in many of the 10-m stretches of shoreline no frogs were found at all. Also notice that the means appear, at least on the surface, to be very different. In fact, one is nearly twice the other. By intuition, we might conclude that a two-fold difference is significant. This poses an interesting question: Why, in reporting the baseline analysis, did Vredenburg conclude that the mean number of frogs did *not* differ in the two groups of lakes? On the way to answering this question, I will remind you that scientific conclusions should not be based only on eyeballing data trends and comparing group means intuitively. Instead, the outcomes of statistical analyses must be considered as well.

In the analysis of Vredenburg's premanipulation study data, the key outcome was a statistic called a *P* value. This statistic indicates the probability of obtaining study results due to chance factors rather than true effects of experimental treatments or conditions. There are many different types of statistical analyses that yield *P* values. For determining whether the differences observed between the means of two groups are statistically significant, a commonly used analysis is a *t* test. As indicated in the preceding excerpt from Vredenburg's results section, a *t* test was used to analyze the baseline study data.

The difference between Vredenburg's baseline means for the fish removal lakes and fish control lakes was 0.042 frogs per 10-m stretch of shoreline. As reported, the *t* test to determine whether this difference was statistically significant produced a *P* value of 0.39. A somewhat simplified but accurate interpretation of this result is as follows: If the baseline study were repeated numerous times on all of the lakes in the Sierra Nevada region that Vredenburg studied, a mean difference of 0.042 frogs per 10-m stretch of shoreline would be obtained 39 times out of 100 due to chance factors. This frequency is fairly high. If chance factors accounted for a difference of 0.042 so often, the observed result likely does not reflect a true difference. Vredenburg thus concluded that the observed difference in frog density was not statistically significant at baseline. In this case, the lack of a significant difference supports the view that Vredenburg's study began on a sound foundation. Indeed, if the numbers of *R. muscosa* in the scheduled fish removal lakes and the fish control lakes differed significantly before trout were removed, the study's outcomes would have been biased.

Note that the phrase *statistically significant* has a highly specialized meaning in scientific research. It refers to the probability that the outcomes observed in a given study are reliable, or that they would be reproduced if the study were conducted on many different samples of the population of interest. A sample is a subset of a larger population, which consists of all individuals or objects that share the characteristics under investigation. Life scientists usually study samples, such as relatively small groups of people with a certain disease rather than everyone in the world who has it. For economic and practical reasons, it's often impossible to study entire populations. The statistical analyses of data collected from a sample are used to make inferences, or reach conclusions, about the larger population. This vital concept is called *statistical inference*.

Now you know the basis for Vredenburg's conclusion that there was no difference between the baseline means for frog density in the scheduled fish removal lakes (0.048 frogs per 10-m stretch of shoreline) and the fish control lakes (0.09 frogs per 10-m stretch of shoreline). Of course, you might justifiably question the conclusion because the mean for the fish control lakes is still nearly two times greater than that for the fish removal lakes. Consider, however, that statistical significance is not determined solely by comparing the means of sample groups. Other properties of data must be included in statistical analyses, specifically properties indicating the extent to which study outcomes accurately represent the entire population of interest. One of these properties is called *variability*, which refers to the spread, or distribution, of study data. The simplest measure of variability to calculate is the range of data values in a study group. Another measure, which is commonly used in life science research, is the *standard deviation*. This more complex statistic is calculated by taking the average deviation, or difference, between the individual data values for each member of a group and the group's mean value.

Note this important point about measures of variability: As they increase, statistical analyses are less likely to reveal significant treatment effects and differences between groups. To grasp this relationship, suppose that we have conducted a study in which one subject group received an experimental treatment and the other, a control group, did not. For a key outcome of interest, we calculated each group's means and ranges of data values. Consider a scenario in which the two groups' means are vastly different and the ranges of data values are narrow. They are so narrow that the individual values for each group do not overlap at all. In this case, the likelihood that the means are truly different would be relatively great, supporting a conclusion that the outcomes were due to a systematic effect of the treatment. Now consider another scenario in which the two group means are still far apart but the ranges of each group's individual data values are very wide. Suppose that the two groups' data values overlap considerably. Given the great amount of variability, it's less likely that the group means are truly different and therefore that the treatment had any systematic effect.

Despite the apparently large difference between the group means for frog density in Vredenburg's baseline study, the author concluded no significant difference. Given the relationship between statistical significance and variability, we might presume that frog density varied considerably in the sample lakes. To find out, let's take another look at the sentence reporting the baseline results.

> In the premanipulation comparison, the number of postmetamorphic frogs in lakes 1–5 did not differ from fish controls [0.048 ± 0.037 ($n = 5$) and 0.09 ± 0.03 ($n = 8$) (mean $\pm$ SE) postmetamorphic frogs per 10 m in removal lakes and fish control lakes, respectively; $P = 0.39$, t test].

Following the mean values, the statistics $\pm \mathbf{0.037}$ and $\pm \mathbf{0.03}$ are the variability measures of frog density in the fish removal and fish control lakes, respectively. These numbers represent a variability statistic called the *standard error of the mean* (SE). Both the definition and calculation of SE are fairly complicated. For our purposes, however, you mainly need to know that if a group's SE is large in relation to its mean, the data are highly variable. This is the case for the SE statistics in Vredenburg's baseline analysis. For example, the SE for the fish removal lakes (±0.037) is nearly as great as the mean value (0.048). The large variability of the data thus contributes to explaining why the apparent difference in mean frog density is not statistically significant.

Another key factor influencing the outcomes of statistical analyses is the number of subjects (or objects, such as lakes) in a study. This number, called the *sample size*, is designated as *n*. In statistical analyses of differences between groups, the likelihood of obtaining significant results is greater as sample size increases.

Recall an important point about the activity at hand: We are focusing on reading to *learn* science. This early stage of the writing process does not involve critically evaluating authors' interpretations of their study results. In addition, you need not yet be concerned with developing your own arguments. These more advanced activities depend on a sound fundamental grasp of what you are reading. Ultimately, however, you are not obliged to accept an author's interpretations. Someone might, for example, raise the criticism that Vredenburg's conclusion is problematic because the variability of frog density was so great. Large variability measures sometimes reflect errors in data collection. In Chapter 3, you will learn how to interpret research data independently and, in the process, construct strong original arguments.

Now let's turn to the results of Vredenburg's trout removal experiment, which the author reported as follows.

> The number of postmetamorphic *R. muscosa* in removal lakes 1 yr after fish removal began was significantly greater than in fish control lakes [Fig. 3A; 0.974 ± 0.12 (*n* = 5) and 0.0811 ± 0.09 (*n* = 8) (mean ± SE) postmetamorphic frogs per 10 m...respectively; *P* < 0.0001, *t* test].

The take-home message is that 1 year into the experiment, the mean number of frogs per 10-m stretch of shoreline was significantly greater in the fish removal lakes (0.974) versus the fish control lakes (0.0811). This interpretation is based on the *P* value for the *t* test used to analyze group differences: $P < 0.0001$. This statistic indicates that if the trout removal experiment were repeated on different samples of the larger population of lakes, the observed difference would occur less than 1 in 10,000 times by chance. Given the extremely low probability that chance factors would have influenced the outcome, we can say that the 1-year results are highly reliable. The *P* value would support the conclusion that the increased *R. muscosa* density in the fish removal lakes, as illustrated in Figure 1.8, was very likely due to the effects of eliminating nonnative trout.

Interpretations of statistical significance do not address the vital matter of *practical* significance. This term refers to the real-world implications of study outcomes. For the 1-year results of Vredenburg's trout removal experiment, an indicator of the practical significance is the change in frog density in the fish removal lakes. From baseline to 1 year, the mean values increased from 0.048 to 0.974 frogs per 10-m stretch of shoreline. This is nearly a 20-fold increase, which might indeed be interpreted as a practically significant restoration of the *R. muscosa* population.

The following excerpt reports the 3-year results of Vredenburg's trout removal experiment. In this case, the data represent frog density in the fish removal lakes versus the fishless control lakes, which had never been stocked with trout and had high numbers of frogs at baseline.

> Three years after removals began, there was no significant difference in postmetamorphic frog counts when comparing removal lakes and fishless control lakes [Fig. 3A; 6.85 ± 1.46 ($n = 3$) and 4.73 ± 0.89 ($n = 8$) (mean $\pm$ SE) postmetamorphic frogs...respectively; $P = 0.24$, t test].

If nonnative trout have indeed been responsible for *R. muscosa* population declines, we would expect that fish removal would lead to an increase in *R. muscosa* numbers. Eventually, the numbers should match those in the lakes that never contained nonnative trout. In other words, we would expect that frog density would not differ significantly in fish removal lakes versus fishless control lakes. This expectation is confirmed by Vredenburg's results. The means for frog density (6.85 and 4.73, respectively) may look different on the surface; however, the statistical analysis supports the interpretation that they do not differ reliably. As indicated in the excerpted text, the probability of detecting the observed difference by chance ($P = 0.24$) was too great for the researcher to conclude that the means were significantly different.

It's certainly worth taking detailed notes about study data and the outcomes of statistical analyses as reported in the results section of research papers. As we consider next, this information is essential for understanding authors' conclusions and arguments in the scientific literature. In addition, the notes will serve as the basis for developing your own conclusions and arguments.

Focusing Questions for Discussion Sections

When you are reading the discussion sections of research papers, your attention should focus on authors' conclusions and the evidence and reasoning presented to support them. In addition, set your sights on learning how the outcomes of studies fit into the larger context of their research fields.

13. What were the author's main conclusions and arguments for them? A discussion section's most fundamental contents are the author's conclusions. These are answers to the questions, or solutions to the problems, that originally inspired reported studies. Well-written discussion sections present conclusions directly, usually in the first few paragraphs or in the last paragraph. To isolate an author's conclusions, first reflect back on the study's specific motivating questions and problems. As an example, consider the two core problems that defined Vredenburg's research issue. The first focused on whether nonnative trout were directly responsible for observed declines in *R. muscosa* populations in the Sierra Nevada Mountain lakes. The second problem involved uncertainty about whether the declines could be reversed if trout were removed from the lakes. In the last paragraph of Vredenburg's discussion section, an excerpt of which is reproduced in Figure 1.9, the author presents a conclusion that directly resolves the two motivating problems.

Notice that Vredenburg's conclusion addresses the effects of nonnative trout on "populations of montane amphibians." The author makes an inference about the larger population based on results obtained from his sample of *R. muscosa* in selected lakes of the Sierra Nevada Mountains.

In the following notes, I have summarized Vredenburg's conclusions along with the supporting results. You may find this note-taking format to be useful for understanding and remembering the take-home messages of published research papers.

Conclusion #1: Vredenburg's study supports the conclusion that nonnative trout play a major role in observed declines in amphibian populations in mountain lakes.

Supporting results for conclusion #1: In the lakes that had trout removed the numbers of R. muscosa frogs and tadpoles increased significantly after only 1 year. But the numbers did not increase in the lakes that did not have trout removed.

Conclusion #2: Over a relatively short time, amphibian populations can be returned to normal levels if trout predators are removed from their communities.

Supporting results for conclusion #2: 3 years after the trout were removed, the population of R. muscosa frogs and tadpoles was not statistically different in the fish removal lakes and the fishless control lakes.

The alarming decline and, in some cases, extinctions of amphibians in protected areas around the world probably do not have a single cause [55]. Understanding mechanisms responsible for declines is important because negative impacts may go unnoticed even though they may be widespread in protected and seemingly pristine areas. Ecologists consider predation to be a significant force shaping amphibian assemblages [2, 56–59] and introduced predators a major threat to worldwide biodiversity. The results of this study show that introduced trout can have dramatic effects on populations of montane amphibians but also demonstrate that amphibian populations have the ability to quickly recolonize habitat and establish large populations in a short period....

FIGURE 1.9 From the discussion section of Vredenburg's (2004) research paper, an excerpt that conveys the author's conclusion. Copyright National Academy of Sciences, U.S.A.

14. How did the author relate the study's outcomes to those from previous studies on the issue? Because the results and conclusions from a single study usually do not fully resolve a research issue, experienced authors place their studies into the larger context of their research fields. The comparisons, which come in the discussion section of research papers, are important to note. Your knowledge of a research field will be incomplete, and perhaps even inaccurate, if it is based only on the study that you are currently reading.

15. What methodological limitations and strengths did the author address? The validity of any study's results and conclusions directly depends on the quality of its methods. Due to many challenges of conducting research, few (if any) studies are methodologically flawless. Accounting for this fact, experienced authors use their discussion sections to acknowledge methodological limitations to their studies. When they believe that their methods are exceptionally sound, scientists write about the strengths. Because your writing project might involve critiquing the studies you are reading, it's important to keep track of authors' discussions of methodological limitations and strengths.

16. What suggestions for future research were offered? Many discussion sections end with suggestions for future research, which may be worth noting. If, for example, you are looking for hot research issues to write about, an author's suggestions for new studies may guide the way. In addition, your paper might require developing original ideas for future research. The suggestions that you read in published articles may spark your creative ideas.

SUMMING UP AND STEPPING AHEAD

This chapter began by introducing a concept known as the rhetorical situation, which comprises all the factors that define a writing project. For any given project, the rhetorical situation includes instructions and guidelines for preparing the manuscript, conventions for the objectives and structure of the paper type, analyses of key characteristics of intended readers, and knowledge about the topic and refined research issue on which the paper is based. This chapter's activities were intended to help you gain a great advantage as you begin to plan a scientific paper. The advantage is a comprehensive and detailed project definition, which will guide you in carrying out challenging activities throughout the writing process. In the next chapter, we will begin applying the products of this chapter's activities, specifically to develop another important aspect of the rhetorical situation: a set of powerful goals for what to do and say in your paper.

CHAPTER 2

Developing a Goal-Based Plan

The Heart of the Matter

To sustain life, our body's cells depend on energy derived from a constant supply of food nutrients and oxygen. In addition, the cells must eliminate the waste products of metabolism, such as carbon dioxide. The body's parts and processes for delivering energy and removing waste are collectively called the cardiovascular system. Its main components are the heart and the vasculature, which includes the arteries, capillaries, and veins. Diseases of the cardiovascular system are among the leading causes of disability and death in Western societies. The most common and deadly of these diseases, called atherosclerosis, is characterized by the accumulation of fatty plaques and blood clots in the arteries that supply blood to the heart muscle and the brain. The buildup of plaque occludes blood flow to these vital organs and, ultimately, to all other organs of the body. If left untreated, the blood clots that form in atherosclerosis can rupture and completely block nutrient and oxygen delivery. The fatal consequences are heart attacks and strokes.

Scientists who study the heart's functions are called cardiovascular physiologists. Some work at the molecular and cellular levels, seeking to understand the basic mechanisms by which the heart muscle contracts and pumps blood. Other researchers in the field study the underlying causes of diseases such as atherosclerosis, high blood pressure, and congestive heart failure. Their investigations have revealed the major risk factors for cardiovascular disease, which include obesity, smoking, high levels of cholesterol, and physical inactivity. In more applied research settings, cardiovascular physiologists conduct research on behavioral and pharmaceutical approaches to preventing and treating heart disease.

This chapter focuses on the heart of the scientific writing process: It's the stage we call *developing a goal-based plan*.

INTRODUCTION

In a nutshell, the defining characteristic of effective writing is that it *says* and *does* the best things. In other words, the success of any document depends on whether it accomplishes the most appropriate goals for its genre and if it ultimately influences readers as the writer intended. For the moment, imagine that the document you are writing isn't a scientific paper at all; instead, let's say that it's a letter of application for your dream job. Your ultimate goal is to convince readers, maybe your future boss and coworkers, to hire you. You will have to write a knockout letter, strong enough to open the door to an interview. The document will be successful if its content accomplishes appropriate communication goals for the genre at hand—that is, the job application letter. Here's one of these goals: **Convince readers that you are qualified for the job**. This goal will guide you in developing ideas about your specific qualifications, including your relevant educational background and previous work experience. Here's another appropriate goal for a job application letter: **Convince readers that you are highly motivated for the job**. To accomplish this one, you will write about the job's intriguing features and how they will inspire your enthusiastic and tireless work. Of course, you won't write about your favorite foods or your roommate's annoying habits, because the conventional goals for job application letters obviously do not include giving irrelevant personal information.

In all endeavors, well-conceived goals sharply focus our attention, fuel our motivation, and direct us in performing highly productive thinking. This is certainly true in the writing process. Skilled authors devote considerable time and effort to developing goals for their projects. Then they apply their goals to carry out various planning, drafting, and revising activities. The result is goal-directed writing, which is a true hallmark of expertise. In a sense, the goal-setting process is really part of defining a writing project, the stage that we covered in Chapter 1. But goal-based planning is so central to successful scientific writing that it warrants a chapter all to itself. You will understand why as this chapter and the rest of the book unfold.

At this stage of the writing process, the objective is to plan your paper by developing goals for its major sections. Our approach to this planning is based on an especially powerful type of goal, which writing instructors call *rhetorical goals*. As introduced in Chapter 1, the word *rhetorical* describes the ideal circumstances in which writers follow genre-specific conventions and effectively address concerns of their audiences. Just around the corner, I will describe the defining features of rhetorical goals in lots of detail. Until then, just to get their gist, have a look at the following few examples of rhetorical goals for scientific papers.

- For the introduction section to a research paper: State the specific purposes of your study, to give your audience advance information for enhancing readability and comprehension.
- For the body of a position paper: Acknowledge and refute viable arguments that oppose your own, to demonstrate that your position is not biased.

• For the conclusion section of an informative review paper: Suggest ideas for future studies to fill knowledge gaps in your research field.

This chapter is not intended to hand you a list of ready-made rhetorical goals for scientific communication. Instead, it will guide you through activities for developing your own goals and for integrating them into well-structured plans. In writing, as well as many other pursuits in life, the goals we develop *independently*, through our concerted efforts and skillful methods, are usually more meaningful and productive than the goals we read about in books and blindly follow. But, of course, it's always helpful to compare your own goals to well-established conventional goals. So to complement this chapter's lessons in independent goal setting, Chapter 8 presents comprehensive sets of rhetorical goals for the primary types of scientific papers.

ABOUT THE PROCESS

Once again, imagine that you are writing a job application letter. So far, you have developed two rhetorical goals: **convince readers that you are qualified** and **convince readers that you are motivated**. Take a minute to ponder *how* you would have actually devised these goals. In other words, what would you have done and thought about in order to conceive them? Certainly, you would have relied on previous experiences in writing job application letters. And you would have used good old-fashioned common sense, reasoning that readers would naturally ask about your qualifications and motivations. But suppose that you don't have much experience with the genre of job application letters. If you get stuck brainstorming more goals, what will you do and think about next? You might ask a career counselor to suggest appropriate writing goals. Or you might buy a how-to book on the genre, seeking excellent examples of application letters that reveal useful goals to adopt. In addition, by thinking more deeply about what your audience is looking for, you will confidently determine what else to do and say in your letter. So the activities for developing goals for job application letters include relying on your previous writing experience and common sense, consulting experts and published educational resources on writing in the genre, analyzing exemplary models, and considering concerns of your audience. These same activities also apply to creating goal-based plans for scientific papers. Our approach covers the eight activities that are illustrated in Figure 2.1 (page 63) and outlined as follows.

Activity 1: Set the framework for your goal-based plan. This first activity covers the preliminaries of goal-based planning, which involve understanding the defining features of rhetorical goals and establishing the structure for integrating them into a written plan.

Activity 2: Rely on your experience in scientific writing. If you have written and read a fair number of scientific papers, you are already familiar with some of their key rhetorical goals. This activity entails brainstorming goals for your current paper by recalling lessons learned through your previous experiences.

Activity 3: Adopt and adapt conventional guidelines. All books and Web sites on scientific writing present conventional guidelines for what to include in different types of papers. This activity is for converting these guidelines into much more powerful rhetorical goals.

Activity 4: Use model papers. As you might imagine, an especially productive approach to developing a goal-based plan is to examine model papers, specifically to identify the goals that their experienced authors set out to accomplish. This activity will guide you through deriving rhetorical goals from published scientific articles.

Activity 5: Apply your task and audience analyses. This activity relies on the products of two activities from Chapter 1—analyzing your writing task and analyzing your audiences. You will learn how to develop rhetorical goals by

- applying key assignment instructions, guidelines, and evaluation criteria for your papers (task analysis); and
- considering the needs, expectations, and values of your intended readers (audience analysis).

Activity 6: Use the helicopter thinking method. At its best, scientific writing requires a method of thinking that psychologists call metacognition. For our purposes, metacognition involves shifting fluidly, as necessary, between two levels of thought processes:

- a lower level that focuses on the concrete details of scientific information; and
- a higher level that focuses on more abstract rhetorical goals.

This method also goes by the less formal name of *helicopter thinking*. The up-and-down flight of a helicopter is a metaphor for the concrete-to-abstract and abstract-to-concrete thinking routines that expert authors apply throughout the writing process. In this activity, you will learn how to use helicopter thinking to develop goal-based plans.

Activity 7: Start drafting. Sometimes we don't know exactly what to do and say until we start doing and saying it. On these occasions, the best way to jump-start a goal-based plan is to jump directly to the drafting process, at least for the moment. Hints for doing so are presented in this activity.

Activity 8: Revise your goal-based plan. Once you have developed a goal-based plan, it's a good idea to double-check it in order to identify any areas for improvement. So this activity is for evaluating and, if necessary, rewriting parts of your plan for the most productive applications.

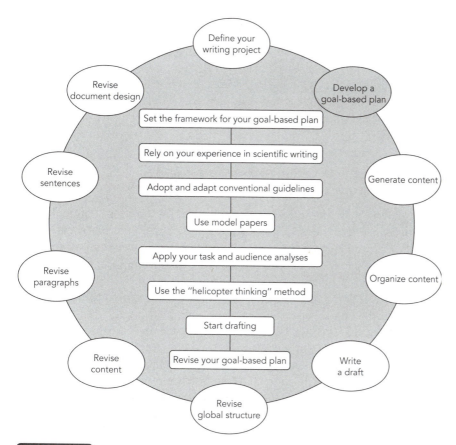

FIGURE 2.1 Process map for *developing a goal-based plan.*

Except for the first and last activities, which should be done in their set positions, the others need not follow any particular order. Perhaps you won't even need to do all eight activities for every writing project. For example, to develop a strong goal-based plan for a review paper, you might rely primarily on your experience—that is, if you know the key rhetorical goals for review papers from having written them before. But if you lack experience in the genre, you should try all of the activities for the best results.

In this relatively early stage of planning, the ideal modes of thinking are exploratory, generative, and unstructured. In other words, this is not the time to worry about creating a perfect linear outline. In Chapter 4, which covers the process of organizing content, you will learn how to convert goal-based plans into well-structured linear outlines. While it's best to concentrate on goal-based planning early in the writing process, don't expect to knock out a complete plan for your paper in just one sitting. In fact, if your project is extensive and complex, you will probably need to refine your plan throughout stages of planning, drafting, and revising.

SETTING THE FRAMEWORK FOR YOUR GOAL-BASED PLAN

The main elements of goal-based plans are the rhetorical goals that have been briefly introduced so far. This section of the chapter begins with more elaborate descriptions of rhetorical goals and their powerful features. Then we will build a framework for incorporating rhetorical goals into comprehensive plans for scientific papers.

Distinguishing between Just-Okay Goals and Powerful Rhetorical Goals

Well-conceived goals are arguably the most powerful tools for communicating successfully in science. But all goals for scientific papers are not equal in the power they afford. Suppose that for a biology class you have set a goal to make an *A* on a major paper due at the semester's end. It's certainly a worthwhile goal, at least for its motivational value. It may very well inspire you to start working on the paper right away and to devote lots of time to the project throughout the semester. If, on a Sunday afternoon, you are trailing off toward your couch and a seemingly inevitable 3-hour nap, the goal to make an *A* might supply a jolt of energy that results in a productive writing session. But considering the most challenging demands in scientific writing, the goal to make an *A* is not especially powerful. Instead, it's a *just-okay* goal. While motivating in a general way, it does not provide specific directions for reaching your final destination. The goal to make an *A* on a school paper reminds me of the comedian Steve Martin's advice for becoming a millionaire: "First," he says, "get a million dollars!"

Here are a few more just-okay goals for scientific papers:

- Be precise and clear in expressing your ideas.
- Hook your readers.
- Present the most accurate information.
- Use correct grammar.
- Avoid needless jargon.
- Impress readers with your advanced analytical and critical thinking skills.

They are useful as general reminders for what writers should do and attend to. But they lack the power for supporting the advanced generative, problem solving, and critical thinking processes that are so fundamental to successful scientific writing. Let's say that you are struggling to express a complex concept, so you set the goal to be precise and clear in expressing your ideas. On its own, this just-okay goal doesn't contribute much to solving the problem at hand—in other words, it doesn't help you figure out exactly *how* to express your ideas precisely and clearly. Now consider the goal to *hook your readers*. In your introduction sections, of course you want readers to be engaged and interested in your topics. But the goal to *hook them* might not sufficiently focus your thinking on what, precisely, to do and say in order to achieve the desired outcome.

Just-okay goals are...well...*just okay*. A much more powerful type of writing goal, as hinted all along, is a rhetorical goal. As a case in point, suppose that I am planning a critical review paper on the effects of dietary fat intake on cardiovascular disease risk. While some published studies support the claim that cardiovascular disease is caused by fat consumption, others indicate that high-fat diets can actually reduce the disease risk. The issue's resolution depends on analyzing the effects of different types of dietary fats—for example, saturated fats, trans fats, and polyunsaturated fats. Saturated fats are generally associated with an increased risk for cardiovascular disease, whereas some forms of unsaturated fats, such as omega-3 fatty acids, appear to promote good heart health. In my review paper, I plan to offer explanations for the inconsistent data and conclusions. In addition, I will use the published research to support suggestions for the optimal types and amounts of dietary fats for preventing cardiovascular disease.

For my review paper's introduction section, one of the rhetorical goals that I have developed is as follows.

> Argue for the importance of my research issue, so that readers are convinced that it is truly worth resolving.

In terms of the cognitive power that it affords, this goal is far superior to its just-okay counterpart, which is the goal to hook your readers. The power of a rhetorical goal comes from its words and phrases that focus the writer's attention on three cues: a what-to-do cue, a what-to-say cue, and an audience-affecting cue. In our sample rhetorical goal, the what-to-do cue is the phrase, *argue for*. It's just two words long, but by no means is it trivial. My goal is not to simply *list* reasons that reflect the importance of my research issue. Likewise, I do not plan to *discuss* the reasons in a general way. Instead, because I want to deeply engage and interest readers in the debate about dietary fat intake and cardiovascular disease, I need to *argue for* why the research issue is important. The what-to-do cue will guide my efforts to construct a convincing argument.

My rhetorical goal also contains a what-to-say cue, which is the phrase *the importance of my research issue*. This is what I need to write about in order to accomplish the goal. By attending to the cue, I will develop specific ideas about why it's important to settle debates about the effects of dietary fat consumption on cardiovascular health. The what-to-say cue inspires a number of useful questions:

- What are the costs—both human and economic—associated with cardiovascular disease?
- Who, specifically, will benefit from knowing how various forms of dietary fat affect cardiovascular health?
- In what ways will people who suffer from heart disease benefit from the issue's resolution? How will the scientific and medical communities benefit?
- For society and science, what are the unfortunate implications of failing to determine the healthy and unhealthy forms of dietary fat?

The answers to these questions will generate a sizable chunk of content, perhaps a well-developed paragraph or two, for part of the introduction section to my critical review paper.

The third element of a well-crafted rhetorical goal statement is an audience-affecting cue, which is a phrase that answers the question, "How do I want my readers to respond to the content that serves the rhetorical goal at hand?" In our sample rhetorical goal, the audience-affecting cue is the phrase *so that readers are convinced that it is truly worth resolving*. I am not aiming for my readers to be informed, enlightened, or intrigued by what I write about the importance of resolving the research issue underlying my review paper. Instead, I want them to be *convinced*. By including explicit audience-affecting cues in your rhetorical goal statements, you will stay focused on the ultimate goal of scientific communication, which is to meaningfully influence your readers' knowledge, viewpoints, arguments, and behaviors.

Rhetorical goals are truly powerful tools for crafting successful content in scientific papers. Some work like spotlights for focusing attention on the most appropriate information and ideas in your mind and in the scientific literature. Others are shovels for digging deep holes, or developing your ideas and arguments extensively. Still others are chisels for shaping content to address concerns of specific audiences. The pages that follow present many demonstrations of rhetorical goals and their applications in scientific writing. For now, just to reinforce your initial understanding of what rhetorical goals are, Figure 2.2 presents a few more examples and highlights their defining features.

For the introduction section of a research paper, thesis, or lab report:
<u>Explain</u> *the rationale for your hypothesis*, **to help readers understand the conceptual foundation on which your study was based**.

For the introduction section of a research proposal:
<u>Argue for</u> *what is novel and unique about your proposed study*, **to convince readers that it will fill important knowledge gaps and extend the scientific field in significant ways**.

For the methods section of a grant proposal:
<u>Justify</u> *the strengths of your proposed study's procedures and analysis*, **to persuade readers to fund your research**.

For the results section of a research paper, thesis, or lab report:
<u>Present</u> *the key data and statistical analyses from your study*, **to prepare readers for the conclusions that you will present (about the results) in the paper's discussion section**.

FIGURE 2.2 Sample rhetorical goals for different sections and types of scientific papers. In these goal statements, *what-to-do cues* are underlined, *what-to-say cues* are italicized, and *audience-affecting cues* are bolded.

For the body of a conceptual review paper:

<u>Explain</u> *the reasons underlying contradictory and debatable findings from previous studies on your research issue,* **to directly answer questions that readers will naturally raise about disagreements in the literature.**

For the discussion section of a research paper, thesis, or lab report:

<u>Explain</u> *how methodological shortcomings may have influenced your study's outcomes,* **so that readers understand the boundaries within which your results and conclusions are valid.**

FIGURE 2.2 *continued.*

Devising Strategies for Accomplishing Rhetorical Goals

The act of setting useful goals naturally leads to more fine-tuned planning, specifically to determine the precise means for achieving them. Consider my rhetorical goal to convince readers that it's worth resolving debates about dietary fat intake and cardiovascular disease risk. On its own, the goal will be quite helpful. But it inspires the following questions: How will I actually go about convincing readers that my research issue is meaningful? What, more specifically, is my plan for what to do and say, as well as for how to influence readers? To answer these questions, I need to develop more refined goals, which we will call *strategies.* Think of a strategy simply as a more specific and detailed way to accomplish a related rhetorical goal. Figure 2.3 presents five strategies that I devised for my sample goal.

Rhetorical Goal: Argue for the importance of my research issue, so that readers are convinced that it is truly worth resolving.

Strategy 1: Highlight the importance of the real-world issue by presenting well-documented statistics on

(a) the number of individuals in our society who suffer and die from cardiovascular disease; and

(b) the economic costs of cardiovascular disease, including costs for health insurance, drug treatments, medical procedures, and lost work income for patients.

FIGURE 2.3 The fundamental unit of a goal-based plan: A rhetorical goal and a set of strategies for accomplishing it. As explained in the text, this example is part of the plan for introducing a review paper on dietary fat intake and cardiovascular disease.

Strategy 2: Discuss the negative influences of cardiovascular disease on quality of life.

Strategy 3: Explain uncertainties and debates about how dietary fat influences cardiovascular health. Demonstrate how the conflicting research on this issue has caused problems for the public and for clinicians, such as doctors and dieticians, who advise patients about dietary approaches to promoting health and preventing disease.

Strategy 4: Discuss the positive implications of successfully resolving the research issue.

Strategy 5: Discuss the negative implications of failing to resolve the research issue.

FIGURE 2.3 *continued.*

Strategies provide helpful directions for generating appropriate content. Take the first strategy in my plan, which involves presenting statistics on the incidence and economic costs of cardiovascular disease. If I don't already know these statistics, the strategy will direct my search for them in the scientific literature. As another example, the second strategy will focus my efforts in brainstorming ideas about how people with cardiovascular disease are affected in their daily lives. Notice that some of the strategies are more specific and detailed than others. While the first strategy is fairly well developed, the last two are phrased in general terms. The specificity and development of your strategies will depend on how far you have progressed in planning your writing project. Early in planning, it's fine to phrase your strategies in general terms. As you progress through planning stages, learning more about your topic and thinking more deeply about how to accomplish your rhetorical goals, you will refine your strategies, making them more specific and detailed. Doing so will ultimately help you generate specific and detailed content for your paper.

A rhetorical goal along with its related strategies, as demonstrated in Figure 2.3, is the fundamental unit of what we are calling a goal-based plan. Next we consider how to integrate this unit into the plan's overall structure.

The Structure of a Goal-Based Plan

In our approach to planning, rhetorical goals and their strategies are organized by the major sections of scientific papers. This format is illustrated in Figure 2.4 (page 69). Let's say that you are writing an IMRAD-structured research paper to report an original study. Your plan would include a number of rhetorical goals for the introduction, methods, results, and discussion sections, respectively. One of your goals for the introduction section, as we have been discussing, is to convince readers that your research issue is important and worth resolving. In addition, your plan for the introduction might include goals for providing background knowledge to help readers understand your research issue, stating your hypotheses, convincing readers that your research is novel, and presenting the specific purposes for your paper. In your plan for the discussion section, you might have goals for presenting your conclusions, relating your study's outcomes to those from previous studies on your research issue, discussing methodological limitations to your study, and suggesting future research to advance knowledge in the field.

As shown in Figure 2.4, each rhetorical goal is accompanied by a set of related strategies. When this framework is fully fleshed out for each major section of a paper, the plan will look like an outline. As I advised earlier, however, goal-based planning should not be constrained by the highly structured thinking that under-lies formal outlining. When you develop a new rhetorical goal, simply add it to its major section without concern about its logical order (in relation to the section's other goals). Similar advice applies to creating new strategies for a given rhetorical

Major Section (e.g., the introduction section to a research paper)
Rhetorical Goal 1
      Strategy 1.1
      Strategy 1.2
      .
      .
      Strategy 1.X

Rhetorical Goal 2
      Strategy 2.1
      Strategy 2.2
      .
      .
      Strategy 2.X
      .
      .
      .

Rhetorical Goal X
      Strategy X.1
      Strategy X.2
      .
      .
      Strategy X.X

FIGURE 2.4 An overall framework for goal-based planning.

goal: In no particular order, add them to your other strategies for accomplishing the goal. Once you have a complete plan, you can concentrate on organizing its elements in an optimal linear order. That's our focus in Chapter 4.

Consider, again, the following rhetorical goal for the introduction section to my critical review paper on dietary fat intake and cardiovascular disease:

> Argue for the importance of my research issue, so that readers are convinced that it is truly worth resolving.

Notice that the goal statement is *general* in the sense that it lacks words or phrases that directly reflect my paper's topic. In fact, the goal would apply *as is* to the introduction section to any and all critical review papers, regardless of their topics. In contrast, several of my strategies for accomplishing this rhetorical goal, as presented in Figure 2.3 (page 68), are *specific*—that is, they refer explicitly to aspects of my paper's topic. A very important bit of advice follows these observations: *You should express your rhetorical goals in general terms and your strategies in more specific terms.* Taking this approach to planning, you will avoid the all-too-common problem of becoming completely overwhelmed by everything that you might do and say in a scientific paper. When rhetorical goals are expressed generally, there are relatively few of them to accomplish in any major section of a standard scientific paper. For example, you would be hard-pressed to think of more than six or seven generally stated rhetorical goals for each major section of an IMRAD-structured research paper or lab report. This manageable number of goals reduces the complexity of the writing process.

To better appreciate the last point, consider an analogy that likens rhetorical goals to dresser drawers for storing your clothes. Imagine that you've just done a huge load of laundry, several weeks' worth. To avoid being overwhelmed by articles of clothing scattered everywhere around your room, and to find exactly what you need quickly, you may rely on an organized system: Underwear goes in the underwear drawer, t-shirts go in the t-shirt drawer, socks go in the sock drawer, and so on. In this scheme, you don't have a separate drawer for each pair of socks—instead, they all go in the one drawer set aside for socks. You also don't put socks in the t-shirt drawer or t-shirts in the sweaters drawer. This system contrasts the chaotic method of hurling a basket of unfolded laundry onto your bedroom floor and then, over the next few weeks, having to search desperately for certain items of clothing when you need them. In no time at all, of course, this troublesome task is compounded by clothes that have disappeared under your bed, behind other pieces of furniture, and perhaps into the jaws of your roommate's dog. In this analogy, the detailed information, ideas, and arguments that you generate for your scientific paper are like articles of clothing. Rhetorical goals are the dresser drawers into which a paper's content is stored, organized, and retrieved for later use in planning, drafting, and revising activities. To extend the analogy, think of strategies as tidy compartments within a dresser drawer. In this scheme, all of the content that comes to mind for your paper has its logical place in your plan. Consider, for instance, all of your good ideas about the importance of your paper's research issue. In your written plan, you will naturally place these ideas with their associated rhetorical goal and the targeted strategies to which they relate. The processes of generating and organizing content in this logical, organized framework are really quite manageable.

Taking Goal-Based Planning to Heart

The preceding dresser-drawer analogy reflects one of many reasons for taking a goal-based approach to planning scientific papers. Another reason is to avoid the frustrating experience of sitting down to draft without a plan. Perhaps you are familiar with this experience. Upon receiving a writing assignment, you read it over and start to draft the paper immediately. Circumventing intermediate stages of planning, you seek a shortcut to your final destination—a finished paper. But the journey goes awry. Getting snagged in the drafting process, and staring at a blank sheet of paper or your empty computer screen, you feel the frustration building. Just short of beginning to pull out your hair, you naturally ask yourself, "What in the world am I trying to do and say here, anyway?" This question, of course, can be translated as, "What are my rhetorical goals and strategies?" To answer these questions, you have no choice but to backtrack to the stage of developing a goal-based plan. So to save time and energy, as well as to avoid counterproductive frustration, it's best to develop your goals and strategies *before* sitting down to draft.

Ultimately, the best reason for taking a goal-based approach to planning is that it produces goal-directed scientific papers, which are the very best kind. The approach ensures that your paper's content will meet the conventional goals for scientific writing and the expectations of its audiences. As an example, part of the introduction section to all scientific papers should aim to accomplish the conventional goal of explaining their motivating questions and problems. Readers will certainly demand such explanations. Our activities for developing goal-based plans include adopting established conventions and analyzing audiences. The outcome of these activities is highly appropriate goal-directed content. This expert method of planning contrasts that of novice writers, who take a topic-based approach to planning. Instead of aiming to accomplish appropriate rhetorical goals in their papers, novice writers focus mostly on talking about their topics.

The experiences of a student named Rudy will serve to demonstrate the remarkably superior outcomes of goal-based versus topic-based planning. Rudy wrote a critical review paper on medications for treating high blood pressure, which is also called hypertension. This condition is a major risk factor for cardiovascular disease and the often-fatal outcomes of heart attack and stroke. Numerous drugs have been developed for treating hypertension. They fall into various categories, which include beta blockers, angiotensin-converting enzyme (ACE) inhibitors, calcium-channel blockers, and diuretics. For patients with severe hypertension, doctors often prescribe two or more drugs that work on different organs and tissues. This method is called combined drug therapy. Rudy's research issue focused on uncertainties about which combinations of antihypertension drugs are most effective and safest for lowering blood pressure. Rudy critically reviewed the literature on combined drug therapy and used published studies to support his own recommendations for optimal combinations.

In the first draft of his introduction section, Rudy took a purely topic-based approach. He reasoned that the overall topic, combined drug therapy, could be divided into subtopics such as high blood pressure, beta blockers, ACE inhibitors, and the other groups of antihypertension drugs. Rudy planned his introduction by making these subtopics the major divisions of his topic-based outline. The result was the draft excerpt that follows (Figure 2.5). Take a few minutes to read Rudy's draft and judge its quality.

High blood pressure is a condition in which the blood pumped by the heart exerts high forces on the walls of arteries, which are the vessels by which blood flows from the heart to the body's organs, tissues, and cells. According to the American Heart Association, approximately 73 million adults in the United States have high blood pressure, or hypertension (http://www.americanheart.org/). In 2004, more than 50,000 Americans died directly as a result of this disease. Hypertension is a major risk factor for heart attacks and strokes, which cause more than 500,000 deaths in the United States every year. Blood pressure is measured with a device called a sphygmomanometer, which consists of an inflatable arm cuff and a numbered dial that indicates pressure values in millimeters of mercury (mmHg). Blood pressure readings are expressed with two values. First, systolic pressure indicates how much force the blood exerts while the heart is contracting. Second, diastolic pressure indicates how much force is exerted while the heart is relaxing in between contractions. Normal values of blood pressure are slightly lower than 120 mmHg (systolic) and 80 mmHg (diastolic). High blood pressure is designated by measures exceeding 140 mmHg (systolic) and 90 mmHg (diastolic).

One treatment for hypertension is a class of drugs called beta blockers. These drugs lower blood pressure by slowing heart rate and reducing the heart's force when beating. The force that the heart produces when it contracts is increased by a naturally oc-curring hormone called epinephrine, which is also called adrena-line. To affect the heart, this hormone must attach to receptor

FIGURE 2.5 An example of topic-based writing from a student's draft.

molecules, called beta receptors, which are located on cardiac cells. Beta blockers are drugs that block the beta receptors so that adrenaline cannot attach to them. Because the drugs block adrenaline, it does not have its normal effect of increasing heart rate and contractile force. Therefore blood pressure is reduced.

High blood pressure can also be treated with another class of drugs, called angiotensin-converting enzyme inhibitors, or ACE inhibitors. These drugs work by relaxing and dilating the peripheral arteries throughout the body. ACE inhibitors block the formation of the hormone angiotensin-II, which naturally constricts the arteries. The drugs reduce blood pressure by relaxing and dilating the arteries.

Doctors prescribe combinations of different medications to lower blood pressure in patients with hypertension. The purpose of this critical review paper is to present research that has been conducted on which combinations are most effective in lowering blood pressure without negative side effects.

FIGURE 2.5 *continued.*

You may have noted that Rudy's draft introduction is logical, coherently structured, and devoid of major grammatical errors. The content, however, has considerable room for improvement. As Rudy planned, his introduction covers subtopics that make up the overall topic of drug treatments for hypertension. The first three paragraphs inform readers about high blood pressure, beta blockers, and ACE inhibitors, respectively. The content is characteristic of topic-based writing because it mostly presents background information, including basic definitions of terms. In his first paragraph, for example, Rudy defined blood pressure and gave background information about how it is measured. Rudy thought that this content was very appropriate because, after all, it relates to his overall topic. In topic-based writing, everything is relevant to include as long as it informs readers about the topic at hand. But we might question whether some, or even much, of the defining and informing is absolutely essential. Does Rudy really need to present the details about sphygmomanometers and about how blood pressure readings are expressed? Are the explanations of how beta blockers and ACE inhibitors work necessary?

In Rudy's first paragraph, note that many of the sentences contain the verb *is* or *are*, which reflects the author's intention to define terms. Of course, definitions can be essential elements of an introduction section. But they become problematic when they take over, leaving little or no room for essential content that accomplishes key rhetorical goals, such as explaining the paper's unresolved research issue and convincing readers that the issue is important. At its core, Rudy's paper is not simply about hypertension, beta blockers, and ACE inhibitors. Instead, it's about the unresolved issue of which drugs to combine in order to effectively and safely lower blood pressure in patients with hypertension. The paper is also about Rudy's critical evaluation of the literature and his suggestions for which combinations of antihypertension drugs are best. However, the first paragraph of Rudy's draft introduction does not introduce the issue and proposed solutions at all.

As originally planned, the second and third paragraphs of Rudy's draft introduction present general information about beta blockers and ACE inhibitors, specifically about how they work to lower blood pressure. In discussing these underlying mechanisms, Rudy did have a goal: to present background information. Whereas this goal might be appropriate for a basic science textbook, its application can lead to superfluous content in the introduction to a critical review paper. The second and third paragraphs of Rudy's draft leave us with no clues about the *relevance* of the background information about how the antihypertension drugs work. We are left asking, "So what?" The final paragraph of Rudy's draft is more promising because it reflects the appropriate goal of presenting a paper's purpose. But the bulk of the draft simply talks about topics with no other purpose than to inform.

After getting feedback on the areas for improvement in his draft, Rudy went back to the drawing board to develop a goal-based plan, an excerpt of which follows.

Rhetorical Goal 1: Present the real-world problems and unresolved scientific issues that led me to write my paper, to give readers a foundation for understanding my critical review of the literature.

Rhetorical Goal 2: Convince readers that my research issue is meaningful to science and society.

Rhetorical Goal 3: State my overall purpose and briefly outline the structure of my paper, to give readers a *heads-up* about what follows the introduction.

Rudy's revised introduction section is reproduced as follows (Figure 2.6). After you read the revision, take some time to compare it to Rudy's first draft.

According to the American Heart Association, high blood pressure, which is also called hypertension, affects approximately 73 million

FIGURE 2.6 An example of goal-directed writing from a student's revision.

adults in the United States (http://www.americanheart.org/). In 2004, more than 50,000 Americans died directly as a result of this disease. Hypertension is a major risk factor for heart attacks and strokes, which cause more than 500,000 deaths in the United States every year. The prognosis for patients with hypertension is good if they seek treatment. Johnson et al. (2) reported that in hypertensive patients who lowered their blood pressure with medication, risks of heart attack and stroke declined by 20% to 40%. However, pharmacological approaches to treating hypertension are not always effective, and many questions remain about the optimal drugs to prescribe for lowering blood pressure.

A major reason that medications for hypertension are not always effective is that the causes of the disease are so complex. Over 50 medications are currently available for treating the disease. Each drug works on specific causes. For instance, drugs in a class called *beta blockers* reduce blood pressure by slowing the heart rate and reducing the heart's contractile force. Other groups of drugs influence the function of different organs and tissues. For instance, ACE inhibitors work by dilating peripheral blood vessels and reducing the resistance to blood flow. Diuretics lower the volume of blood in the body, reducing how much pressure the blood creates against the arteries when the heart contracts. One problem that doctors face is determining which antihypertension drug to prescribe. For instance, a beta blocker that lowers blood pressure in one patient may not work effectively in another patient, who might respond better to an ACE inhibitor.

When doctors first diagnose hypertension they usually prescribe a single drug. This approach, which is called "monotherapy," is effective in lowering blood pressure to healthy levels in only 50% to 60% of cases (3). For individuals who do not respond optimally to monotherapy, doctors prescribe a combination of two to four drugs. This approach is called "combined drug therapy." While the combined approach has been reported to be effective, there

FIGURE 2.6 *continued.*

are several unanswered questions about its use. A major question deals with which drugs to combine for the optimal effects of lowering blood pressure. Should a beta blocker be combined with an ACE inhibitor or a diuretic? Which drugs work optimally when combined with calcium-channel blockers? Also at issue is the optimal dose for each combined drug. Should the doses be kept at the same levels that are used for monotherapy? Or should the doses of each drug be lowered, and if so, by how much? Finally, there are concerns about dangerous side effects of combining certain drugs.

Research has been conducted to provide answers to these questions about optimal approaches to combined drug therapy for hypertension. In this paper, I will critically review studies that have addressed optimal combinations, correct dosing, and negative side effects. On the basis of this review of the literature, I will recommend certain drug combinations and doses for the best outcomes in hypertensive patients.

FIGURE 2.6 *continued.*

In both versions of Rudy's introduction, the first paragraph presents statistics on the number of Americans who suffer from hypertension and who have died from the disease. Notice, however, that the statistics serve different purposes in the two versions. In the draft, the statistics come across mostly as background information. In contrast, the statistics in the revision serve the higher goal of introducing the research issue that inspired Rudy to write his critical review paper. The last four sentences of the revised first paragraph convey the overall issue, which can be summarized as follows: Although drugs for lowering blood pressure can reduce the risk of potentially fatal outcomes caused by hypertension in some cases, the drugs do not always work, and doctors face many questions about which ones to prescribe.

While much of Rudy's first draft defines terms and concepts, his revision mostly defines questions and problems, which is highly appropriate content for introduction sections. Even when Rudy presents definitions in his revision, he does so for the important purpose of helping readers understand the unresolved research issue that motivated his paper. In the second paragraph of his revision, for example, Rudy defines how the different antihypertension drugs work. It's not simply background information about the underlying mechanisms, as it was in the first draft. Instead, the revised content contributes to explaining a problem that Rudy identifies in the first sentence of the second paragraph: The problem is that blood pressure drugs are not always effective, because hypertension is a complex disease. In the third paragraph of the revision, Rudy presents the specific research issue underlying his critical review paper by raising direct questions about the best approaches to combined drug therapy. As we have discussed, Rudy's draft does not directly present the research issue.

The striking improvement in Rudy's revised introduction is attributable to his revised approach to planning, specifically to his taking a goal-based approach. You'll appreciate this point by looking again at the three rhetorical goals that Rudy developed for his plan. Clearly, the revised introduction reflects Rudy's application of these goals. In the revision process, Rudy detected and eliminated unnecessary information and ideas from the first draft. For example, he realized that it was not necessary to explain how blood pressure is measured, because the explanation did not serve any appropriate rhetorical goals for his paper.

RELYING ON YOUR EXPERIENCE IN SCIENTIFIC WRITING

Let's revisit the essential question at hand: *How* do you develop effective goal-based plans? A good way to begin is to contemplate what you have already learned about goals for communicating in science. So this activity simply involves brainstorming rhetorical goals and strategies by reflecting on your previous experiences in reading and writing scientific papers. Suppose that you are planning a lab report, a type of paper that you have written several times before. To develop your plan, ask yourself, "What do I know about the best things to do and say in the different sections of lab reports?" You may be surprised by how many useful goals and strategies come to mind.

ADOPTING AND ADAPTING CONVENTIONAL GUIDELINES

The conventions for the content of scientific papers are highly standardized. Most educational resources on the subject present these conventions directly. For instance, all good books on writing IMRAD-structured papers convey the following common guidelines:

- The introduction section should present the hypotheses that motivated the study being reported.
- The methods section should include detailed descriptions of the procedures for collecting and analyzing the study's data.
- The discussion section should acknowledge limitations to the study methods.

Because conventional guidelines for scientific writing are so well established and universal, it makes good sense to include them straightaway in plans for your papers. However, there is a potential downside to this approach if it does not involve *adapting* conventional guidelines to meet the specific demands of your rhetorical situation. In other words, you should avoid a paint-by-numbers approach to planning your paper. Take the common direction to describe study procedures and analyses in the methods sections of research papers. It's certainly advice on the right track. But as it stands, the conventional guideline is somewhat limited. For one reason, it lacks audience-affecting cues. So writers who adopt the guideline *as is* may fail to shape their descriptions of study methods to achieve desired outcomes for intended audiences. If you were planning to write a methods section, what outcomes would you have in mind for your readers? Do you intend for them to be able to replicate your experimental methods themselves? Or do they just need a general overview of your methods in order to get the gist of how you collected and analyzed data? Are you trying to convince your audience that your methods were the strongest ones for resolving your study's research issue? Or, is such justification and argument unnecessary? Without audience-affecting cues that address these questions, the conventional guideline to simply describe study methods might lead to underdeveloped and perhaps inessential content. Let's say that you actually intend to convince readers that your methods were stronger than those used in previous studies on the research issue. In this case, you might transform the conventional guideline into a well-developed rhetorical goal.

> Argue for the strength of your study's methods compared to those of related previous studies, so that readers are convinced that your approaches to resolving the research issue were superior.

Some assignments challenge writers to develop completely original rhetorical goals, or ones that are not easily derived from established conventions. This is especially true for school writing assignments that professors design to test students' knowledge of basic concepts presented in class lectures and readings. Imagine that your assignment is to write a lab report about a class experiment on the effects of caffeine on blood pressure. In class lectures, your professor has stressed the importance of learning the basic concept that blood pressure is the product of cardiac output and the resistance to blood flow in the peripheral vasculature. Anticipating that your professor will be looking for students to relate the findings from the class experiment to this basic concept, you develop the following nonconventional but effective rhetorical goal.

> Explain your findings by relating them to the key scientific concepts, so that your audience is convinced that you know your stuff.

My point here is that sometimes writers must adapt established conventions or go beyond them in order to develop their strongest plans.

USING MODEL PAPERS

When you are determined to master a physical skill, like a technically demanding sport, one of the best learning methods is to observe and imitate the movements of experts. This approach is called observational learning, or *modeling*. If you play golf, for example, you can improve your game by imitating the swinging actions of professional golfers. Modeling is also a great method for developing goal-based plans for scientific papers. In this activity, we develop goal-based plans by analyzing the content of published articles, specifically to identify the rhetorical goals and strategies that their authors sought to accomplish.

For a demonstration of the activity, we will examine parts of the discussion section from a published research paper titled, "Regular Exercise, Hormone Replacement Therapy and the Age-related Decline in Carotid Arterial Compliance in Healthy Women" (Moreau et al., 2003). The authors, a team of cardiovascular physiologists led by Dr. Kerrie Moreau at the University of Colorado, reported a study that focused on the carotid arteries, which are the vessels through which blood flows from the heart to the brain. These arteries must be compliant, or flexible, for the brain to receive all of the oxygen-rich blood that it needs. As we age, however, our carotid arteries naturally become less compliant. This condition, which can increase blood pressure and impede normal blood flow, is a risk factor for stroke. Moreau et al. found that in postmenopausal women who exercised regularly and received hormone replacement therapy (HRT), the carotid arteries were more compliant than those of age-matched peers who did not exercise and take the supplemental hormones. The complete first paragraph of Moreau et al.'s discussion section is presented in Figure 2.7. Incidentally, in Moreau et al.'s model passage, and in many other writing samples throughout the book, I have numbered the sentences for easy reference to them in my evaluations of their distinguishing features.

To derive rhetorical goals and strategies through analyzing published articles, you must first understand what you are reading. So we will begin by reviewing the key ideas in Moreau et al.'s paragraph. The authors present the two main findings of their research, which emerged from two separate but related studies. The first was a cross-sectional study in which carotid arterial compliance was compared in three groups of postmenopausal women: regular exercisers, sedentary women who were

> **(1)** The primary findings of the present study are as follows. **(2)** First, carotid arterial compliance is greater in post-menopausal women who use HRT and those who perform regular aerobic exercise than in their sedentary estrogen-deficient peers. **(3)** Second, a short-term moderate aerobic exercise program restores carotid arterial compliance in healthy, previously sedentary, HRT-supplemented postmenopausal women to levels observed in premenopausal women, consistent with an additive effect. **(4)** These results suggest that both HRT and habitual exercise have beneficial effects on carotid arterial compliance, and that a short-term aerobic exercise program can reverse the age-related reduction in carotid arterial compliance in HRT-supplemented postmenopausal women.

FIGURE 2.7 From Moreau et al. (2003). Reproduced with permission from Oxford University Press.

taking a form of HRT that contained the hormone estrogen, and sedentary women who were not taking HRT. Sentence 2 summarizes the main finding from this study, which revealed greater arterial compliance in the exercise and HRT groups compared to the sedentary control group. The second study was an experiment in which one group of postmenopausal women, subjects who were already on HRT, participated in a 3-month aerobic exercise (walking) program. The researchers measured carotid arterial compliance in these women before and after the exercise intervention. Sentence 3 presents the main finding from the intervention study: Previously inactive and HRT-supplemented postmenopausal women who complete a 3-month aerobic exercise program will experience a marked improvement in carotid arterial compliance. Indeed, the compliance of their arteries will improve to a level that is commonly observed in healthy premenopausal women. In sentence 4, Moreau et al. interpret the two main findings to support their overall conclusion about the favorable effects of exercise and HRT.

In Figure 2.7, it's easy to see that Moreau et al. sought to present their study findings as support for their overall conclusions. Through modeling, we can derive the following rhetorical goal for discussion sections of research papers:

Apply the relevant results from your study to develop your conclusions and to convince readers to accept them.

Upon deriving rhetorical goals and strategies from a model paper, you must decide whether to adopt and adapt them for your paper. For the goal at hand, this decision is a no-brainer. In well-written discussion sections of research papers, authors always present their main findings and conclusions. The content is essential for meeting readers' needs and expectations. In addition, the conclusion becomes an important element of a scientific argument that is supported with evidence and reasoning throughout the rest of the discussion section. In their introduction, Moreau et al. posed the question about how exercise and HRT affect carotid arterial compliance in postmenopausal women. At the start of the paper's discussion section, readers will surely ask, "What were the key results that indicate the effects? How did the researchers interpret those results to reach a direct conclusion that answers the motivating research question?" The first paragraph of Moreau et al.'s discussion section effectively responds to readers' pressing questions.

Let's do some more modeling. Figure 2.8 presents the second paragraph of Moreau et al.'s discussion section. As you read the paragraph, keep in mind that your purpose is not to acquire scientific knowledge or to evaluate the quality of the writing. Instead, the objectives of this activity are to (a) identify the rhetorical goals and strategies that underlie the content, (b) decide whether to adopt them for your own paper, and (c) adapt them to meet your specific needs.

In Figure 2.8, Moreau et al. were clearly aiming to relate their study's outcomes to those from previous studies on exercise, HRT, and the carotid arteries. In sentence 2 the authors cite one of their own previous studies on the effects of exercise

(1) Little information exists on the effects of regular exercise on carotid arterial compliance, particularly in women. (2) To our knowledge the only published study in women is from our laboratory in which carotid arterial stiffness (measured via aortic pulse wave velocity and augmentation index) was lower in endurance-trained compared with sedentary postmenopausal females [14]. (3) We recently demonstrated that endurance-trained older men have a higher carotid arterial compliance compared with age-matched sedentary controls, and that the age-related reduction in carotid arterial compliance is partially reversed in middle-aged and older men following a 13-week intervention consisting of moderate-intensity aerobic exercise [2]. (4) The results of the present study are consistent with and extend these findings in healthy men. (5) First, our cross-sectional group comparisons allowed us to determine if chronic HRT use and habitual exercise are similarly associated with enhanced carotid artery compliance among healthy postmenopausal women; HRT use has previously been associated with reduced arterial stiffness [15, 16]. (6) We found that the age-related reduction in carotid arterial compliance is attenuated in both postmenopausal women who use HRT and those who participate in regular and vigorous endurance exercise. (7) Second, our intervention study demonstrated that carotid arterial compliance was restored to pre-menopausal levels after 3 months of exercise in post-menopausal women who were using chronic HRT, whereas the same exercise program only partially restored carotid compliance in healthy middle-aged and older men [2]. (8) The enhanced carotid arterial compliance associated with habitual exercise was independent of changes in body composition, diet and traditional CVD risk factors, suggesting a primary effect of regular exercise on the arterial wall. (9) Importantly, the exercise program, which consisted of moderate intensity walking is consistent with that recommended by health care organizations for general health maintenance [17,18]. (10) Our findings suggest a possible interactive effect of HRT and exercise on carotid arterial compliance; this may explain the more complete restoration of arterial compliance in postmenopausal HRT users than we observed with this exercise program alone in men [2]. (11) Specifically, although the increase in arterial compliance with the exercise intervention was similar to what we observed in men [2], the fact that baseline levels of carotid compliance were higher (presumably due to chronic HRT use) allowed our women to improve to values seen in premenopausal females. (12) Our results provide initial insight into the potential benefits of using a mutifactorial risk intervention approach [19] to restore the age-related loss in large artery compliance in women. (13) However, in order to accurately determine the separate and interactive effects of HRT and exercise on carotid arterial compliance, randomized placebo controlled intervention trials need to be performed.

FIGURE 2.8 From Moreau et al. (2003). Reproduced with permission from Oxford University Press.

on carotid arterial *stiffness*, which is the converse of arterial compliance, in exercise-trained versus sedentary postmenopausal women. As summarized, stiffness measures were lower in the physically active subjects, reflecting favorable effects of regular exercise. Sentences 3 and 4 relate Moreau et al.'s current study to one that they conducted on exercise and carotid arterial compliance in middle-age and older men. Sentence 3 indicates that a 13-week exercise program reversed the males' age-related reduction in carotid arterial compliance; notice, however, that the exercise only "partially" reversed the loss of compliance in these subjects. As Moreau et al. acknowledge in sentence 7, this result differs from that observed in their study on HRT-supplemented postmenopausal women, who experienced a complete reversal due to exercise. In sentences 10 and 11, the authors speculate on why the effects of exercise were more advantageous in the women versus the men. Specifically, Moreau et al. reason that the women exercisers benefited from an interactive effect of HRT (which men do not take). Two more features of the model paragraph are worth noting. First, in sentences 8 and 9, the authors seek to convince readers that the greater arterial compliance they observed was due specifically to exercise rather than to potentially confounding variables such as changes in body composition, diet, and other risk factors for cardiovascular disease. Second, in sentences 12 and 13, they suggest future studies to directly test their speculation that the more advantageous effects of exercise on women (compared to men) may have been due to interacting effects of HRT.

Moreau et al.'s goal to compare their study's outcomes to those from related previous studies is very appropriate for discussion sections of research papers. By presenting substantiating evidence from other studies, the authors address questions that readers will ask about how the current study fits into the larger context of the research field. In addition, they strengthen arguments for their conclusions about the beneficial effects of exercise and HRT on the carotid arteries. If you were planning an IMRAD-structured paper with a discussion section, you would do well to include a rhetorical goal and a set of strategies based on modeling Moreau et al.'s paragraph in Figure 2.8. Your plan might look something like this:

Rhetorical Goal: Relate your study's outcomes to those from previous studies on the issue, to give readers the "big picture" and to convince them that your conclusions are viable.

Strategy 1: Directly compare your key results and conclusions to those from other studies, noting similarities and differences.

Strategy 2: Discuss how your results and conclusions relate to those from studies in which the subject population differed.

Strategy 3: Justify the validity of your results and conclusions by pointing out strengths in your methods for controlling against confounding variables.

Strategy 4: If your results and conclusions contrast those of previous studies, speculate on underlying reasons for the differences.

Strategy 5: Suggest new research to test your speculations about the differences between your study's outcomes and those of previous studies.

To get the most out of this modeling activity, try it on numerous published papers of the type that you are writing. Comparing the models, you will identify similar rhetorical goals and strategies that reflect well-established conventions and that generate especially effective content. These would be worth adopting and, if necessary, adapting for your plan. You might also identify problematic content that reflects misguided goals and strategies. For example, the discussion sections of some published research papers do little more than summarize their results sections. The authors simply restate their results without using them to support conclusions or to make comparisons with the results of related studies. Through our modeling activity, you would be mindful to avoid the goal to simply restate your study's results in the discussion section of an IMRAD-structured paper.

If you read published articles in a wide variety of journals covering different life science disciplines, you will find interesting variations on the rhetorical goals that authors apply. In the introduction sections to articles in psychology journals, authors commonly devote a considerable amount of content to presenting the theoretical frameworks on which their research is based. As another example, in some research papers published in biology journals, authors present a brief overview of their study results and their conclusions in the introduction section. Traditionally, this content is reserved for discussion sections. But consider a case in which a study's results are stunningly groundbreaking. Or perhaps they completely contradict the outcomes from previous studies. The excitement generated by these results may warrant breaking conventions by bringing the findings to the audience's attention immediately. Indeed, readers may respond quite favorably to the unconventional approach. Modeling this technique, you would adopt the rhetorical goal to summarize any remarkable results and conclusions in the introduction sections to your papers that report original research. In response to the content that this goal generates, your audience will be engaged and motivated to keep reading for more details.

APPLYING YOUR TASK AND AUDIENCE ANALYSES

As emphasized in Chapter 1, many activities in the writing process are ideally guided by task and audience analyses. This is certainly true for developing goal-based plans. To demonstrate the application of audience analysis, I will pose a scenario in which we have conducted a 6-month clinical trial on a new drug for lowering cholesterol. Our subjects were 500 patients with abnormally high levels of LDL-cholesterol, the harmful form that contributes to fatty plaque formation in the coronary arteries. Over the course of our study, LDL-cholesterol decreased by an average of 20 mg/dL of blood in subjects who took the new drug. The improvement was statistically significant. Encouraged by these results, we plan to report them in a research paper that we will submit to the *New England Journal of Medicine*. The primary readers of this leading scientific journal are biomedical researchers and practicing physicians. At the moment, we are planning the discussion section of our research paper. To develop this section's rhetorical goals, we will reflect back on our audience analysis, focusing on the following characteristics that we attributed to the practicing physicians in our audience.

- Our readers are doctors who subscribe to biomedical journals to learn about the effectiveness of new medications for the diseases they treat. From research papers on new cholesterol-lowering drugs, this audience is especially interested in the *clinical* significance, not just the statistical significance, of study results.
- Our readers will expect comparisons between the effects of our new drug and the drugs they currently prescribe for patients with high cholesterol.
- Our readers will likely ask whether our tested drug's effects depend on individual differences. For example, does the drug affect women differently than it affects men? Does the cholesterol-lowering response to the drug depend on the severity of the disease across individuals?
- Given their clinical concerns, our readers will demand reports of any side effects and contraindications that we observed in subjects who took the cholesterol-lowering drug.

Through audience analysis, we derive the following rhetorical goals for our research paper's discussion section.

Rhetorical Goal 1: Discuss the clinical significance of our study's results, to aid readers in deciding whether the treatment is practically worthwhile.

Rhetorical Goal 2: Compare the effects of the new drug treatment to those of currently prescribed treatments, to guide readers in deciding whether to change their prescriptions.

Rhetorical Goal 3: Explain all potential negative side effects and contraindications of the drug treatment, so that readers can accurately weigh the benefits versus the risks.

Rhetorical Goal 4: Discuss any results from our study that indicate individual differences in response to the new drug treatment, to help readers determine which patients will benefit most from it and which patients should avoid it.

It's a fairly simple but extremely powerful process: Figure out what to do and say by considering the needs, expectations, and values of your audience.

Now suppose that our paper was published in the *New England Journal of Medicine* to rave reviews. To extend our research on the new cholesterol-lowering drug, we need additional funding. So we are planning to write an NIH grant application to propose a study on the drug's long-term effects on patients' risks of heart attack and stroke. To develop part of our goal-based plan, we will apply the task analysis activity that was presented in Chapter 1. Recall that this activity involves attending to the instructions, guidelines, and evaluation criteria that are included in writing assignments. On the NIH Web site, we have found several documents that contain this information. One of them presents the primary criteria that peer reviewers use to evaluate NIH grant applications. An excerpt from this document is presented in Figure 2.9 (page 86) (the entire document is on the NIH Web site at http://grants.nih.gov/grants/guide/notice-files/NOT-OD-09-025.html).

In applying task analysis, we transform key instructions, guidelines, and evaluation criteria for a writing project into the elements of a goal-based plan. Consider, for instance, the evaluation criteria reflected by the questions in the item labeled ***Investigator(s)*** in Figure 2.9. These are questions that peer reviewers ask about the training, experience, accomplishments, collaborative abilities, and leadership skills of the project directors (PDs) and principal investigators (PIs) who author NIH grant applications. These criteria are extremely important for making decisions about whether to fund research proposals. The NIH most definitely does not want to spend money on projects that are strong on paper but that cannot be executed successfully because the scientists lack the necessary knowledge, experience, and skills. If we were actually writing an NIH grant application to extend the research on our hypothetical cholesterol-lowering drug, we would have to ask, "What must we do and say to meet the peer reviewers' evaluation criteria regarding our ability to carry out our proposed study?" Applying this task analysis, we would develop a rhetorical goal and a set of strategies as follows.

Rhetorical Goal: Present relevant details about our scientific training, experience, accomplishments, and skills, in order to convince readers that we are well qualified to execute our proposed study successfully.

Strategy 1: Present details about the educational background and training of the members of our research team.

Strategy 2: Highlight our specific experience and skills in conducting clinical trials on drugs for treating cardiovascular disease.

Strategy 3: To convince reviewers that we have a strong track record in the field, summarize our previously published research on cholesterol-lowering drugs.

NIH Criteria for the Evaluation of All Research Applications

The mission of the NIH is to support science in pursuit of knowledge about the biology and behavior of living systems and to apply that knowledge to extend healthy life and reduce the burdens of illness and disability. As part of this mission, applications submitted to the NIH for grants or cooperative agreements to support biomedical and behavioral research are evaluated for scientific and technical merit through the NIH peer review system.

Overall Impact. Reviewers will provide an overall impact score to reflect their assessment of the likelihood for the project to exert a sustained, powerful influence on the research field(s) involved, in consideration of the following five core review criteria, and additional review criteria (as applicable for the project proposed).

Core Review Criteria. Reviewers will consider each of the five review criteria below in the determination of scientific and technical merit, and give a separate score for each. An application does not need to be strong in all categories to be judged likely to have major scientific impact. For example, a project that by its nature is not innovative may be essential to advance a field.

Significance. Does the project address an important problem or a critical barrier to progress in the field? If the aims of the project are achieved, how will scientific knowledge, technical capability, and/or clinical practice be improved? How will successful completion of the aims change the concepts, methods, technologies, treatments, services, or preventative interventions that drive this field?

Investigator(s). Are the PD/PIs, collaborators, and other researchers well suited to the project? If Early Stage Investigators or New Investigators, do they have appropriate experience and training? If established, have they demonstrated an ongoing record of accomplishments that have advanced their field(s)? If the project is collaborative or multi-PD/PI, do the investigators have complementary and integrated expertise; are their leadership approach, governance, and organizational structure appropriate for the project?

Innovation. Does the application challenge and seek to shift current research or clinical practice paradigms by utilizing novel theoretical concepts, approaches or methodologies, instrumentation, or interventions? Are the concepts, approaches or methodologies, instrumentation, or interventions novel to one field of research or novel in a broad sense? Is a refinement, improvement, or new application of theoretical concepts, approaches or methodologies, instrumentation, or interventions proposed?

Approach. Are the overall strategy, methodology, and analyses well reasoned and appropriate to accomplish the specific aims of the project? Are potential

FIGURE 2.9 NIH criteria for evaluating grant applications.

problems, alternative strategies, and benchmarks for success presented? If the project is in the early stages of development, will the strategy establish feasibility and will particularly risky aspects be managed?

If the project involves clinical research, are the plans for (1) protection of human subjects from research risks, and (2) inclusion of minorities and members of both sexes/genders, as well as the inclusion of children, justified in terms of the scientific goals and research strategy proposed?

Environment. Will the scientific environment in which the work will be done contribute to the probability of success? Are the institutional support, equipment and other physical resources available to the investigators adequate for the project proposed? Will the project benefit from unique features of the scientific environment, subject populations, or collaborative arrangements?

FIGURE 2.9 *continued.*

Strategy 4: Give examples of how the outcomes of our previous research have been productively applied in the scientific and medical communities.

Strategy 5: Describe the integrated, complementary nature of our research team's training and skills. Demonstrate that our collaborative skills are the essential ones for carrying out the proposed study.

In the appropriate section of our grant application, this rhetorical goal and its strategies would guide us in generating substantial content that is right on target for the task at hand.

This activity—developing a goal-based plan through audience and task analysis—is especially useful for school writing projects. If your professor is your primary audience, you will naturally ask what he or she is looking for. What qualities of student writing does your professor value? What are his or her criteria for giving excellent grades? Suppose that one of your professors regularly talks about the importance of making well-rounded arguments. It's never good enough, she says, to present only the evidence and reasoning that support *your* side of a debatable issue. Strengths and weaknesses of other sides must also be addressed. This standard for excellence, derived through audience analysis, can be transformed into a new rhetorical goal:

Acknowledge and refute claims that oppose yours, so that readers are convinced that you are not inherently biased against the alternative arguments and that your argument is the strongest one.

USING THE *HELICOPTER THINKING* METHOD

As introduced earlier, helicopter thinking is an advanced cognitive skill that involves flexibly shifting attention and thought processes from abstract to concrete levels, and vice versa. The method is diagrammed in Figure 2.10. Among its other powerful applications in the writing process, helicopter thinking is an especially productive method for developing goal-based plans. For a demonstration of this process, I will recall the experience of a student named Elaine, who wrote a lab report for an exercise physiology class. The paper was based on an experiment intended to measure the cardiovascular responses to exercise. One of Elaine's classmates, a recreational runner, volunteered to be the study's subject. In the experimental protocol, he ran on a treadmill at progressively faster speeds until he was too fatigued to continue. For the first 4 minutes of the exercise bout, the treadmill speed was set at 4 miles per hour, which is a slow jog. At the 4-minute mark, the rate was increased to 6 miles per hour. Then, at regular 4-minute intervals thereafter, the rate continued to increase by 2 miles per hour. By 16 minutes into the test, the subject was running at the brisk pace of 12 miles per hour. Succumbing to fatigue, he called for the test to stop at around the 18-minute mark.

At 2-minute intervals throughout the treadmill test, the subject's heart rate and oxygen consumption were measured. The latter variable indicates how much oxygen the body uses to produce energy for physiological processes such as muscle contraction during exercise. Elaine's class measured oxygen consumption with instruments that sampled the volume and concentration of oxygen that the

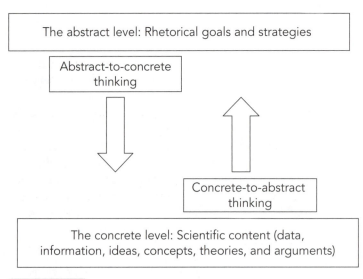

FIGURE 2.10 A model of the *helicopter thinking* method for developing goal-based plans.

exercising subject breathed in and out through a mouthpiece. The results were expressed as the volume of oxygen consumed in milliliters (ml) relative to the subject's weight in kilograms (kg) over successive 1-minute periods. Figure 2.11 presents a graph of the results; the values for heart rate and oxygen consumption are represented on the left and right vertical axes, respectively.

Elaine used the graph in Figure 2.11 to guide her helicopter thinking, which resulted in several excellent rhetorical goals and strategies for her lab report. One mode of helicopter thinking involves shifting from concrete to abstract levels of attention and thought processes. As illustrated in Figure 2.10, the concrete level comprises specific data, information, ideas, concepts, theories, and arguments. When these details of scientific content come to mind, skilled writers sensibly ask, "Do they belong in my paper?" The best answer is that the content indeed belongs *if it contributes to achieving appropriate rhetorical goals.* Just thinking about the details of scientific content can lead to the development of such goals. For example, examining the graph of her study's results, Elaine pondered the similar pattern of changes in heart rate and oxygen consumption. As the subject ran faster, both variables increased linearly. Elaine's thoughts inspired questions about the underlying relationships between heart rate and oxygen consumption during exercise:

- Are the increases in heart rate and oxygen consumption directly driven by the muscles' needs for energy?
- What physiological factors make the heart beat faster as running speed increases?
- Do the exercising muscles produce chemical or neural signals that increase heart rate?
- Can the heart somehow sense the need for oxygen in the muscles?

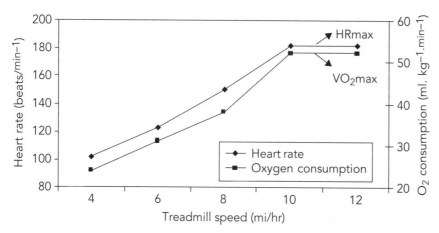

FIGURE 2.11 Results from a hypothetical lab experiment on the influence of exercise on heart rate and oxygen consumption.

In her graph, Elaine also noticed the plateau that occurred for both heart rate and oxygen consumption. Exercise physiologists call these plateaus HRmax and VO$_2$max, respectively. When the treadmill speed increased from 10 to 12 miles per hour, there was no corresponding increase in oxygen consumption and heart rate. Focused on this observation, Elaine wondered how the subject had the energy to continue running, at least for a short time, at 12 miles per hour. More questions came to Elaine's mind:

- What physiological processes caused the subject to fatigue and stop the test?
- Would the results have differed if the subject had been a highly trained endurance athlete rather than a recreational jogger?
- Would the results have differed if the subject had been female?

As Elaine was immersed in thinking about these concrete aspects of her study's results, her mental helicopter was still on the ground, so to speak. To start the blades turning for liftoff, she began thinking about how the concrete science might relate to the more abstract rhetorical goals for her paper. In other words, she asked, "What goals would I be accomplishing by writing about my analysis of my study's results?" Several of Elaine's observations and questions address underlying mechanisms, or the factors that caused the results to turn out as they did. Elaine reasoned that it would indeed be appropriate to write about these mechanisms in the discussion section of her lab report. So she developed the following rhetorical goal and strategies.

Rhetorical Goal: Explain the underlying mechanisms that could have accounted for my study's results.

Strategy 1: Describe the similar linear pattern of the results for heart rate and oxygen consumption.

Strategy 2: Explain how the increased demand for oxygen in the muscles might have caused the increase in heart rate.

Strategy 3: Focusing on the plateau effects in my graph, explain how the subject was able to continue exercising at a higher intensity, at least for a short time, even though heart rate and oxygen consumption were not increasing.

To achieve the ambitious goal and strategies, Elaine realized that she would need to learn more about cardiovascular responses to exercise. But she also realized that her plan would guide the way to the essential knowledge.

Elaine practiced some more helicopter thinking, focusing on her observations and questions about the subject who performed the exercise test. Recall that he was a recreational jogger rather than a trained endurance athlete. In addition, the study did not include female subjects. These concrete details prompted

Elaine to reflect on higher-level ideas about potential methodological limitations to her study and about how to avoid them in future studies. This helicopter thinking produced the following part of Elaine's plan for her lab report's discussion section.

Rhetorical Goal: Suggest ideas for new studies to overcome the methodological limitations to my study, in order to guide future researchers in the field.

Strategy 1: Discuss the restrictions of studying only one male subject who represented only recreational runners. Explain why my study's results cannot be applied to the entire population of runners.

Strategy 2: Suggest a new study to determine the effects of increasing exercise intensity on heart rate and oxygen consumption in trained endurance athletes, including a sufficient number of male and female subjects.

Strategy 3: State and support a hypothesis about the effects of increasing exercise intensity on heart rate and oxygen consumption in female runners.

Applied to goal-based planning, helicopter thinking might seem straight-forward and intuitive. Of course you will figure out what to do and say in your paper by reflecting on the concrete science that defines your topic. But when you are immersed in contemplating the finer points of scientific knowledge, it's not always so natural and easy to elevate your thinking to the more abstract level of rhetorical goals. The shift requires a deliberate, conscious effort to reflect on whether and how the science might contribute to developing and ultimately achieving your plans.

STARTING TO DRAFT

It just makes good common sense: Before beginning any complex and challenging task, you should spend a sufficient amount of time planning. This essentially involves setting goals and figuring out how to accomplish them. As revealed through many studies on writing, the time that authors spend planning is highly correlated with the quality of their final papers. Planning is also the antidote to the frustrating experience of writer's block. There is an occasion, however, when the best course of action is to interrupt the planning process and, at least for the moment, to jump right into drafting: It's when we get completely stuck trying to figure out our rhetorical goals and strategies. The point is that sometimes we just don't know what to do and say until we start doing and saying it. If you find yourself in this situation, try quickly drafting a paragraph or two in order to get your rough ideas on paper. In the drafting process, you may find that well-targeted rhetorical goals and strategies begin to emerge in your mind somewhat spontaneously. When this happens, make sure that you don't completely abandon your goal-based planning to go headlong into drafting. Instead, draft as much as necessary to crystallize key rhetorical goals and strategies, and then return to developing your plan until it is complete.

REVISING YOUR GOAL-BASED PLAN

A considerable amount of time and effort go into developing effective plans for scientific papers. Without a doubt, however, the time and effort are well spent. As demonstrated in upcoming chapters, a strong goal-based plan greatly enhances efficiency and productivity throughout the writing process. On the other hand, a poorly conceived plan can turn the process into a nightmare. To ensure that your plans are most helpful, it's worth taking the necessary time to strengthen them through revision. This last section of the chapter presents guidelines for doing so. In addition to revising your goal-based plans independently, it's always helpful to get feedback from collegial reviewers, including peers and professors. So you might share the following guidelines with your plan's reviewers to solicit their well-targeted comments and advice for improvement.

Check for Whether Your Rhetorical Goals Are Appropriate for the Major Sections in Which You Have Placed Them

This guideline is for ensuring that you are doing and saying the best things in the best places in your paper. For example, in the results sections of IMRAD-structured papers, an appropriate rhetorical goal involves presenting your study data and statistical analyses. The results section, however, is not the place to develop arguments in support of your overall conclusions; by convention, this is a goal for the discussion section. To determine whether you have placed your rhetorical goals and their strategies in the appropriate major sections, you must know the conventions for structuring scientific papers. One way to learn the conventions is to study the structural elements of published articles, identifying rhetorical goals by the major sections in which authors usually aim to accomplish them. Another useful resource is this book's Chapter 8, which presents conventional rhetorical goals organized by the major sections of scientific papers.

Check Your Rhetorical Goals for Their Content-Generating Potential

A well-crafted rhetorical goal should generate a substantial amount of content—at least a well-developed paragraph, if not several unified paragraphs in succession. If any goals in your plan fail to stir up brainstorms of useful content, check the clarity and specificity of their what-to-say cues. As described earlier, these are

phrases in goal statements that sharply focus attention on appropriate information, ideas, and arguments to develop. The following goal, intended for the body
of a critical review paper, has a weak what-to-say cue.

> Discuss the problems with previous studies on my topic.

The phrase *the problems with previous studies on my topic* is too vague. It might
confuse the writer about what sorts of problems to discuss and what to say about
them. To improve the goal, the writer must think more deeply about what he truly
wants to say. Does he want to present a laundry list of any old problems? Or does
he want to demonstrate the influence of a certain type of problem, such as limitations to the previous studies' methods? Here's the result of this revision:

> Discuss how specific methodological limitations have compromised the outcomes
> of previous studies on my topic.

The what-to-say cue in this revised goal will generate especially appropriate and
well-developed content.

Check Your Rhetorical Goals for Their
Audience-Affecting Potential

In the preceding example of a rhetorical goal with a strong what-to-say cue,
notice that there is no audience-affecting cue. That is, the goal statement does
not focus the writer's attention on how he wants readers to respond. This is not
a problem when our intentions for readers are obvious to us. In the preceding
scenario, however, suppose that the writer has a special reader-focused reason for
explaining how methodological limitations have compromised previous studies
on his topic. The writer intends for the content to preface very specific suggestions for future research. To stay focused on this intention, he revised the goal
statement as follows.

> Discuss how specific methodological limitations have compromised the outcomes
> of previous studies on my topic, to prepare readers for my proposals for future
> research to resolve the limitations.

As you might imagine, the content generated by applying this fully revised goal
will be far superior to that produced by the writer's original goal, which was "to
discuss the problems with previous studies on my topic."

Check Your Strategies for Their Detail and Depth

Imagine that I am writing an informative review paper on the causes of atherosclerosis, the disease in which fatty plaques form in the inner linings of blood
vessels. A number of hypotheses have been proposed to explain how atherosclerosis develops in the cardiovascular system. One is that LDL-cholesterol,
which our bodies naturally produce and we also consume in our diets, contributes to plaque formation. A second hypothesis involves the influence of chronic

inflammation caused by reactions of the immune system. A third hypothesis is that free radicals, the unstable molecules that form as a natural byproduct of metabolism, damage the coronary vessels and thereby contribute to the development of atherosclerosis. In the body of my review paper, I will present evidence from published studies that support the three hypotheses. Here's a draft of my plan so far:

> **Rhetorical Goal:** Summarize the published research that supports the hypotheses of interest, to convince readers that they are plausible.
>
> **Strategy 1:** Present studies on cholesterol.
>
> **Strategy 2:** Present studies on inflammatory chemicals and processes.
>
> **Strategy 3:** Present studies on free radical damage.

The plan's rhetorical goal is definitely on target, but my three strategies obviously lack detail and depth. This would not be a major problem if I had the specific plans stored in mind. But I don't, so I will need to flesh out my strategies on paper. What, precisely, will I do and say in presenting the studies on cholesterol, inflammatory chemicals and processes, and free radical damage? After contemplating answers to this question, I realize that my strategies should reflect detailed plans for summarizing the specific studies that support the three hypotheses about how atherosclerosis develops. Here, then, is a revised version of my first strategy:

> **Strategy 1:** Present studies in which Sutherland et al. and Hudson et al. directly tested the hypothesis that diets high in cholesterol contribute to the development of atherosclerosis in rodents. Describe the key methods of these studies, including (a) the dietary manipulations to increase cholesterol intake and (b) the procedures for measuring fatty plaque formation and hardening of the arteries. Synthesize the study's results showing that high-cholesterol diets induced atherosclerosis.

Comparing my original strategy to this revision, you will appreciate how well-developed and specific plans can ultimately make the processes of generating content and drafting much more manageable and productive.

Check Your Strategies for Whether They Are Logically Related to Their Rhetorical Goals

Every strategy in your plan should be clearly relevant to its associated rhetorical goal. In other words, each strategy should be a logical means of accomplishing the goal. For example, in the plan for my review paper on atherosclerosis, my original three strategies are indeed logically related to their rhetorical goal. Each strategy is a means of accomplishing the goal to summarize the published research that supports hypotheses about how the disease develops. But let's say that, without thinking carefully enough, I happened to add a fourth strategy:

> Suggest future research on the effects of high-cholesterol diets on the development of atherosclerosis.

This one is not relevant at all because it would not serve the goal at hand. So an appropriate revision would be to delete the errant fourth strategy or to move it to a part of my goal-based plan that focuses on suggesting future research. To identify potentially irrelevant strategies, raise the following question about each strategy in your plan: Will it directly help me accomplish its associated rhetorical goal?

SUMMING UP AND STEPPING AHEAD

At the start of this chapter, I hailed goal-based planning as the heart of the scientific writing process. Now you know that goal-based plans are constructed with rhetorical goals, along with detailed and specific strategies for accomplishing them, for each major section of a scientific paper. You also know how to develop effective rhetorical goals and strategies through activities such as using model papers, applying task and audience analyses, and helicopter thinking. Along the way, this chapter has hinted at how expert authors use their goal-based plans to guide key activities throughout the writing process. One of the most productive applications comes in the stage of generating content, or developing the information, ideas, and arguments that will eventually compose a draft. Up next, we focus on the goal-directed process of generating content.

CHAPTER 3

Generating Content

Nutrition for Health and Performance

Regarding nutrition, we are all scientists of sorts. Our bodies are laboratories for daily experiments in biochemistry, metabolism, and psychology. We naturally raise questions and pose hypotheses about how various nutrients will affect us. And we reach conclusions about the foods we eat based on how they make us feel, look, and perform. Our everyday intuitive studies in nutrition can certainly be useful, although they are quite limited compared to taking the scientific approach. As a case in point, imagine that a friend has gone on a popular but controversial low-carbohydrate diet. Eight weeks and 12 lost pounds later, she is a true believer in cutting carbs. It's worth asking, however, whether your friend's weight loss can be directly attributed to carbohydrate restriction *per se*. What if she happened to start exercising upon starting the diet? Or what if the weight loss was actually due to unhealthy consequences of the diet, such as muscle wasting or dehydration? To answer these questions with confidence, we would need to conduct well-controlled studies in nutrition science.

These days, the growing field of nutrition science is especially interesting because it abounds with unanswered questions and unsettled controversies. This is especially true for research on dietary approaches to weight loss and promoting optimal health. To weigh in successfully on the debates, one must apply advanced critical thinking and reading skills—the very skills that are central to this chapter's activities for generating content in scientific writing.

INTRODUCTION

Suppose that for a nutrition class you have an assignment to write a 15-page review paper on a topic of choice. You have decided to write about the effects of

Vitamin C on the common cold. The results of many published studies indicate that high doses of Vitamin C significantly reduce the duration and severity of cold symptoms. In contrast, numerous other studies have revealed no treatment effect at all. It's a perfect issue for a position paper, the specialized form of review papers that make original arguments on debatable issues. So far, your planning has been extremely efficient and productive. You have astutely analyzed the task at hand, taking stock of the key instructions, guidelines, and evaluation criteria in your assignment. Your search for literature has been prolific, and you are making good progress in learning the science underlying your paper's issue. You have a clear sense of your readers' needs, values, and expectations. To top off your promising preparation, you have developed a comprehensive plan of well-targeted rhetorical goals and strategies.

So far, so good, indeed. But now you face a new challenge, clearly the most fundamental of all in scientific writing: You must figure out what you will actually say in your paper—the points to make, the ideas to develop, the concepts to convey, and the arguments to advance. How will you go about filling those 15 pages with highly appropriate, well-developed, and convincing content? When storms of ideas build in your mind, how will you decide which to refine and develop? Reading journal articles, books, and Web sites on your research issue, how will you identify the most relevant content for your paper? By what principles and practices will you interpret data from studies on your issue? How will you form strong independent conclusions? To go beyond paraphrasing what others have written about your issue, what's your approach to constructing convincing original arguments? These questions are central to this chapter, which covers a stage of the writing process that we call *generating content*.

ABOUT THE PROCESS

As mapped in Figure 3.1, our approach to generating content involves the following five activities.

Activity 1: Brainstorm independently and collaboratively. Some of the best content for your paper may already be stored in your mind, either as well-formed prior knowledge or as rough ideas and images. To stir up and shape this content, you will undertake the activity of brainstorming. In our approach, brainstorming can be a solo or collaborative process.

Activity 2: Read for relevance. In the early stages of planning scientific writing projects, we read published literature mainly to acquire topic knowledge. In the current stage of generating content, our intention shifts from reading to *learn* to reading to *write*. This activity is for identifying relevant content in the published literature—specifically, the content that will help you successfully achieve your rhetorical goals and strategies.

Activity 3: Interpret study data. To accomplish many key rhetorical goals in scientific writing, authors must skillfully interpret study data. This activity

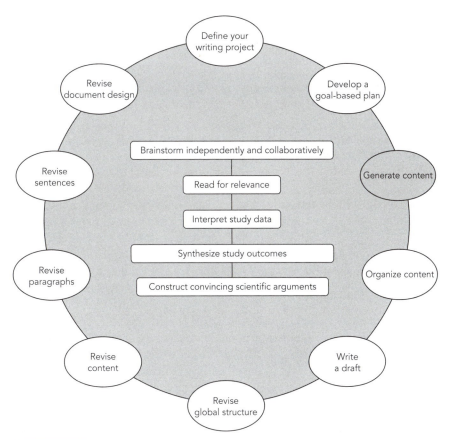

FIGURE 3.1 Process map for *generating content.*

relies on specialized knowledge about statistics, and it involves making well-informed judgments about the practical importance of research outcomes.

Activity 4: Synthesize study outcomes. As defined in Chapter 1, synthesis entails piecing together the outcomes of numerous studies on a research issue. In some cases, it's a matter of identifying common patterns in studies that have produced similar outcomes. A more difficult challenge is solving the puzzles of conflicting study outcomes. This activity will guide you through an organized approach to generating truly superior content through synthesizing research.

Activity 5: Construct convincing scientific arguments. The most valuable content in scientific communication advances original arguments to resolve debatable issues. To construct convincing scientific arguments, you must understand their structure and successfully apply critical thinking skills for generating their content. As you will learn in this activity, these skills involve evaluating arguments in the published literature.

At its best, generating content is a goal-directed process. All five of this chapter's activities are closely guided by the rhetorical goals and strategies that make up goal-based plans. To appreciate this point, recall a key take-home message from the previous chapter: Whether certain information and ideas are worth writing about depends on whether they will help you accomplish appropriate rhetorical goals for your paper. Another defining feature of this chapter's activities is that they rely on helicopter thinking, the powerful mode of cognition that I described in Chapter 2. The current chapter presents many demonstrations of helicopter thinking applied to generating useful content.

As diagrammed in the outer ring of our process map (Figure 3.1), generating content is an intermediate stage between developing a goal-based plan and the stages of organizing content and writing a draft. Organizing content entails structuring information, ideas, and arguments in relatively large units of discourse, specifically a paper's major sections and subsections. Before you can determine the best global organization for your paper's content, you must first generate it. At the current stage of the writing process, it's usually counterproductive to get hung up on detailed matters of global organization. It's also problematic to interrupt the process of generating content by struggling to transform ideas into the structured sentences and paragraphs of a draft. The processes of generating content and drafting engage competing modes of thinking. In generating content, we bring information into the mind and use it to construct images, ideas, concepts, and arguments. Drafting works in the other direction. We draft by translating the content in our minds into a structured form, so the content flows from our minds to our papers. Often, the best results come from concentrating on generating content and drafting separately. So the product of this chapter's activities is content generated in rough note form. However, our approach places one important constraint on structure: As you will see, the notes produced through this chapter's activities are organized by the rhetorical goals and strategies of goal-based plans.

SOLO AND COLLABORATIVE BRAINSTORMING

Brainstorming generally refers to spontaneous and intense thinking. Our specialized definition of brainstorming is conceiving and developing useful ideas without relying on external knowledge. The main distinction is that this activity does not directly involve reading scientific literature (unlike the other activities in this chapter). You might creatively stir up ideas by brainstorming knowledge that is already stored in your mind; or you can tap into in the minds of colleagues and coauthors. We will call these two methods *solo brainstorming* and *collaborative brainstorming*, respectively.

Solo Brainstorming

Solo brainstorming begins with a rhetorical goal, its related strategies, and a simple but powerful question: *What do I know and what can I say to achieve this plan?* This question will begin turning your helicopter thinking blades, so to speak.

Concentrating on one rhetorical goal at a time, you can focus deeply on brainstorming helpful ideas. You might be surprised by how prolifically your ideas will flow. Brainstorming is especially powerful for working with rhetorical goals that call for creative thinking rather than relying strictly on established knowledge in published literature. One example is the goal to convince readers that your research issue is meaningful to society; another is the goal to propose ideas for future studies on your research issue.

For a demonstration of solo brainstorming, consider the approach of a student named Angela whose project was an informative review paper on dietary approaches to weight loss in obese children. For part of her introduction section, Angela developed the rhetorical goal and strategies presented in Figure 3.2.

It's a good idea to first identify goals and strategies in your plan that are especially fitting for brainstorming. Perusing her plan, Angela realized that she would not be able to brainstorm ideas for Strategy 1.1 because, off the top of her head, she did not know the statistics on the incidence and economic costs of obesity-related diseases in children. Angela also realized that brainstorming would not work for Strategy 1.2. Although she had heard that obese children are at a high

Rhetorical Goal 1: Explain why my topic is meaningful to society, so that my audience sees the value of my paper and wants to keep reading it.

Strategy 1.1: Present statistics on the number of children and adults who are obese in the U.S. and on the economic costs of treating obesity-related diseases.

Strategy 1.2: Discuss the negative psychological outcomes, including risks for low self-esteem and depression, that are linked with obesity in children.

Strategy 1.3: Discuss the negative physical health outcomes, including type II diabetes and cardiovascular disease, that are linked with childhood obesity.

Strategy 1.4: Explain why I'm focusing my paper on dietary approaches to weight loss specifically in obese children. Stress the importance of my topic by making distinctions between dietary and health-related concerns for obese children versus obese adults.

FIGURE 3.2 An excerpt from a student's goal-based plan for a review paper on diet and obesity in children.

risk for depression and low self-esteem, Angela had not yet studied the research on these associations. Angela had, however, previously learned about how childhood obesity affects physical health, which is the focus of Strategy 1.3. Angela did not consider herself an expert on the topic, but at least she could get started on developing her ideas through brainstorming. In addition, Angela would brainstorm to generate content for Strategy 1.4. She developed this strategy to show readers that compared to obese adults, obese children have special dietary needs and health concerns. Angela envisioned that drawing distinctions between obese adults and obese children would highlight the meaningfulness of her research issue. To achieve Strategy 1.4, Angela would rely at least partly on her own creative reasoning.

Through brainstorming, Angela produced a set of notes that she integrated into her goal-based plan, shown as follows (Figure 3.3).

Rhetorical Goal 1: Explain why my topic is meaningful to society, so that my audience sees the value of my paper and wants to keep reading it.

Strategy 1.3: Discuss the negative physical health outcomes, including type II diabetes and cardiovascular disease, that are linked with childhood obesity.

Note 1.3.1: Previous research shows that many obese children (as young as 5) already have early signs of cardiovascular disease—ones commonly found in obese adults. Including high blood pressure, high LDL cholesterol, and fatty plaques in the coronary arteries. Obesity is at root of these health problems. Changes in diet can reduce body fat.

Why my topic is important: doctors and parents need to know about the research on optimal dietary approaches to help obese children lose body fat and avoid cardiovascular disease.

Note 1.3.2: Obese children have a high risk of developing type II diabetes. An excessive amount of body fat causes insulin resistance, an underlying problem linked to type II diabetes. Insulin is the hormone responsible for transporting glucose from the blood into the body's cells, which use glucose for energy. When the body's cells become resistant to insulin, glucose builds up in the blood. This condition, called hyperglycemia, damages the nerves and the linings of blood vessels.

Why my topic is important: Effective dietary approaches to reducing obesity in children will prevent type II diabetes and the serious neural and vascular diseases that are linked to it.

FIGURE 3.3 Brainstorming notes integrated into a goal-based plan.

Note 1.3.3: Health risks associated with lack of exercise among obese kids. Research shows they tend to be sedentary, likely to be sedentary when they become adults, too. Sedentary adults get diseases caused by physical inactivity—heart disease, some types of cancer. So due to lack of exercise, obese kids could grow up to be physically inactive obese adults. Bad for health!

Why my topic is important: The effect of obesity on exercise behavior in childhood could lead to poor health in adulthood.

Strategy 1.4: Explain why I'm focusing my paper on dietary approaches to weight loss <u>specifically</u> in obese children. Stress the importance of my topic by making distinctions between dietary and health-related concerns for obese children versus obese adults.

Note 1.4.1: Some studies show that adults can lose a large amount of weight on certain diets (like low-calorie and low-carbohydrate diets) without any negative health risks. But the effects of these diets might be different in children for a few reasons. One is that there are differences in metabolism between adults and children. Another reason is that children have different nutritional needs, especially for vitamins and minerals that promote normal growth and development.

Why my topic is important: Metabolism and nutritional needs different between children and adults. We can't be sure that research on adults applies to children. So we need to look specifically at research on the effects of dietary changes on obese children.

TO-DO LIST

1. For note 1.3.1: Find actual statistics from research to back up the idea that obese children already have signs of cardiovascular disease.

*2. For note 1.4.1: Find articles on differences in metabolism between children and adults. Important: Explain **how** the differences in metabolism between children and adults might lead to different responses to certain dietary changes. Give specific examples. Compared to adults, will children on certain diets lose more weight or less weight? Compared to effects on adults, how might certain diets affect health of children? Get help on developing this idea from Professor Harris!*

FIGURE 3.3 *continued.*

If you don't already have a tried-and-trusted approach to taking notes for your papers, you might find Angela's system to be especially helpful. Its most powerful feature is that the notes are structured within a goal-based plan. Each note fleshes out a specific strategy for accomplishing a rhetorical goal. Take Note 1.3.1, which records what Angela previously learned about alarming markers of cardiovascular disease in obese children. The note also reflects Angela's ideas about the importance of her research issue from the perspective of doctors and parents. The content of Note 1.3.1 is right on target for achieving Angela's strategy to discuss the negative physical health outcomes that are associated with childhood obesity. In addition, the note's content logically contributes to accomplishing Angela's rhetorical goal to explain the meaningfulness of her research issue. This approach to generating and noting content naturally translates into well-organized and goal-directed papers.

Another advantage of Angela's note-taking system is that it enables efficient approaches to organizing content and drafting. As Angela continues to generate content through brainstorming and other activities, she will build on her existing notes and create new notes. Later, when Angela concentrates on organizing her plan's content, she can simply cut and paste her notes to arrange them in a linear outline. When she sits down to write a draft, Angela will be able to include some of her notes verbatim. An example is Note 1.3.2, which concerns risks of developing type II diabetes in obese children. The entire note is written in complete sentences that make up a fairly coherent paragraph. When Angela was brainstorming the ideas for this note, they fortunately came to mind in complete sentences. Angela recorded the ideas as such, figuring that they would be perfect for her draft. In contrast, some of Angela's notes are informal, written in sentence fragments and nonscientific language. This is true of Note 1.3.3. Angela took this note quickly because she did not want to interrupt her brainstorming activity to translate the rough ideas into scientific language and complete sentences.

At the bottom of Angela's plan, have a look at the to-do list. It contains Angela's notes to herself about tasks for continuing to generate content for her paper. The list's first item is a reminder to find statistics on the incidence of cardiovascular disease markers in obese children. This reminder refers to Note 1.3.1, which Angela recognized as incomplete. When she brainstormed the idea that obese children are already at risk for cardiovascular disease, Angela did not have the supporting statistics at hand. So the first item in her to-do list is to obtain those statistics from the research literature. The second item reminds Angela to expand an underdeveloped idea in Note 1.4.1. Her idea is that metabolic functions might differ in adults and children, so the research on diets that promote weight loss in adults might not apply to children. Through brainstorming alone, Angela was not able to fully develop the idea. She could not come up with examples of differences in metabolic functions between children and adults. Angela realized that she needed to learn more about that topic. So the to-do note reminds Angela to seek literature and help from her professor.

In all activities for generating content, you must decide how extensively to develop your notes. Suppose that for accomplishing one of your rhetorical goals, you already know exactly what to say and how to say it. In this case, it doesn't make sense to spend time fleshing out the content in note form. But if you are uncertain about what to say, your best bet is to take well-developed notes before drafting. You might not choose to use the formal numbering system demonstrated in Angela's notes if it doesn't suit your needs. But the general approach of integrating your ideas into a goal-based plan is well worth the effort.

At unexpected times and in the oddest places, great ideas may arise in spontaneous brainstorms. It's an experience that skilled writers relish and are always prepared for. Imagine that you're standing in line at the grocery store checkout on a Sunday afternoon, gazing at tabloid covers. The farthest thing from your mind is a lab report due on Wednesday. Out of nowhere, a torrential brainstorm sets into motion. It starts with a small, vague notion of how to interpret perplexing results from the class experiment that you conducted earlier in the week. Within a matter of seconds, an elegant solution comes to mind. You're in line at the supermarket, about to unload your shopping cart. So you assure yourself that such a potent solution will be easy to recall later. But, of course, when you eventually sit down in front of your computer to draft, the idea is long gone and forgotten. Experienced writers are prepared to note great ideas wherever and whenever they come to mind. The old-fashioned method is to jot them down on a pocket-sized notepad. These days, we can use digital technologies, such as voice recorders, PDAs (personal digital assistants), and cell phones. When I tell my students to leave themselves a voicemail message to save great ideas that arise at unexpected times and in odd places, they look at me as if I'm from another planet. But those expressions are certainly better than the pained ones I see when students try unsuccessfully to recall promising spontaneous brainstorms.

Collaborative Brainstorming

Brainstorming is often defined as a group activity, which certainly characterizes its use among scientists. Because most journal articles and grant applications are coauthored, collaborative writing is the standard approach in science. In collaborative brainstorming, coauthors identify their rhetorical goals and then work together to develop their ideas through shared knowledge and creative team-oriented thinking. Even if you don't have coauthors, you can still benefit from collaborative brainstorming by sharing ideas with peers and perhaps even experts on your paper's topic. It's perfectly acceptable for student writers to contact established scientists on campus, or even around the world via the Internet, to raise questions and share ideas. Whether the scientists will agree to collaborate depends on how students present themselves and their requests. The worst thing you can do is to show up uninvited on the office doorstep of an elite scientist intending to shoot the breeze. To prepare for a prearranged meeting, make sure that you

- have read the scientist's published papers and that you know the details of his or her research,
- can effectively demonstrate your interest in the scientist's work,

- have made considerable efforts to develop your ideas independently,
- have a well-defined and goal-focused writing project,
- have specific questions and comments that are directly relevant to the scientist's research interests, and
- are not trying to get the expert to simply give you ideas and write your paper for you.

Whether you are brainstorming with a classmate or a famous researcher, make sure to note who comes up with which ideas. To meet codes of ethics and honesty in scientific writing, your paper should appropriately credit your collaborators if you include their original ideas.

READING FOR RELEVANCE

Although brainstorming is surely a productive way to begin developing ideas for scientific papers, it has its limits. At some point, even the most knowledgeable and creative scientists must turn to the published literature to build on the content they have generated through brainstorming. Reading skillfully requires purposefully focusing attention on the most relevant information, ideas, and arguments. In our goal-directed approach to writing, the relevance of reading material depends on how much it helps writers accomplish their rhetorical goals and strategies. Of course, all reading activities in the scientific writing process are interdependent. To determine the relevance of what you are reading, for example, you must first make the effort to understand the material. In addition, it's most efficient to first determine the relevance of what you are reading before spending the time and mental energy critically evaluating it. Here we focus on reading for *goal* relevance; later in the chapter we will cover several activities that engage more critical reading.

Reading for goal relevance is pure helicopter thinking. You can begin at the highest level of your goal-based plan by focusing on individual rhetorical goals, using them to guide your reading and to identify content worth developing for your paper. Working in the other direction, you begin with concrete information, ideas, and concepts in the literature. In this approach, you will pause every now and again, maybe after reading a paragraph or a section, to reflect on the following question: *Is this material relevant to what I'm planning to write about in my paper?* The answer depends on your rhetorical goals and strategies. So you will elevate your thinking to focus on these elements of your plan. If the reading material is indeed valuable for accomplishing one or more of your rhetorical goals, you will take notes accordingly.

To demonstrate reading for relevance, let's suppose that you and I are coauthoring an NIH grant application. We are proposing a new study on the effects of omega-3 fatty acids on mental depression. Dietary sources of omega-3 fatty acids include vegetable oils and fish. In contrast to unhealthy forms of dietary fat such as saturated fatty acids and trans fat, omega-3 fatty acids have been associated with positive physical health outcomes. However, researchers have not fully determined the influence of omega-3 fatty acids on mental health, including their potential role for treating clinical depression. One of the articles that we have collected is a review paper titled "Omega-3 Fatty Acids and Major Depression: A Primer for the Mental Health Professional" (Logan, 2004). The paper was written by Dr. Alan C. Logan, a leading expert on dietary supplements and mental health. A three-paragraph excerpt from Logan's article is presented in Figure 3.4. As indicated by the section heading, Logan summarizes the outcomes of epidemiological studies on omega-3 fatty acids and depression. (Epidemiology is a branch of the life sciences in which researchers study behaviors, genetic characteristics, and environmental factors that influence the incidence and prevention of diseases in large populations.) Take a few minutes now to read the three-paragraph excerpt, just to understand the content.

Epidemiological Data

A number of epidemiological studies support a connection between dietary fish/seafood consumption and a lower prevalence of depression. Significant negative correlations have been reported between worldwide fish consumption and rates of depression [10]. Examination of fish/seafood consumption throughout nations has also been correlated with protection against post-partum depression [11], bipolar disorder [12] and seasonal affective disorder [13]. Separate research involving a random sample within a nation confirms the global findings, as frequent fish consumption in the general population is associated with a decreased risk of depression and suicidal ideation [14]. In addition, a cross-sectional study from New Zealand found that fish consumption is significantly associated with higher self-reported mental health status [15].

Not all studies support a connection between omega-3 intake and mood. A recent cross-sectional study of male smokers, using data collected between 1985 and 1988, indicated that subjects reporting anxiety or depressed mood had higher intakes of both omega-3 and omega-6 fatty acids [16]. In a large population-based study of older males aged 50–69, there was no association between dietary intake of omega-3 fatty acids or fish consumption and depressed mood, major depressive episodes, or suicide [17].

The epidemiological studies which support a connection between dietary fish and depression clearly do not prove causation. There are a number of cultural, economic and social factors which may confound the results. Most significantly, those who do consume more fish may generally have healthier lifestyle habits, including exercise and stress management. Despite the limitations, the epidemiological data certainly justify a closer examination of omega-3 fatty acids in those actually with depression.

FIGURE 3.4 An excerpt from a published review paper on the effects of omega-3 fatty acids on depression (Logan, 2004). Reproduced with permission.

From the first paragraph of Logan's review, we learn that numerous epidemiologic studies have indicated a "connection" between consuming fish and other seafood, which are rich in omega-3 fatty acids, and relatively low incidences of depression and other psychiatric diseases. In the excerpt's second paragraph, Logan highlights conflicting studies, including one that revealed that dietary omega-3 intake was actually associated with a greater incidence of depressed mood. In the third paragraph, Logan points out that whereas a number of epidemiological studies may indicate that omega-3 fatty acids protect against depression, the studies do not prove a cause-and-effect relationship. The main reason is that the studies have not adequately controlled for extraneous variables; these are factors that interfere with a study's independent variables, creating confusion about the actual causes of observed outcomes.

To evaluate the relevance of Logan's article to writing our grant application, we must refer to our goal-based plan, excerpts of which are presented in Figure 3.5. The plan includes two rhetorical goals for different subsections of our grant application; for the sake of simplicity in this demonstration, each goal has just one strategy. Rhetorical Goal 1 involves presenting previous studies on our research issue, specifically ones that have produced conflicting results. This is an especially important goal for a grant application because reviewers will respond negatively to proposals for new studies on research issues that are not unresolved. As indicated in Strategy 1.1, our approach to achieving this goal is to present the contrasting findings from specific studies on our research issue. Later in our plan, Rhetorical Goal 6 involves arguing that our proposed research methods are superior to those used in previous studies. Strategy 6.1 reflects our intention to justify our proposed methods aimed at controlling for extraneous variables, which are also called confounding variables.

Rhetorical Goal 1: Present the contrasting findings from previous studies on our research issue, to convince readers that the issue is still unresolved and therefore worthy of future study.

Strategy 1.1: Present the contrasting findings from epidemiologic studies on omega-3 fatty acids and depression.

Rhetorical Goal 6: Explain and justify our proposed experimental methods, to argue that they will solve methodological problems from previous studies on the issue.

Strategy 6.1: Justify our methods intended to control for possible confounding variables.

FIGURE 3.5 An excerpt from a goal-based plan for a hypothetical grant application on omega-3 fatty acids and depression.

Take a few minutes to reread the three paragraphs from Logan's review paper—this time reading for goal relevance. After each paragraph, pause to reflect on whether the content is relevant to our grant application. Specifically, ask whether the information and ideas might help us accomplish any of the rhetorical goals and strategies in our plan. To understand my reflections and the notes that I took on Logan's article, which are presented as follows, you will benefit from doing this activity before reading on.

On its own, Logan's first paragraph is not directly relevant to our rhetorical goals. The paragraph's take-home message is that numerous studies have linked omega-3 intake, through fish and seafood consumption, to a relatively low incidence of depression and to good mental health. If we were to emphasize this idea exclusively, the reviewers of our grant application would surely deny our request for funding. Why would a grant agency fund our proposed study if many previous investigations have already revealed favorable effects of omega-3 fatty acids? Of course, Logan's first paragraph becomes more obviously relevant in the context of his second paragraph. There, the author writes about study outcomes that disagree. This content is definitely applicable to our first rhetorical goal in Figure 3.5. If we can demonstrate that the previous research on omega-3 fatty acids and depression reveals inconsistent and uncertain findings, our grant application's reviewers will be more inclined to recognize that our proposed research is truly necessary. So we should definitely note the relevant ideas in Logan's first two paragraphs. My notes for Rhetorical Goal 1 and its associated strategy are presented in Figure 3.6. The notes simply paraphrase Logan's review of the conflicting epidemiological studies. (Normally, I take notes in sentence fragments and chicken scratch. But for your reading ease and comprehension, the notes in Figure 3.6 are written in fairly complete sentences.)

Rhetorical Goal 1: Present the contrasting findings from previous studies on our research issue, to convince readers that the issue is still unresolved and therefore worthy of future study.

Strategy 1.1: Present the contrasting findings from epidemiologic studies on omega-3 fatty acids and depression.

Note 1.1.1: As reviewed by Logan (2004), the outcomes of epidemiologic studies do not agree (these are studies 10-15 in Logan's reference list). Some studies have revealed that large populations of individuals who regularly consume omega-3 fatty acids (in fish and seafood) have relatively low rates of different types of depression, including seasonal affective disorder, bipolar disorder, and postpartum depression.

FIGURE 3.6 Notes integrated into a goal-based plan. As explained in the text, the notes were generated through the activity *Reading for Relevance.*

Note 1.1.2: Other epidemiologic studies have not revealed lower rates of depression in people who regularly consume omega-3 fatty acids (these are studies 16 and 17 in Logan's reference list). In a cross-sectional study, male smokers who reported high intakes of omega-3 fatty acids were more likely to have anxiety and depression than counterparts who consumed less omega-3 fatty acid. Another study revealed no connection at all between omega-3 fatty acid intake and depression in 50-59 year-old men.

TO-DO LIST:

For notes 1.1.2 and 1.1.2: Need to read the original research articles to verify Logan's review and to obtain the actual study data (get citations 10-17 in Logan's reference list).

FIGURE 3.6 *continued.*

At the bottom of my notes in Figure 3.6, the to-do list includes an important reminder to obtain the original research papers on which Logan based his review of literature. As discussed in Chapter 1, review papers are secondary sources of scientific knowledge. As such, they summarize information and ideas from primary sources, which you know are peer-reviewed research papers. To verify Logan's interpretations of the studies that he summarized, we need to read the research papers that originally reported the studies. From these papers, we will generate more content to flesh out the notes in Figure 3.6. We will look for details about the methods and results of the conflicting epidemiological studies on omega-3 fatty acid and depression. In our grant application, to serve Rhetorical Goal 1, these details will better help us convince readers that the previous research is contradictory.

For more practice at reading for goal relevance, let's focus on Logan's third paragraph (Figure 3.4). Its main point is that epidemiological studies cannot prove that omega-3 fatty acid intake directly *causes* low rates of depression. The reason, says Logan, is that various factors "may confound the results" of such studies on this issue. As examples of confounding variables, Logan lists cultural, economic, and social factors as well as exercise and stress management. For our grant application, are Logan's ideas about these possible confounding variables relevant? Consider that one of Logan's points is that people who eat a lot of fish, which is high in omega-3 fatty acid and is generally considered a healthy food, may also exercise regularly and control stress effectively. Logan does not explain exactly how these factors may confound the results of epidemiological studies, but his message is implied. Regular exercise and stress management are known to be associated with a low risk of depression. In an epidemiological study, it's quite possible that the subjects who report consuming large amounts of omega-3 fatty acid intake also participate in exercise and stress reduction behaviors. If the incidence of depression is low in these subjects, the researchers might conclude that omega-3 fatty acids played a role. But unless the researchers control for the subjects' exercise and stress management, their conclusion might very well be flawed.

The preceding interpretations of Logan's third paragraph are indeed relevant to Rhetorical Goal 6 in our goal-based plan, which is shown again in Figure 3.7. This is the goal to explain and justify our proposed experimental methods. A key strategy for accomplishing this goal, Strategy 6.1 in the plan, is to justify our proposed methods aimed at accounting for confounding variables. Under this strategy in Figure 3.7, I have noted my ideas for justifying our methods for controlling against possible confounding effects of exercise. As you read the notes, consider how they would serve as the basis for appropriate content for our hypothetical grant application.

Reading for relevance is another example of a writing process activity that ain't rocket science but is nevertheless extremely challenging and equally powerful. The greatest challenge comes with having to continually elevate your thinking—that is, to rise above what you are reading in order to reflect on how it might,

Rhetorical Goal 6: Explain and justify our proposed experimental methods, to argue that they will solve methodological problems from previous studies on the issue.

Strategy 6.1: Justify our methods intended to control for possible confounding variables.

Note 6.1.1: Previous epidemiologic studies have shown that people who consume large amounts of omega-3 fatty acids in fish have a low incidence of depression. But these studies have not adequately controlled for subjects' possible exercise participation. Regular exercise can directly reduce symptoms of depression through its effects on the brain. In people who consume omega-3 fatty acids in fish, low rates of depression might not be due to their fish intake at all. Instead, these individuals might be experiencing the positive effects of their exercise participation. Our proposed study will control for the possible confounding effects of exercise by screening subjects (only sedentary subjects will be accepted into the study) and by having subjects regularly report on their physical activity levels throughout the study.

TO-DO LIST:
1. *Make sure to give credit to Logan (2004) when we write about possible confounding variables in previous epidemiological studies on omega-3 intake and depression.*
2. *Need to track down research on exercise and depression. Include statistics and citation information from these studies in our proposal.*

FIGURE 3.7 More notes, generated through *Reading for Relevance*, integrated into a goal-based plan.

or might not, relate to what you are writing. The preceding demonstration was somewhat simplified because it involved only three paragraphs from just one published article, and it was based on only two rhetorical goals. Imagine a more realistic scenario in which you are reading something like 20 articles on your topic and your plan consists of 15 rhetorical goals, each with many well-developed strategies. In this case, the cognitive demands would be substantial. But the effort would pay great dividends in terms of the quantity and quality of the content that you would generate. Even in our simple example, we developed a fairly large chunk of appropriate content from a small excerpt of a published article.

INTERPRETING STUDY DATA

One of the core activities for generating original content in scientific papers is interpreting study data. The data might be your own, collected from a class experiment or a thesis project; or they might be presented in published journal articles on your research issue. Research-derived data are simply numbers that serve no real purpose until we interpret them, which entails explaining their meaning and messages, assessing their significance, and using them to reach conclusions and support arguments. The skill of interpreting study data is fundamental to accomplishing many vital rhetorical goals for scientific papers. Here are a few examples:

- In the introduction section to research proposals, authors interpret the results of previous studies on their research issues in order to present the current state of knowledge and to convince readers that the issues are unresolved.
- In the body section of position papers, authors interpret the data from published research to support their claims.
- In the discussion section of IMRAD-structured papers, such as research papers and lab reports, authors interpret data to compare their studies' outcomes to those of related studies in the published literature.

Our approach to the activity at hand emphasizes interpreting study data critically and independently. When reading published articles, you are by no means obliged to accept authors' interpretations of their data unconditionally. The best practice is to interpret the results for yourself and to compare your conclusions to the authors'. Along the way, we will discuss the purposes and advantages of taking this independent approach.

Imagine that you are writing a critical review paper on dietary approaches to lowering blood pressure. You will devote a major section of your paper to evaluating published research on low-sodium diets. Dietary sodium, an element of table salt and salty foods, causes the body to retain water. A consequence is an increase in plasma volume, which is the watery component of the blood. When plasma volume increases so does the amount of blood that fills the heart and that the heart must then pump. Through this mechanism, high-sodium diets can elevate blood pressure and cause hypertension. So it makes sense that most doctors advise their patients with hypertension to restrict sodium intake. However, studies on this issue do not collectively support the universal prescription of low-sodium diets. In fact, the published literature reflects quite a bit of uncertainty about whether low-sodium intake effectively lowers blood pressure in everyone with hypertension. Through critically reviewing the literature, you intend to offer useful recommendations for the cases and conditions that warrant dietary sodium restriction.

One of the published research papers that you have collected on your issue reports a clinical trial that the authors called *SALT*, or Study About Low-blood-pressure Treatments. The SALT study included 200 men and women with physician-diagnosed hypertension. The subjects were randomly assigned to one of two diets:

(1) 100 subjects consumed a low-sodium diet (LS), reducing daily sodium intake to 50% of normal values; and
(2) 100 subjects consumed a control diet (C) that maintained their normal amounts of sodium intake.

The researchers precisely controlled for every conceivable extraneous variable, so we can interpret the results without concerns about confounding methodological flaws. Once a week over the 12-week study period, the researchers measured systolic and diastolic blood pressure. Systolic pressure, which indicates the force that develops as the heart contracts, is the top number in blood pressure readings (such as 120/80). Diastolic pressure, which indicates the force generated as the heart muscle relaxes, is the bottom number. To simplify our demonstration of how to interpret study data, we will focus only on the SALT study's results for systolic blood pressure, which are graphed in Figure 3.8. The graph shows the two study groups' changes in systolic blood pressure relative to baseline values, which are designated by the zero line.

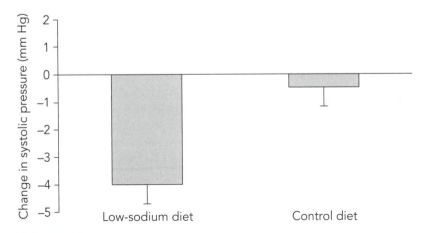

FIGURE 3.8 Results from the hypothetical SALT study on sodium restriction and blood pressure.

In their research paper's discussion section, the SALT authors interpreted their data as follows.

> The change in systolic blood pressure was statistically greater for subjects on the low-sodium diet (mean change = −4.0 mm Hg) versus subjects on the control diet (mean change = −0.5 mm Hg); $P < 0.0001$. This highly significant effect is the basis for our unconditional endorsement of sodium-restricted diets for patients suffering from hypertension.

Acknowledging the authors' conclusion, but recognizing that you are not obliged to accept it categorically, how will you interpret the SALT study's data for your critical review paper? Like many data interpretations in the published literature, this one can be challenged by questions and viable alternative interpretations. It's definitely worth asking, for example, what exactly do the authors mean by "statistically greater" and "this highly significant effect"? What does the P value less than 0.0001 tell us about the significance? Are you really convinced that the results indeed support the "unconditional endorsement" of low-sodium diets for treating hypertension? You will need to answer these questions skillfully in order to generate effective content for your paper. The skill at hand involves critically evaluating the statistical and practical significance of research outcomes.

Interpreting the Statistical Significance of Study Data

As introduced in Chapter 1, the term *statistical significance* has a specialized meaning in research. Recall that it refers to the reliability, or reproducibility, of research outcomes. Take the SALT study's key outcome, which the researchers calculated by subtracting the mean change in systolic blood in the control group (−0.5 mm Hg) from the mean change in the low-sodium diet group (−4.0 mm Hg). The resulting *difference score* is −3.5 mm Hg. The SALT authors interpreted this outcome as statistically significant. By definition, this interpretation is an *inference*, which I defined in Chapter 1 as a conclusion about a population that is based on results obtained from a sample, or a subset of the population. More specifically, the SALT authors' interpretation is an inference about the probability of obtaining a −3.5 mm Hg difference score, or greater, if the study were repeated many times under certain assumptions that will be clarified as our discussion continues. The SALT team had to rely on inference because they studied a relatively small sample of 200 subjects. The researchers could not possibly have determined the effects of low-sodium diets on the world's entire population of hypertensive people. In the United States alone, this population is over 70 million strong.

Whereas inference is a well-established approach to reaching conclusions, all scientific inferences lend themselves to a type of bias called sampling error. To illustrate this concept, let's say that from the world's entire population of hypertensive individuals we randomly selected numerous samples of 200 subjects. Suppose that we formed 1,000 samples in all. Using these samples, we conducted 1,000 separate experiments that repeated the original 3-month SALT study. We randomly assigned 100 subjects in each sample to a low-sodium diet and the other 100 subjects to a control diet with normal levels of sodium. For each of the 1,000 experiments, we calculated difference scores for systolic blood pressure by subtracting the mean change in the control group from the mean change in the low-sodium group. We now have 1,000 difference scores. As you might expect, all are not exactly −3.5 mm Hg, the difference score obtained from the SALT study's sample. Instead, the scores vary. In one experiment, for example, systolic blood pressure decreased 5.3 mm Hg more in the low-sodium group than in the control group. In another experiment, the difference score was only −.8 mm Hg. In six more experiments, the difference scores were −6.9, −3.7, −4.8, −2.9, −3.5, and +.2 mm Hg.

Now imagine that with tremendous ambition, along with a healthy dose of magic dust, we conducted the SALT study on everyone in the world who has hypertension. In the overall population, we found a difference score of −4.2 mm Hg. The deviation between this true difference score and the others reported in the previous paragraph reflects sampling error in the repeated sample-based experiments. Any number of factors can cause sampling error. Suppose, for example, that in one of our 1,000 experiments a large number of subjects assigned to the treatment group just happened to have a unique genetic predisposition that made them somewhat resistant to sodium's effects on blood pressure. In this experiment, the difference score was only −0.3 mm Hg. If we based our interpretations on this value only, we would conclude that a low-sodium diet is not very effective at treating hypertension. But consider that the study sample does not represent the larger population, in which the sodium-resistant genetic predisposition is rare. So our conclusion, influenced by sampling error, would be problematic.

To determine the statistical significance of a study's data, we want to know whether they represent the true conditions in the population of interest. (By *conditions*, I mean the true effects of treatments or the true differences between groups.) The problem, however, is that we usually do *not* know the true population conditions—if we did know the conditions, we would not need to conduct studies on them. Following conventional practices in research and statistics, we must make several assumptions about what's happening in populations. A fundamental assumption, called the null hypothesis, is a tentative and testable belief that nothing especially interesting is going on. For example, in a study to determine whether two variables are related, the null hypothesis holds that no relationship truly exists in the population. Or in a study to determine the effects of a medical treatment, the null hypothesis is that the treatment is actually no more effective than a placebo pill. The null hypothesis is meant to establish an unbiased foundation on which research is conducted. In addition, it is the standard for guiding decisions about the viability of competing research hypotheses. These are scientists' statements about the study outcomes that they anticipate based on previous research and established knowledge and theory.

At the start of their study, the SALT research team acknowledged the null hypothesis that low-sodium diets have no effect on blood pressure. That is, they began by operating on the assumption that the difference score in the population is zero. The SALT team also established their competing research hypothesis, stating that low-sodium diets do reduce blood pressure and that their study would yield a difference score less than zero. By convention, the null hypothesis is tested through statistical analyses such as *t* tests, ANOVA (analysis of variance), multiple regression, and chi-square tests. As explained in Chapter 1, analyses that guide inferences about study samples produce statistics called *P* values. Earlier I defined a *P* value as the probability of obtaining study results due to chance factors. That definition, while generally accurate, is missing a few important details. The complete definition, presented as follows, will help you interpret study data more astutely. (By the way, to avoid passing out, you might want to take a deep breath before you start reading the full definition.)

> A *P* value is the probability of obtaining an observed sample-derived study result, or a more extreme result, if the study were repeated many times on additional samples of the population, assuming that the null hypothesis is true.

To grasp this definition, consider again the observed sample-derived result from the SALT study: Systolic blood pressure decreased 3.5 mm Hg more in the low-sodium group versus the control group. Working on the presumption of the null hypothesis, we would initially attribute this difference score to chance factors rather than to a true, systematic effect of low-sodium intake. Suppose that the study's statistical analysis produced a high *P* value, say 0.48. Here's the interpretation: In repeated SALT experiments on different samples of the population, a difference score of −3.5 mm Hg or greater would result 48 times out of 100, assuming that the null hypothesis is true. Given that the result would occur so frequently under this assumption, we would be inclined to attribute the −3.5 mm Hg difference score to chance factors, such as sampling errors. We would then accept the null hypothesis—that the true difference score is zero—as more viable than the research hypothesis. Recall, however, that the SALT study's statistical analysis actually yielded a *P* value less than 0.0001. This means that in fewer than 1 out of 10,000 repeated SALT experiments, a difference score of −3.5 mm Hg or greater would result if the null hypothesis were true. Given that this result is so extremely unlikely to occur, we would be inclined to reject the null hypothesis and to conclude that the result reflects a true, systematic effect of reduced sodium intake. In this case, we would also say that the result is *statistically significant*, although a more precise interpretation is to say that it is *reliable*, or reproducible.

Following a long-standing tradition in statistics, many scientists begin their studies by setting a criterion *P* value, or a cutoff above which they will interpret findings as nonsignificant. This criterion is called *the level of significance*, or *alpha* (α). In life science studies, common values for α are 0.05 and 0.01. Let's say that before their study, the SALT researchers set α at 0.05. They used this value to guide decisions about whether to accept or reject the null hypothesis. If their statistical test had produced a *P* value of 0.049 or less, following tradition the researchers would have rejected the null hypothesis and concluded that their results were statistically significant, because $P < 0.05$. If, however, the calculated P value had been 0.051 or greater, the decision would have been to accept the null hypothesis and to conclude that the results were not significant. The traditional practice of setting α has been criticized in recent years. Many forward-thinking statisticians and scientists recognize that criterion P values can be quite arbitrary and therefore unnecessary. A more progressive approach is to report calculated P values and let readers decide for themselves whether the results are reliable or not.

We began this discussion by looking at the SALT authors' interpretation of their study data. Here's the text again:

> The change in systolic blood pressure was statistically greater for subjects on the low-sodium diet (mean change = -4.0 mm Hg) versus subjects on the control diet (mean change = -0.5 mm Hg); $P < 0.0001$. This highly significant effect is the basis for our unconditional endorsement of sodium-restricted diets for patients suffering from hypertension.

Once more, I raise the question, how would *you* interpret the study's results? Recall that in this scenario you are planning to write a critical review paper on dietary approaches to lowering blood pressure. Applying what you have learned about statistical inference, hypothesis testing, and P values, you would accurately view the SALT results as statistically significant. The low P value indicates that the treatment group's reduction in systolic blood pressure was likely due to a systematic effect of the low-sodium diet rather than to sampling error. Nonetheless, a few other questions linger about the results. It's worth critically evaluating the phrase "highly significant effect." We should also ask about the validity of the authors' "unconditional endorsement of sodium-restricted diets for hypertensive individuals." Is this conclusion actually justified by the results? The following guidelines will help you answer these questions and thereby generate especially insightful content for your critical review paper. The guidelines apply to interpreting the outcomes of all studies in which conclusions depend on statistical inference.

1. Don't jump to the conclusion that statistically significant results are *practically* significant, or that they have any real-world importance. The next section of this chapter elaborates on this vital guideline for interpreting research outcomes.

2. Don't be wowed by infinitely small P values. How low must a P value go before you decide to reject the null hypothesis? Would you be any more convinced to reject it if a calculated P value were 0.0001 (1 in 10,000 chance of making an incorrect decision) compared to 0.01 (1 in 100 chance of making an incorrect decision)? Consider that once a reasonable level of significance has been reached, the number of zeros following the decimal point in a P value becomes fairly inconsequential. So the SALT authors' interpretation of a *highly* significant effect is a bit misleading.

3. Consider sample size and the variability of study data. Just to make a point, let's alter the SALT study scenario. Suppose that the researchers were short on funding and could afford to test only 20 subjects. In the end, the mean changes in systolic blood pressure were −7.5 mm Hg and −0.5 mm Hg in the low-sodium and control groups, respectively. On the surface, the −7.0 mm Hg difference score suggests that a low-sodium diet is a very effective treatment for lowering blood pressure. But the researchers interpreted the results as nonsignificant because their statistical analysis yielded a high P value of 0.23, well above the criterion α P value of 0.05. The researchers thus concluded that hypertensive patients should seek treatments other than dietary sodium restriction.

Now imagine that the study included 1,000 subjects and that the very same difference score, −7.0 mm Hg, was obtained. All other factors being equal, the resulting P value would be considerably lower. It would almost certainly fall well below $P = 0.05$. This outcome would support the conclusion that a low-sodium diet reduces blood pressure reliably. The key point is that as sample size increases, the probability of obtaining statistically significant results also increases. This makes sense because as a study's sample size increases, more members of the population are included and therefore sampling error decreases. A related guideline holds for the effects of variability, which refers to the dispersion, or range, of study data. As discussed in Chapter 1, variability is commonly assessed by measures such as the standard deviation and the standard error of the mean. As measures of variability decrease, which reflects less sampling error, P values also decrease.

So in our altered SALT study scenario, the one in which only 20 subjects participated, you would actually be justified in advancing an interpretation that differs from the authors'. You could very well argue that a 7 mm Hg reduction in blood pressure would have been revealed as a reliable result *if the study had included more subjects*. But given the study's lack of statistical power, the best conclusion is that the results cannot be used to offer valid advice about whether individuals with hypertension should reduce their sodium intake. This example is by no means unrealistic. Especially in studies on humans, it can be difficult and expensive to recruit large numbers of subjects. In some research fields, small sample sizes are quite common. Interpretations of their results should thus account for sample size and variability.

Interpreting the Practical Significance of Study Data

Let's have one more look at the results from the SALT study: The reduction in systolic blood pressure was 3.5 mm Hg greater in the low-sodium diet group than in the control group. The P value was less than 0.0001. Based on these results, the authors reached the following conclusion:

> This highly significant effect is the basis for our unconditional endorsement of sodium-restricted diets for patients suffering from hypertension.

Before you jump on board with this conclusion, consider that interpretations of statistical significance in no way, shape, or form imply *practical* significance, which refers to meaningful real-world implications of research results. Based only on the difference score and *P* value obtained from the SALT study, the endorsement of a low-sodium diet for treating hypertension is, to say the least, problematic. Statistical significance indicates nothing about the efficacy of a treatment or the importance of an observed difference between study groups. This is an extremely important point, one that scientists and science students occasionally overlook in their papers.

To offer the soundest advice about whether patients with hypertension should restrict sodium intake, the SALT researchers needed to interpret the practical significance of their data, which in this case concerns the clinical, health-related implications. What difference does a reliable (statistically significant) reduction in blood pressure make if it doesn't actually improve a patient's health status? To convincingly support their practical advice, the SALT researchers needed to raise and answer the following questions.

- Does a 3.5 mm Hg reduction (relative to control conditions) in systolic blood pressure improve cardiovascular function to a degree that would have noticeable effects on the quality of life in people with hypertension?
- Could a 3.5 mm Hg reduction in systolic blood pressure lower the risk for suffering heart attacks and strokes, the life-threatening outcomes that are associated with high blood pressure?
- If a 3.5 mm Hg reduction in systolic blood pressure is not enough to improve quality of life and prevent heart attacks and strokes, how much more must it decrease?

To reach valid interpretations of the practical significance of a study's results, you must turn to knowledge in your research field. To accurately interpret the SALT study's results, for example, you would need to know a good bit about cardiovascular physiology. Specifically, you must understand how high blood pressure can damage the arteries that supply oxygen-rich blood to the heart muscle and brain. In addition, you would need evidence from sound research conducted to determine the magnitudes of blood pressure reductions that are necessary to improve cardiovascular function and to prevent heart attacks and strokes. If you were writing a critical review paper on dietary approaches to reducing blood pressure, you could take the SALT authors to task for overlooking this essential support for their conclusion. Better yet, you could seek sources of literature that would inform your strong independent interpretation and conclusion.

In this section, we interpreted data from just one study, which by no means fully resolves the issue at hand. In reality, some studies have produced a fair amount of support for claims that hypertensive individuals should reduce their dietary sodium intake. But other studies have indicated no benefit. In recent years, many new studies have been conducted to determine whether individual differences between hypertensive patients might explain the contrasting findings.

For example, researchers have sought to determine whether the effects of dietary sodium restriction depend on an individual's sensitivity to salt, initial blood pressure, and medication status. Reflecting science at its best, each new study has inspired numerous others in the field. To finally derive individually targeted recommendations regarding sodium restriction, researchers and clinicians will be challenged to compile and interpret all of the studies' results. To do so, they will apply the advanced critical thinking skill of synthesis, which we cover next.

SYNTHESIZING STUDY OUTCOMES

As introduced in Chapter 1, the process of synthesis in science is somewhat like putting together the pieces of a puzzle. More precisely defined, synthesis is a critical thinking skill that writers apply to constructing complete and accurate representations of the knowledge that defines their fields. The pieces of any scientific puzzle are the conclusions, results, and methods from the related studies on a focused research issue. Synthesis is necessary to generate successful content for achieving a number of key rhetorical goals for scientific papers. In general, these are goals that involve explaining relationships between study outcomes and developing arguments to resolve contrasting findings and conclusions.

In scientific papers, the antithesis of synthesis is fragmented content that fails to present a clear, composite picture. Take the case of a review paper that summarizes 20 studies on its topic. In the paper's body, the author devotes each paragraph to briefly describing the methods, results, and conclusions of just 1 of the 20 studies. The first paragraph begins, "Jones conducted a study on X and found Y." The second paragraph begins, "Smith conducted a study on X and found Z." The third paragraph begins, "Williams conducted a study on X and found Y." The remaining 17 paragraphs follow suit, reviewing one study at a time. This *dis*integrated format might be acceptable if review papers were intended simply to list the methods and outcomes of separate research studies. However, scientific audiences generally have higher standards for the overall goals of these documents. Experienced readers will demand synthesis. If the 20 studies in this scenario all have similar methods, results, and conclusions, readers will complain that the author is being redundant by presenting the disjointed paragraph-by-paragraph summaries. Or consider the more common scenario in which related studies on a research issue are characterized by agreements as well as debatable disagreements. Readers will justifiably expect a composite picture that reveals the consistencies, explains the mismatches, and offers resolutions to the debates. So readers will be understandably frustrated if the author presents the studies in a fragmented form.

The experience of a student named David will serve to demonstrate how to generate effective content through synthesizing study outcomes. David is writing a position paper on the effects of low-carbohydrate diets on weight loss. The traditional stance on dieting is that the only way to lose weight is to consume less energy, or fewer calories, than the body expends from day to day. According to this view, the composition of dietary macronutrients—which are carbohydrates, fats, and proteins—does not matter as long as the dieter cuts calories. For health reasons, however, mainstream nutrition experts recommend calorie-cutting diets that are relatively high in carbohydrates and low in fats. High-fat diets have been linked to cardiovascular disease and cancer. In striking contrast to the conventional view on dieting, supporters of popular low-carbohydrate diets argue that they cause substantial weight loss without requiring calorie restriction. This debatable argument is based on the premise that low-carbohydrate diets shift the body's metabolism to burn more fat. We will follow David as he takes an exemplary approach to synthesizing studies on this issue in order to develop an original argument in his position paper. Through synthesis and by applying critical thinking skills to evaluate the studies' methods (skills that we will cover later in the chapter), David has decided to argue that low-carbohydrate diets are indeed effective for promoting healthy weight loss.

Focusing on Rhetorical Goals that Require Synthesis

Like all activities for generating the content of scientific papers, synthesis is ideally a goal-directed activity. So a fitting way to begin is to focus attention on your rhetorical goals that, to be achieved successfully, demand synthesis. One such goal, which applies to the introduction section to all scientific papers, is to present the current state of knowledge derived through published studies on your research issue. Another example, this one applying to the discussion section of IMRAD-structured papers, is to relate your study findings and conclusions to those from previous studies on your issue. Any rhetorical goal that entails resolving conflicts and debates in the scientific literature requires synthesis for successful outcomes. Given the debatable nature of David's research issue on low-carbohydrate diets and weight loss, his plan definitely calls for synthesis. Among other rhetorical goals for his position paper, David has developed two essential ones for constructing a well-synthesized argument. One of the goals is to support his claim for the effectiveness of low-carbohydrate diets by presenting data from published studies. The other key goal is to acknowledge and refute opposing claims.

Creating a Summary Chart to Guide Synthesis

Setting out to solve an intricate jigsaw puzzle, you will first empty its numerous pieces onto a large table, perhaps in your kitchen or living room. Then you will look for complementary shapes and angles in the randomly organized pieces, identifying ones that naturally fit together. When the puzzle pieces do not match up, you will try to figure out why by examining them more closely and testing different configurations. In the end, this process solves the puzzle and reveals a clear composite picture. To begin solving scientific puzzles through synthesis, we also lay out their many pieces onto tables. These tables are called summary charts. They are ideal for organizing notes about the methods, results, and conclusions of published studies. Most important, summary charts guide the process of synthesizing study outcomes. Table 3.1 presents an excerpt from a summary chart that David constructed to guide his synthesis of studies on low-carbohydrate diets and weight loss. David's complete summary chart reviews many published studies on his research issue. To simplify our demonstration of synthesis, however, we will concentrate on the three studies summarized in Table 3.1.

Summary charts facilitate efficient comparisons of study methods, results, and conclusions. The organization of information in rows and columns helps writers identify similarities and differences across numerous studies, which is the core thinking skill in synthesis. An ideal way to generate a summary chart's content is to apply the *focusing questions* that were presented in Chapter 1 to guide the activity of reading to learn science (Figure 1.5, page 41). Recall, for example, one of the focusing questions for reading a research paper's discussion section: *What were the author's main conclusions and arguments for them?* Applying this question to reading published research papers on his issue, David noted the studies' results and the authors' conclusions in the last two columns of his summary chart.

Following the standard format for summary charts, David noted citation information about the published papers in his chart's first column. The second column includes notes about the studies' participants. From the methods section of each article, David isolated this information by asking the focusing question: *What key characteristics describe the study's subjects and materials?* It's important to note all subject characteristics that might significantly influence the outcomes of studies on your research issue. Among other characteristics, David noted the number of women and men who participated in the studies on low-carbohydrate diets and weight loss. Knowing that metabolic processes differ across the sexes, David figured that low-carbohydrate diets might conceivably affect weight loss differently in women and men. He reasoned that if the results and conclusions varied across the studies, a contributing explanation might be differences in the numbers of female versus male participants.

The *Subjects* column of David's summary chart also includes information about the study participants' mean (M) weight and body mass index (BMI) at baseline. BMI is a measure of body composition, calculated as an individual's weight (in kilograms) divided by height (in meters squared). Higher BMI values generally indicate greater amounts of body fat. David included the subjects' initial weight and BMI values because he reasoned that these factors might influence the extent to which low-carbohydrate diets affect weight loss. In synthesizing study outcomes, David will examine whether the amount of weight that subjects lost was related to their initial weight and body fat composition.

The third column of David's summary chart, titled *Methods*, includes essential details about the studies' experimental procedures and variables. It's especially important to note the independent variables of research, including details about the treatments that the researchers administered. For David's research issue, the key independent variable is the carbohydrate composition of the subjects' diets. So David noted the prescribed amounts of carbohydrates, along with the fat and protein intake, for the diet groups in each study. In his synthesis,

TABLE 3.1 Summary Chart of Studies on Low-Carbohydrate Diets and Weight Loss

Study	Subjects	Methods	Results	Conclusions
Brown (2006)	74 female community-based volunteers Age: M = 36.6 yrs Weight: M = 127.8 kg BMI: M = 42.3 Other subject characteristics: sedentary lifestyle; 85% of the subjects had preexisting cardiovascular disease and type 2 diabetes	Prescribed diets: Subjects were randomly assigned to 1 of 2 diets 1. Low-CHO diet: Subjects were given prepared meals in which CHO accounted for 10% of total daily calories; PRO and fat intake were unrestricted 2. Conventional diet: Subjects were given prepared meals consisting of 55% CHO, 30% fat, and 15% PRO; individualized calorie intake was targeted at 500 calories per day under subject's normal consumption Study duration: 6 months Control variables: Subjects were instructed to maintain their sedentary lifestyle	Total weight loss after 6 months 1. Low-CHO group: M = 8.7 kg 2. Conventional group: M = 3.6 kg *Subjects on the low-CHO diet lost 142% more weight *Difference was statistically significant ($P < 0.001$)	Over a 6-month period, low-carbohydrate diets are superior to conventional diets for promoting weight loss

Abbreviations: BMI = body mass index; CHO = carbohydrate; M = mean; PRO = protein.

(continued)

TABLE 3.1 *continued*

Study	Subjects	Methods	Results	Conclusions
Franklin (2004)	6 female and 44 male community-based volunteers Age: M = 38.3 yrs Weight: M = 132.4 kg BMI: M = 45.3 Other subject characteristics: sedentary lifestyle; 92% of the subjects had preexisting cardiovascular disease and type 2 diabetes	Prescribed diets: Subjects were randomly assigned to 1 of 2 diets 1. Low-CHO diet: Subjects were given prepared meals in which CHO accounted for 16% of total daily calories; PRO and fat intake were unrestricted 2. Conventional diet: Subjects were given prepared meals consisting of 60% CHO, 25% fat, and 15% PRO; individualized calorie intake was targeted at 500 calories per day under subject's normal consumption Study duration: 4 months Control variables: Subjects were instructed to maintain their sedentary lifestyle and avoid dieting drugs	Total weight loss after 4 months 1. Low-CHO group: M = 5.9 kg 2. Conventional group: M = 2.3 kg *Subjects on the low-CHO diet lost 157% more weight *Difference was statistically significant ($P < 0.001$)	Over a 4-month period, low-carbohydrate diets are superior to conventional diets for promoting weight loss

Abbreviations: BMI = body mass index; CHO = carbohydrate; M = mean; PRO = protein.

| Gonzalez (2004) | 38 female and 4 male community-based volunteers
Age: M = 37.0 yrs
Weight: M = 92.8 kg
BMI: M = 32.9
Other subject characteristics: sedentary lifestyle; subjects were admitted into the study only if they had no markers of cardiovascular disease and type 2 diabetes | Controlled diets: Subjects were randomly assigned to 1 of 2 diets
1. Low-CHO diet: Subjects were instructed to prepare their own meals in which CHO accounted for 30% of total daily calories; PRO and fat intake were unrestricted
2. Conventional diet: Subjects were instructed to prepare their own meals consisting of 50% CHO, 28% fat, and 22% PRO; individualized calorie intake was targeted at 500 calories per day under subject's normal consumption
Study duration: 2 months
Additional treatments: Subjects received nutritional counseling and participated in 45 minutes of light exercise per day | Total weight loss after 2 months
1. Low-CHO group: M = 2.1 kg
2. Conventional group: M = 2.3 kg
*Difference was NOT statistically significant (P = 0.35) | Low-carbohydrate diets do NOT lead to greater weight loss than conventional high-carbohydrate, low-fat diets |

Abbreviations: BMI = body mass index; CHO = carbohydrate; M = mean; PRO = protein.

David will assess whether the amount of weight that subjects lost was related to the specific amounts of carbohydrates, fats, and proteins that they consumed. David also recorded information about how the diets were administered. Notice, for example, that in the Brown and Franklin studies, the subjects were given prepared meals that contained the researcher's prescribed amounts of carbohydrates, fats, and proteins. In the Gonzalez study, however, the subjects prepared their own meals according to the researchers' instructions. David insightfully speculated that the studies' results might have been influenced by how strictly the researchers were able to control the subjects' diets. By noting information about the methods for administering the diets, David will be able to easily test his speculation through synthesis. The *Methods* column of the summary chart includes other key details about the studies' design and procedures, including the duration of each study and extraneous variables that the researchers sought to control.

All summary charts should include the key results and conclusions of the studies they review. As David noted in the last column of Table 3.1, both Brown and Franklin concluded that weight loss is significantly greater in individuals on low-carbohydrate versus conventional high-carbohydrate, calorie-cutting diets. In contrast, Gonzalez concluded that low-carbohydrate diets do *not* cause greater weight loss than conventional diets. The agreements and conflicts in these conclusions pose the challenges of synthesis. David's summary chart is an essential tool for successfully meeting these challenges. (Incidentally, following a useful convention for writing position papers and other forms of literature reviews, David will include his entire summary chart in the body of his paper.)

Synthesizing Studies with Similar Conclusions

Suppose that for a position paper one of your rhetorical goals is to back your central claim with the outcomes of published studies. You will need to compile the supporting evidence from studies that yielded similar conclusions that align with your argument. For David, who will argue that low-carbohydrate diets are effective and safe for promoting weight loss, this process begins with the concurring conclusions from Brown's and Franklin's studies.

After identifying the studies on your research issue that have produced similar conclusions, you should check for how closely their supporting data line up. For example, as indicated in the *Results* column of David's summary chart, the weight-loss advantage afforded by a low-carbohydrate diet was fairly similar in Brown's and Franklin's studies. Compared to subjects on the conventional diets, subjects on the low-carbohydrate diets lost 142% and 157% more weight in the Brown and Franklin studies, respectively. David has learned that many additional published studies on this issue have yielded similar results: Over periods of approximately 4–6 months, subjects on low-carbohydrate diets tend to lose approximately two to three times more weight than subjects on conventional diets. This pattern in the studies' results, revealed through synthesis, is important for David to write about in his position paper. The pattern supports his claim for the effectiveness of low-carbohydrate diets (especially if David can successfully argue that the magnitude of weight loss observed in the studies is clinically significant).

When the conclusions and supporting data of different studies on your research issue agree, you might expect to find similarities in their methods as well. It's worth synthesizing the methodological pieces of scientific puzzles because doing so affords greater explanatory and predictive power. As indicated in the *Methods* column of David's summary chart, for example, the diets that Brown and Franklin prescribed were similar in their composition of carbohydrates, fats, and proteins. In the low-carbohydrate diets, carbohydrates accounted for 10% and 16% of total calorie intake in the two studies, respectively. In addition, both studies did not restrict the caloric intake from protein and fat. These observations will help David develop specific guidelines for the macronutrient composition of low-carbohydrate diets.

Considering the matching puzzle pieces in Brown's and Franklin's studies, we can see part of a picture forming. It's really quite an interesting picture, because it challenges traditional views on optimal dietary approaches to losing weight. These views do not predict the greater amounts of weight loss experienced by subjects on low-carbohydrate diets. In reality, however, the picture is not so clear and complete. The reason is that other studies on David's issue have produced contrasting results and conclusions.

Synthesizing Studies with Contrasting Conclusions

Our approach to this truly challenging but especially powerful aspect of synthesis takes the following four steps:

Step 1: Isolate any contrasting conclusions from studies on your research issue.

Step 2: Form your own conclusions by independently interpreting the contrasting studies' results. (This step involves the skills for interpreting data, as presented earlier in the chapter.)

Step 3: Compare your conclusions to those of the studies' authors; if your conclusions disagree with those in the published literature, develop your synthesis by explaining and justifying your interpretations.

Step 4: If your conclusions agree with those in the published literature, look for differences in the studies' methods as possible explanations for their contrasting outcomes.

David took this approach to synthesizing the contrasting conclusions from the Gonzalez study versus the Brown and Franklin studies. As noted in the last column of David's summary chart, Gonzalez claimed that low-carbohydrate diets do *not* cause greater weight loss than conventional diets. The supporting data for this conclusion indicated no significant difference in mean weight loss between subjects on the low-carbohydrate diet versus the conventional diet. Because we have already covered the process of independently interpreting study data, we will skip over Steps 2 and 3 in the preceding list. Let's say that David agreed that Brown, Franklin, and Gonzalez interpreted their results accurately.

So here's our situation in a nutshell: The conclusions differ across studies on the very same research issue, the supporting results also conflict, and the authors' interpretations of the results are sound. In this case, synthesis requires systematically comparing the studies' methods to identify similarities and differences. Upon identifying differences, you will consider whether they might actually have accounted for the studies' contrasting results and conclusions. This process of synthesis is often powerful enough to fully resolve complex scientific puzzles, so it results in truly valuable content for scientific papers. The following questions guide a productive approach to synthesizing the contrasting outcomes of studies through comparing their methods.

- Across the contrasting studies, were there any differences in relevant characteristics of the subjects, such as their ages, sex, initial physical and health conditions, lifestyle behaviors, and so on?
- Did levels of the studies' independent variables differ? If the research issue concerned a drug treatment, for example, did the researchers administer similar or different doses of the drug across the studies?
- Were the dependent variables similar or different? In other words, across the studies did the researchers measure identical outcomes?
- Did the studies differ in setting, length, and other aspects of their design?
- Were there differences in the researchers' approaches to controlling for potential confounding variables?

David applied these questions to guide his comparisons of the methods used by Brown, Franklin, and Gonzalez. He identified a number of relevant similarities and differences. For example, as noted in the *Subject* column of David's summary chart, the participants' mean ages were similar across the studies, ranging from 36.6 to 38.3 years. So David can rule out *age* as a factor that explains Gonzalez's contrasting conclusion (i.e., the conclusion that low-carbohydrate diets do not afford a weight-loss advantage). There were differences in the distribution of female and male subjects across the three experiments. Compared to Franklin's study, Gonzalez's study included relatively more women. However, David figured that this difference in subject characteristics would not adequately explain the difference in the two studies' outcomes. The reason is that Brown's study included relatively more women than Franklin's study, but the studies' outcomes agreed.

David noted interesting differences in the weight and body composition values across the studies' subjects. For example, Gonzalez's subjects (mean weight = 92.8 kg) weighed considerably less than Brown's subjects (mean weight = 127.8 kg) and had lower BMIs (32.9 and 42.3, respectively). In addition, Gonzalez's subjects were healthier because the researcher screened out individuals with markers of cardiovascular disease and type 2 diabetes. Off the top of his head, David did not know how initial values for body weight and health might influence the effectiveness of low-carbohydrate diets. So in his goal-based plan's to-do list, David noted a reminder to search for literature on these possible associations.

For a bit more practice at synthesizing contrasting study outcomes, let's compare the diets in Brown's and Gonzalez's studies. Both studies address the effects of dietary carbohydrate restriction on weight loss. But there is no absolute standard for what actually defines the *low* in low-carbohydrate diets. In published studies on this issue, you will see considerable variation in subjects' prescribed carbohydrate intake. Could these variations account for contrasting conclusions about the effects of low-carbohydrate diets on weight loss? In David's summary chart, you will see that the mean daily carbohydrate intake for subjects on the low-carbohydrate diet differed considerably across Brown's and Gonzalez's studies. The values were 10% and 30% of total daily calorie intake, respectively. Through reading articles about the physiological mechanisms by which low-carbohydrate diets promote weight loss, David learned that the greatest advantages result when carbohydrate intake is restricted to very low levels. He thus reasoned that the lack of a significant treatment effect in Gonzalez's study might have been due to the relatively high carbohydrate content of the low-carbohydrate diet. In his goal-based plan, under the rhetorical goal to present supporting evidence for his claim for the effectiveness of low-carbohydrate diets, David took notes on the differences in carbohydrate intake across the studies; in addition, he noted the mechanisms by which these differences might have influenced the studies' results and conclusions.

There are a number of other relevant and insightful comparisons of the methods, results, and conclusions from the three studies in David's summary chart. David will make these comparisons to generate content that successfully accomplishes key rhetorical goals for his paper. Applying the advanced skills of synthesis, as described and demonstrated in this section of the chapter, you will have the same positive experience and productive outcomes in your writing projects.

CONSTRUCTING CONVINCING SCIENTIFIC ARGUMENTS

In the order of their presentation, this chapter's content-generating activities have posed progressively more challenging critical thinking and reading skills. For example, synthesis is more cognitively demanding than brainstorming, and the content generated through synthesis is more sophisticated. In this section, we cover the most advanced content-generating activity of all in scientific writing, which entails the skills for constructing strong, original scientific arguments. The occasion to argue in science arises when you establish your own view on an issue that someone else might question or debate. Your view might be an assertion about the importance of your paper's research issue, a statement to justify your choice of an experimental method, a conclusion that you derive through interpreting data, or the position that you are taking in a debate. In response to a statement of your view, engaged readers will naturally ask, "What knowledge, evidence, and reasoning support it?" This question invites scientific argument. It's definitely an invitation worth accepting. The greatest advancements in science are realized through constructing convincing arguments, especially ones that resolve debatable research issues. In addition, the skills of scientific argument enable writers to go well beyond simply saying what someone else has already said and what everyone already knows.

The customary approach and etiquette to arguing in science differ considerably from how many people argue in other walks of life. A scientific argument, for instance, is nothing like a derisive political debate or the fighting words exchanged between roommates about whose turn it is to clean the toilet. When elite scientists argue, they are neither contentious, nor confrontational, nor excessively competitive. Instead they are generally collegial, cooperative, and even complimentary to opponents. This is not to say that scientists lack the qualities of forcefulness, decisiveness, and passion in making their arguments. They demonstrate all of these qualities. However, strong scientific arguments are shaped by the ultimate goals of science, which are *not* to win, to prove the other side wrong, or even to prove one's own side right. Instead, the ultimate goals of science and scientific argument are to resolve debatable research issues, to discover truths in nature, and to gain knowledge for the most beneficial applications in life.

Setting up the Structure of a Scientific Argument

A number of rhetorical goals for all types of scientific papers engage writers in constructing arguments. Whatever the occasion, all scientific arguments have a classical structure, which can be described by four main elements: claims, lines of support, warrants, and counterarguments.

Claims

The assertions, conclusions, theses, and position statements that scientific authors seek to convince readers to accept are the *claims* of scientific arguments. Consider the following examples of common types of claims in scientific papers.

- Conclusions based on interpreting research results: This study's data support the conclusion that Vitamin C supplementation is effective in treating the common cold.
- Assertions about the underlying mechanisms of research observations: Vitamin C speeds recovery from the common cold by enhancing immune system function.
- Views on the comparative value of a treatment: For reducing the severity of cold symptoms, Vitamin C is more effective than the herb Echinacea.
- Position statements concerning practical applications of study results: Based on the findings of the present study, we recommend that adults take 500 mg of Vitamin C per day at the onset of cold symptoms.

Lines of Support

A unified synthesis of evidence and reasoning to back a claim is called a line of support (or a line of argument). Some lines of support are *data driven*, or based on statistics derived from research. Others are *concept driven*, or based on existing knowledge, established theory, or creative new ideas and hypotheses. Consider, for example, a data-driven line of support for the claim that Vitamin C effectively treats the common cold. The author might synthesize statistics on subjects' ratings of cold-symptom severity and the number of days that subjects needed to recover from their colds. A conceptual line of support for this claim would involve explaining the underlying physiological mechanisms by which Vitamin C reduces the severity of cold symptoms and promotes recovery. The strongest scientific arguments generally rely on both data-driven and concept-driven lines of support.

Warrants

To support the claim for Vitamin C as an effective treatment for the common cold, suppose that an author presents a data-driven line of support that involves a type of immune system cell called neutrophils. The data indicate that the proliferation and activity of these cells increase in subjects taking Vitamin C. A reasonable response from readers is, "What's the connection between the proliferation and activity of neutrophils and the effectiveness of Vitamin C in reducing cold symptoms?" An equivalent question is, "What's the *warrant* for this argument?" Warrants are assumptions, definitions, ideas, concepts, and theories that explain how lines of support are related to argument claims. In some cases, the connections between claims and their supporting evidence and reasoning are obvious to readers; so authors need not develop explicit warrants. If, however, readers might question the connections, it's worth drawing the warrants. In the preceding example, this would involve explaining the physiological processes by which (a) Vitamin C promotes the proliferation and activity of neutrophils and (b) neutrophils influence the progression and symptoms of the common cold.

Counterarguments

By definition, any scientific argument can be contested by counterarguments, which comprise alternative and opposing claims, lines of support, and warrants. As an example, for the argument that Vitamin C supplements speed recovery from the common cold, a counterargument is that the treatment is no more effective than a placebo pill. Proponents of counterarguments (to your argument) might back their claims with directly contrasting data-driven and concept-driven lines of support. Or, they might raise criticisms about the logic of your claims and supporting evidence and reasoning.

The major structural elements of scientific arguments—claims, lines of support, warrants, and counterarguments—are logically related to the genre's main rhetorical goals. For example, an appropriate goal for any argument is to present data-driven lines of support in order to convince readers to accept a claim. Another standard goal is to acknowledge and refute all viable counterarguments to your own argument. Chapter 8 presents a number of rhetorical goals for constructing arguments in different sections and types of scientific papers. Here we focus on the underlying skills for generating the content of scientific arguments. This partly involves applying the content-generating activities that we have covered so far in the chapter. For example, you might develop a concept-driven line of support for your argument through brainstorming. Or, to build a data-driven line of support, you will interpret and synthesize study outcomes. The remainder of this chapter describes and demonstrates two especially advanced skills for generating strong content for scientific arguments. The skills involve evaluating the strengths and weaknesses of arguments and research methods in the published literature.

Evaluating Published Scientific Arguments

Part of generating content for an original scientific argument involves figuring out what to say about scientific arguments in the published literature. This, of course, demands critical reading. Like all critical reading activities, this one ideally operates on the principle that we should not unconditionally accept what we read in the published literature. Reflecting on this principle, student writers may understandably respond, "How can we possibly critique the advanced and complex arguments of veteran scientists?" It's certainly not a simple task, but students can succeed remarkably well if they know (a) the right questions to ask about scientists' arguments and (b) how to skillfully apply the questions to identify strengths and weaknesses of authors' claims, lines of support, warrants, and responses to counterarguments. Our systematic and comprehensive approach takes the following five steps. (By the way, our approach to evaluating arguments in the published literature is easily adaptable to evaluating arguments in oral presentations.)

> **Step 1: Determine whether what you are reading is actually a scientific argument.**
>
> **Step 2: Reflect on the rhetorical goals that you seek to accomplish through evaluating the argument.**
>
> **Step 3: Raise key diagnostic questions for identifying strengths and weaknesses in the argument.**
>
> **Step 4: Answer the diagnostic questions, and thereby evaluate the argument, by applying think-ahead and think-through strategies.**
>
> **Step 5: Note your critical evaluations in your goal-based plan, or develop them directly as draft material.**

The first step is to decide whether what you are reading is an argument or not. In certain types of articles, such as informative review papers, much or even all of the content serves rhetorical goals only for informing and describing, rather than for arguing. In contrast, position papers are nothing but arguments. In some papers, argumentative content is interspersed with informative and descriptive content. For example, arguments are made (a) in the introduction sections to all types of scientific papers, where authors assert that their motivating research issues are important; (b) in the methods sections of grant applications, where authors seek to justify their proposed study procedures and analyses; and (c) in the discussion sections of research papers, where authors seek to convince the audience to accept their conclusions. Any sentence (or set of sentences) that reflects a questionable or debatable claim indicates an argument, which should set the reader's evaluative process in motion. Once you know that you are reading an argument, it's worth taking the time to identify its major structural elements—its central claim, lines of support, warrants, and counterarguments. The evaluative process addresses each of these elements.

Step 2 in our approach to evaluating arguments is to continually reflect on your rhetorical goals. Keep in mind that the ultimate reason for evaluating arguments in the published literature is to generate content for your own arguments. So when you identify strengths and weaknesses in published arguments, do the helicopter thinking—that is, consider how your evaluation might serve the rhetorical goals and strategies that you have developed for your paper. Ultimately, if your evaluation is relevant, you should note it in your goal-based plan or develop your ideas as draft material (Step 5).

Step 3 engages diagnostic thinking, similar to a physician's routine for assessing a patient's condition during a comprehensive examination. For assessing strengths and weaknesses in scientific arguments, we will apply five diagnostic questions, which are listed in Figure 3.9 and described in detail just around the corner. As designated in Step 4 of our approach, we answer the diagnostic questions by applying two critical thinking methods, which we will call the *think-ahead* and *think-through* strategies. They happen to be pivotal to a number of the book's upcoming activities that involve critically assessing content in scientific papers—activities that include evaluating research methods, revising content in your own drafts, and reviewing peers' drafts. So let's take a moment to talk about how the think-ahead and think-through strategies work.

As its name implies, the think-ahead strategy involves getting into mind what you are looking for *before* you go looking for it. It's the method that you would

---

1. How relevant are the argument's lines of support to its claim?
2. If an argument's lines of support are not obviously relevant, how successfully has the author developed the necessary warrants?
3. How convincing are the data-driven lines of support for the claim?
4. How convincing are the concept-driven lines of support for the claim?
5. How successfully does the author acknowledge and refute viable counterarguments?

FIGURE 3.9 Diagnostic questions for evaluating published scientific arguments.

use in buying a new car—that is, if you are a savvy consumer. You won't just show up at the dealership with no idea of what you are looking for. Instead, you will have thought ahead, ideally preparing a list of desirable criteria. Your think-ahead list might include details about performance standards that your new car must meet or surpass, the price range that you are set on, the minimum gas mileage that you desire, and your safety requirements. You might even have a list of criteria for disqualifying certain cars, such as *anything shaped like a box and painted bright orange*. With your list of think-ahead criteria guiding your evaluation of prospective new vehicles, you will be confident that you are covering all bases. In addition, you will not be overwhelmed by the loads of information, the endless choices, and the many so-called bargains that the salesperson is hurling at you.

In brainstorming ideal criteria for your new car, you will consider their implications. Take the criterion for the minimum gas mileage that you desire. These days many cars that get good gas mileage are hybrids that run on gasoline and electricity. They are generally more expensive than conventional cars that use gasoline only. Mulling over the implications of buying a hybrid versus a conventional car, you will reason along these lines: If I buy a conventional car that costs X amount of dollars and is not very fuel efficient, then every month I will be faced with spending approximately Y for gas. If I buy a hybrid car I will have to spend an additional Z amount of dollars, but then I will save approximately A on gas. This sort of if-then reasoning reflects our think-through strategy. It's simply a matter of thinking through the implications of your think-ahead criteria.

Of course, in thinking ahead you might overlook at least a few useful criteria to guide your new car purchase. So you might arrive at the dealership with 15 items on your list, only to add 5 or 10 more as you talk with the salesperson and take a few test drives. Likewise, your critical evaluations of scientific arguments will not be completely proactive. But for the best outcomes it's still extremely helpful to do as much proactive thinking as possible. Our think-ahead and think-through strategies are demonstrated in the following presentation of diagnostic questions for evaluating scientific arguments.

Diagnostic Question 1: How Relevant Are the Argument's Lines of Support to Its Claim?

In this diagnostic question, relevance refers to the logical relationships between a claim and its supporting evidence and reasoning. To demonstrate how to evaluate scientific arguments for the quality of relevance, I will pose a hypothetical argument for the latest and self-proclaimed greatest popular diet. It was devised by a Dr. Willie Cashin from Peach County Georgia, where they grow some of the South's most delicious peaches. Not surprisingly, Dr. Cashin's creation is a peaches-only diet: peaches for breakfast, peaches for lunch, and more peaches for dinner. Dr. Cashin calls it the South Peach Diet. In an article intended to make the diet's case, Dr. Cashin presents the claim that it "promotes positive changes in countless aspects of health." In one line of support for his claim, the author synthesizes the outcomes of published studies on the South Peach Diet and weight loss. Calculating mean weight-loss values from the various studies over 6-month periods, Dr. Cashin determined that clinically obese subjects on his diet lost 11.2 kg, while counterparts on nutritionally balanced conventional diets lost only 5.3 kg. At first glance, these results seem quite impressive. On principle, however, we will not accept them unconditionally. Guided by our first diagnostic question we ask, "Are these results actually *relevant*, or logically related, to the author's specific claim that the South Peach Diet leads to countless *positive health outcomes*?"

Instead of attempting to answer the preceding question straightaway, we will set Dr. Cashin's article aside for the moment. Doing so, we can concentrate our efforts on the powerful think-ahead and think-through strategies. Again, the think-ahead strategy entails bringing to mind what you are looking for before you go looking for it. In this example, we need to brainstorm the most logical evidence and reasoning that would support Dr Cashin's claim for positive health outcomes associated with the South Peach Diet. We would reasonably expect Dr. Cashin to present supporting data from studies in which subjects underwent standard medical examinations before and after going on the diet. So our think-ahead list would include items reflecting cardiovascular health outcomes such as *evidence of favorable changes in concentrations of cholesterol and triglycerides (fat) in the blood*. Another logical line of support would be diet-induced changes in concentrations of glucose and insulin, which are markers for type 2 diabetes. And if the South Peach Diet truly promotes countless positive health outcomes, as Dr. Cashin claims, we might even expect research-based evidence showing that it reduces the risk of infectious illnesses.

To brainstorm productive think-ahead criteria, you really must know a lot about the science that underlies the arguments you are evaluating. Take our criterion that Dr. Cashin should present data on changes in blood glucose and insulin. To establish this criterion and judge its relevance to Dr. Cashin's claim, you would have to know about the metabolic mechanisms by which diet can influence glucose and insulin concentrations, as well as how changes in these substances influence the risk of developing type 2 diabetes.

If you were to study the science of nutrition and obesity, you would learn that one source of weight loss is the body's fat mass, which is composed primarily of triglyceride molecules stored in adipose cells. If a diet causes weight loss primarily from fat, then the outcomes usually favor good health, especially in clinically obese individuals. However, weight can also be lost from fat-free sources, including water and muscle mass. If a diet causes weight loss primarily through dehydration (water loss) and muscle wasting, then its effects are most definitely *not* healthy. This if-then analysis is an example of our think-through strategy. The upshot of this critical thinking exercise is that we will not add *evidence of substantial weight loss* to our list of relevant lines of support for Dr. Cashin's claim. On its own, weight loss does not necessarily indicate a positive change in health. Instead, we should be looking for Dr. Cashin to present data showing that obese subjects on the South Peach Diet experience substantial *fat* loss (relative to their total weight loss). Incidentally, some popular diets do cause extreme amounts of water and muscle loss. In addition, in many published studies on popular diets researchers unfortunately do not report measures of fat versus fat-free weight loss.

The immediate objective for brainstorming think-ahead criteria is to guide critical evaluations of scientific arguments. But recall that this activity serves the higher goal of generating content for original arguments. Imagine that you are writing a critical review paper on popular diets. What would you say about the relevance of Dr. Cashin's argument so far? In response to the author's synthesis of studies on weight loss, a fair criticism is that weight loss, *per se*, does not reflect a positive change in health. You might raise the point that the weight loss could have been due to muscle wasting because peaches are deficient in protein. You would be justified in saying that Dr. Cashin's line of support should have included changes in body fat, which better reflect health outcomes. Developing these ideas with scientific knowledge and evidence, you would construct a convincing argument of your own. We might even say that it would be a peach of an argument!

Diagnostic Question 2: If an Argument's Lines of Support Are Not Obviously Relevant, How Successfully Has the Author Developed the Necessary Warrants?

Let's extend the previous example by supposing that Dr. Willie Cashin has actually presented relevant evidence to support his claim for the health-promoting effects of the South Peach Diet. He has synthesized results from studies revealing that obese subjects on the diet lost substantial amounts of fat. The results indicated that 99.5% of their total weight loss came from fat, leaving negligible amounts of water and muscle loss. In response to this line of support, some readers may ask for explanations: *Why* is fat loss good for health? How does excess fat cause health problems? The answers to these questions are not intuitive. For some audiences, particularly readers who are not experts on nutrition and obesity, the warrants need to be developed to connect diet-induced fat loss to positive health outcomes. Recall that warrants are assumptions, definitions, ideas, concepts, and theories that explain how an argument's lines of support are actually connected to its claim. In this case, unless Dr. Cashin can safely assume that readers already know the physiological mechanisms by which fat loss can improve health, he would need to develop the warrants convincingly.

This diagnostic question is especially powerful for generating high-quality content for students' original arguments because a common problem in published arguments is that authors fail to sufficiently develop essential and convincing warrants. Instead, they assume that readers will automatically accept numerical data, on their own, as valid support for claims. Based on a comprehensive critical evaluation of research on the South Peach Diet, let's say that you are not convinced that it promotes optimal health. In a position paper of your own, to support your claim that obese individuals should avoid the diet, you will set the rhetorical goal to acknowledge and refute counterarguments. One of them is Dr. Cashin's argument that is backed by the fat-loss data. If he does not convincingly draw the warrants—especially one that *soundly* explains how the magnitude of fat loss observed could actually result in positive health outcomes—you would be wise to write about the oversight in your paper.

Diagnostic Question 3: How Convincing Are the Data-Driven Lines of Support for the Claim?

In many scientific arguments, the strongest support for claims is hard, cold data derived from impeccably conducted research. Evaluations of strengths and weaknesses in data-driven lines of support should be based on the following criteria.

- Whether the author has presented a *critical mass* of data—in other words, a sufficient amount of support to move even the most skeptical readers to accept the claim.
- Whether the data have been derived from sound research methods.
- Whether the data are statistically significant.
- The extent to which the data are practically significant.
- How effectively the author has synthesized contrasting data from various studies on the issue.

We have already covered the skills of interpreting the statistical and practical significance of study data. It's easy to see that these skills are fundamental to evaluating published scientific arguments. For especially insightful interpretations of the significance of study data, you might try using the think-ahead and think-through strategies. Let's say that you are interpreting data backing a claim for a popular diet's positive effects on health outcomes. To be convinced that the data were indeed statistically and practically significant, what would you be looking for ahead of time? Specifically, what criterion P values would you set as acceptable levels of significance? For interpreting data that the author presents on changes in body composition, exactly how much fat loss will you deem as practically significant for optimal health? If the data include markers of cardiovascular disease and type 2 diabetes, what magnitudes of change will you look for to indicate healthy outcomes for subjects on the diet? These are by no means trivial questions. If you do not raise and answer them ahead of time, you might react blindly to the author's claims and supporting data—if you react at all. But if you do raise the questions, you will have definite standards to guide your critical judgments of the argument.

Concerning assessments of the critical mass of an argument's supporting data, there are no set-in-stone rules. The best evaluations demand problem solving. Arguments backed by a sparse amount of data from relatively few studies should not automatically be dismissed. If a study is absolutely flawless in its methods, and especially if its sample comprehensively represents the population of interest, its data alone might convincingly support an author's claim. This is not the case, however, when study methods are imperfect and samples are incomplete. So it's appropriate to question an author's claim that is not backed by data from numerous studies. In raising your criticism, avoid simply saying, "The argument is weak because it's not backed by enough data." The challenge is to explain *why* the sparse data might actually be problematic. This usually requires explaining potential sampling errors and other methodological flaws.

By and large, we should expect scientists to present data-driven support from most, if not all, of the relevant studies that support their claims. We should also expect effective syntheses of the studies. So when you are evaluating arguments based on data from numerous studies, look for whether the authors have successfully covered the research field. Also look for whether they have adequately synthesized the research by piecing together studies with similar as well as contrasting results and conclusions.

Diagnostic Question 4: How Convincing Are the Concept-Driven Lines of Support for the Claim?

In strong arguments, authors complement data-driven support with nonnumerical factual information, structural and mechanistic explanations, theories, and other forms of conceptual knowledge and reasoning. To identify strengths and weaknesses in such concept-driven lines of support, you will need to raise questions about their accuracy, validity, logic, and development. Below is a set of refined diagnostic questions for these purposes.

- To what extent does the author's conceptual support fit into the framework of current knowledge in the research field?
- Has the author presented convincing documentation that leading scientists in the field agree with the conceptual support?
- Does any research exist to confirm the conceptual support?
- Is the conceptual support logical—that is, are you convinced by the author's reasoning?
- To what extent has the author presented sufficient details and depth of explanation in the conceptual support?

Your answers to these questions will be the basis for deciding whether to accept an author's published argument; in addition, they will help you develop ideas to explain your own position in the argument.

Diagnostic Question 5: How Successfully Does the Author Acknowledge and Refute Counterarguments?

Because scientific arguments are by definition multisided, we must evaluate them by how effectively their authors account for counterarguments. This entails thinking ahead about opposing arguments to the one that an author is making. What are the claims of the counterarguments? What lines of support and warrants back them? If the author fails to acknowledge viable counterarguments, you are entitled to raise the oversight as a criticism. In addition, you should base your evaluation on how convincingly the author refutes the counterarguments by pointing to shortcomings in their claims, lines of support, and warrants. In applying this diagnostic question, you are entitled to play the role of devil's advocate, saying to the author, "Unless you can convince me that viable counterarguments are not as strong as your argument, I might as well line up with them."

Evaluating Research Methods

Earlier in the chapter, the hypothetical SALT study (on the effects of a low-sodium diet on blood pressure) served to demonstrate the activity of interpreting study data. To concentrate on the activity, we assumed that the SALT study was flawless in its methods. In reality, of course, this is hardly ever the case in science, especially in life science research. In addition to outright procedural and analytical mistakes, perfection in research methods may be limited by technological challenges in studying life's complexities, by shortcomings attributable to excessive costs of money and time, and even by necessary measures taken to protect the well-being of subjects. So the validity of study results and conclusions is largely determined by the quality of the methods used to derive them. It follows that the strongest arguments in scientific papers account for strengths and weakness in study procedures and analyses. If you are basing an argument on data from a methodologically flawed study, the data and your argument will naturally be flawed as well. In contrast, if the study's methods were strong and you can explain why, the supporting data for your claim and ultimately your entire argument will be strong as well.

The process of generating content through evaluating research methods is central to accomplishing a number of rhetorical goals for scientific argument. In the discussion section of research papers and lab reports, for example, writers argue for the validity of their conclusions by presenting strengths of their study procedures and analyses. In grant applications, a key goal is to convince readers that proposed study methods will be optimal for planned research projects. In addition, to resolve debatable issues in position papers, writers argue for methodological strengths of studies supporting their claims, and they argue that the studies supporting counterarguments have problematic methods.

Just imagine that you are writing a position paper on a controversial issue in nutrition or, for that matter, in any other scientific field. Let's say that the arguments in the published literature line up on two sides of a fence, each side supported by a substantial amount of evidence and reasoning. How will you go about determining the strongest position? And once you establish your position, how will you convince readers that the support for your argument is superior to the support for opposing arguments? The solutions depend on a skillful approach to identifying strengths and weaknesses in the methods from which an argument's supporting evidence was derived.

Our approach to evaluating research methods engages the same sort of goal-directed, diagnostic thinking that underlies evaluating arguments in the scientific literature. The steps in the process involve (a) reflecting on the rhetorical goals that you seek to accomplish through evaluating a study's research methods, (b) raising key diagnostic questions for identifying strengths and weaknesses in the methods, (c) answering the diagnostic questions by applying the think-ahead and think-through strategies, and (d) noting your critical evaluations in your goal-based plan, or developing them directly as draft material.

Figure 3.10 presents eight diagnostic questions for evaluating major aspects of study design, procedures, and analyses in published research papers. If your future holds advanced training in science, you will learn additional analytical questions to guide the activity at hand.

Diagnostic Question 1: Were the Study's Subjects Screened and Selected for the Appropriate Characteristics?

In well-executed studies, subjects are screened and selected for specific characteristics that enable scientists to successfully answer their research questions and test their hypotheses directly. For example, to make sound inferences about the results obtained from a study's sample, the researchers must select subjects who possess all relevant characteristics that represent the larger population of interest. Accordingly, to develop think-ahead criteria to guide your application of this first diagnostic question, you should brainstorm the qualities that distinguish targeted populations. What age ranges should the subjects fall into? What physical features and behaviors would ideally describe the sample's subjects? Should the study include both sexes? In addition to thinking proactively about characteristics that the subjects should possess, brainstorm qualities that the researchers should have screened to avoid extraneous variables and confounding outcomes.

After setting your sights on appropriate subject characteristics for a study, pick up the research paper that reports it and turn to the methods section. Focusing on the subsection devoted to describing the study's subjects, use your think-ahead criteria as a checklist. If the researchers screened the subjects to eliminate the influence of extraneous variables, note this methodological strength accordingly. If the subjects lacked essential qualities that represent the study's targeted population, note the weakness. To complement the proactive think-ahead method,

1. Were the study's subjects screened and selected for the appropriate characteristics?
2. Were subjects assigned to groups and conditions without bias?
3. Did the study include a sufficient number of subjects?
4. How appropriate was the study design for resolving the research issue?
5. How valid and comprehensive were the study's independent variables?
6. How valid and reliable were the study's dependent variables?
7. During the course of the study, how effectively did the researchers control for extraneous variables?
8. How appropriate and accurate were the study's statistical analyses?

FIGURE 3.10 Diagnostic questions for evaluating research methods.

be mindful of appropriate subject characteristics that you glean from reading articles on your research issue. If you have overlooked these characteristics, add them to your think-ahead list. And, as always, reflect on whether and how your critical evaluation serves the rhetorical goals and strategies that you have set for your paper.

Diagnostic Question 2: Were Subjects Assigned to Groups and Conditions without Bias?

Before any treatments are administered in an experiment, subjects assigned to treatment and control groups should be alike in all characteristics that could influence the outcomes of interest. Once the study begins, the groups should be distinguished only by factors associated with the targeted independent variables. Take a study to determine the effects of dietary calcium supplementation on bone mineral density (BMD) in young women. The independent variable, dietary calcium intake, should be the only condition that distinguishes the treatment group from the control group. If all other conditions are equal, the researchers can confidently conclude that any differences in outcomes are due to the treatment rather than to extraneous variables. One approach to avoiding bias caused by preexisting differences between treatment and control groups is to screen subjects for extraneous characteristics. Another method is to assign subjects to groups randomly. In theory, this ensures that subjects in the treatment and control groups will be matched for all relevant variables before the experiment begins. But, in practice, screening and random assignment do not always work perfectly.

To determine whether bias has been introduced, begin by brainstorming subject characteristics that should have been initially matched across treatment and control groups. Think ahead about extraneous variables that could give one of the groups an unfair advantage or disadvantage. To evaluate the study on calcium supplementation and BMD in young women, for example, I will look for whether the treatment and control groups were initially matched for diet and amount of exercise. Both of these variables can influence BMD independently of calcium supplementation. Weight-bearing exercise, for example, can maintain bone density in young women. What if a large number of subjects in the treatment group just happened to participate in more weight-bearing exercise than subjects in the control group? In this scenario, the treatment group would have a biased advantage. If the results indicated that BMD was maintained at higher levels in the treatment group, the researchers' conclusions about the effects of calcium supplementation would be limited, if not flawed.

After thinking ahead about variables that should have been initially matched across a study's groups, go looking for these variables in the research paper you are evaluating. In a research paper on calcium supplementation and BMD, for example, the results section might begin with a table comparing the treatment and control groups for their initial BMD, age, weight, diet, and exercise levels. These tables commonly include means and P values from statistical tests to determine whether the groups were initially different. If the groups were matched appropriately for key variables, note that as a methodological strength. If you find significant preexisting differences, think through their consequences. For any differences that could lead to biased outcomes, note them as methodological weaknesses.

Diagnostic Question 3: Did the Study Include a Sufficient Number of Subjects?

In Chapter 1, I talked about the relationship between sample size, or the number of subjects in a study (designated by the letter n), and statistical significance. As n increases, the probability of observing group differences and treatment effects by chance decreases. It's helpful to keep this relationship in mind when you are evaluating conclusions from studies conducted on a relatively small number of subjects. If, based on an n of only 10 or 20 subjects, an author concludes that a result is not significant, you would be justified in raising an alternative interpretation: The study lacked a sufficient sample size to detect reliable outcomes.

There are no set-in-stone rules for determining whether a study's n is too small. In some circumstances, studies with relatively few subjects can yield perfectly valid results and conclusions. The subjects would have to be screened appropriately, randomly assigned to groups, and representative of the population. In addition, the experimental design would have to be flawless. However, when all other factors are equal across studies, put more trust in the results and conclusions of those with relatively large sample sizes. In addition, look for whether the researchers reported a statistical calculation called a power analysis. Statistical power is defined as the probability of detecting reliable differences between groups or true effects of treatments. This level of confidence concerning the significance of study results is determined partly by sample size. As n increases, so does power. A power analysis is a valid procedure for estimating appropriate sample sizes. The actual calculations for power analysis are too complex to present here. However, the procedure's outcomes are straightforward enough to interpret and guide evaluations of study methods. Because power analyses yield probabilities, their values are expressed on a scale of 0–1. Values of 0.80 or greater are generally considered to reflect strong power. Suppose, for example, that an author reports that a study had a power of 0.90. The interpretation is that reliable group differences or treatment effects would be detected in 90% of repeated instances of the study. This interpretation indicates that the study included a sufficient number of subjects.

Diagnostic Question 4: How Appropriate Was the Study Design for Resolving the Research Issue?

As introduced in Chapter 1, the term *study design* refers to the overall organization and execution of research projects. Observational and experimental study designs, the two most general types, can be further divided into numerous subtypes. For instance, some observational studies have a cross-sectional design. The defining characteristic of cross-sectional studies is that all of their data is collected at one slice, or point, in time. Take a cross-sectional study to determine the relationship between dietary calcium supplementation and BMD. The researchers would recruit subjects who already supplement their diets with calcium as well as subjects who do not. At one slice in time, the subjects would report how much calcium they take, and they would undergo a procedure for measuring BMD. The researchers would then compare the BMD scores across the two groups. This cross-sectional design differs from a longitudinal design, another subtype of observational study. In a longitudinal study on calcium supplementation and BMD, the researchers would follow subjects over months or years, correlating their intake of the supplement with changes in BMD scores.

Study design can strongly influence the conclusions that can be drawn from research. Cross-sectional designs, for example, cannot be used to reach cause-and-effect conclusions. Suppose that our hypothetical cross-sectional study revealed that BMD was significantly greater in the group of subjects who took calcium. From this result, the researchers cannot conclude that calcium supplementation *caused* the favorable outcome. It's quite possible that the subjects practiced other healthy lifestyle behaviors that were responsible for their dense and strong bones. The only way to determine whether calcium supplementation directly increases BMD is through a well-controlled experimental design. The experiment would begin with people who had never supplemented their diets with calcium. The subjects would be randomly assigned to treatment and control groups. If the researchers successfully control for all possible confounding variables, the results can be interpreted to indicate a cause-and-effect relationship.

An ideal way to apply this diagnostic question is to begin by brainstorming what you would deem as the ideal design features for the study you are critiquing. Based on the research questions and hypotheses, should the study have been an experiment, or would an observational design have been more appropriate? How long should the study have lasted? What would be the optimal setting? How frequently should data have been collected? After you have thought through the implications of conducting the study with ideal design features, apply your think-criteria to evaluate the actual study design. Because the perfect design is hardly ever completely feasible in life science research, you will almost always find some separation between your think-ahead criteria and the actual design features. That separation is what you will note and, if it helps you accomplish your rhetorical goals, write about in your paper.

Diagnostic Question 5: How Valid and Comprehensive Were the Study's Independent Variables?

Suppose that you are reading a research paper that reports an experiment to determine the effects of a high-protein diet on athletic performance. The independent variable is protein intake, which is usually expressed in daily grams of dietary protein relative to an individual's body weight in kilograms. Nutrition experts recommend that normally active adults consume 0.8–1.0 g of dietary protein per kilogram of body weight each day. One criterion for evaluating a study's independent variables is their validity, which refers to how well they represent the treatment or condition that the researchers set out to investigate. In the study that you are reading about, the researchers assigned athletes to a daily diet containing 1.5 g of protein per kilogram of body weight. It's certainly a greater amount of protein than that recommended for normally active individuals. So you might think that the experimental treatment is valid because it represents what appears to be a high-protein diet. But in a study on this issue, *how* high should high be? Is 1.5 g really a high-protein diet? Would this amount of protein be sufficient to induce changes in muscle mass and strength? Would 1.2 g do the trick? Or should the protein intake have been as high as 2.5 g, or even higher? These are important questions that address the validity of the independent variable.

The best answers to the preceding questions would come from a study that included a wide and resolute range of dietary protein values. We would credit such a study with strong methods because the levels of the independent variable are comprehensive and they would account for possible "dose–response" effects. In research in fields such as nutrition, medicine, exercise, and education, the outcomes often depend on the specific amounts, or doses, of treatments. It's very common for dose–response effects to be nonlinear. For example, very low as well as very high levels of regular exercise are associated with poor immune function. But a moderate level of regular exercise enhances immune function.

For this diagnostic question, the think-ahead strategy involves brainstorming criteria for valid and comprehensive independent variables. Which specific treatments or conditions should the researchers have assigned in order to resolve the issue at hand? How should the researchers have gone about accounting for possible dose–response effects? Exactly how comprehensive and resolute should the levels or doses of treatments have been? Thinking through the implications of strengths and weaknesses of independent variables in the studies you are reading about, you will develop effective ideas and arguments for your paper.

Diagnostic Question 6: How Valid and Reliable Were the Study's Dependent Variables?

For a demonstration of how to apply this diagnostic question, consider a study to determine whether a nutritional supplement called glucosamine sulfate is an effective treatment for osteoarthritis, the disease in which cartilage deteriorates in the body's joints. Cartilage is the elastic, shock-absorbing tissue that protects the ends of bones from rubbing against each other. Cartilage deterioration can cause considerable pain and debilitation. In this study, the independent variable is glucosamine sulfate supplementation, and the dependent variables are laboratory and clinical measures of osteoarthritis. If the study is truly sound, its dependent variables will be valid, which means that they logically and accurately measure the outcomes of interest.

Even if you are not an expert on treatments for osteoarthritis, the think-ahead technique and common sense will help you identify outcomes that the researchers should have measured. As you might imagine, valid tests of an osteoarthritis treatment must account for changes in the physical properties of cartilage. The reason is that the condition of cartilage deterioration defines the disease. So you will look for whether the researchers directly measured changes in the composition, thickness, and resiliency of the subjects' cartilage. Tools for assessing these properties include x-rays and magnetic resonance imaging (MRI). If the researchers in our hypothetical study did measure the effects of glucosamine sulfate directly on joint cartilage, they would have a basis for reaching valid conclusions about the treatment. This is so because their dependent measures were valid. Suppose, however, that the researchers collected data only on the subjects' ranges of joint motion and self-reported pain levels. Of course, increased joint motion and decreased pain would be important clinical outcomes. But if the researchers specifically intended to determine the effects of glucosamine sulfate on *osteoarthritis*, these dependent variables are limited. The researchers could not conclude with certainty whether the treatment actually caused the outcome that they intended to study. It's possible, for example, that the improved clinical outcomes were due to glucosamine sulfate's effects on other joint tissues besides deteriorated cartilage.

Dependent variables should also be evaluated for their reliability, which refers to the consistency and dependability of instruments and procedures used to measure study outcomes. To trust the results from a study on glucosamine sulfate and osteoarthritis, for example, we need to be assured that the x-ray and MRI machines worked properly each time they were used. So, in the methods section of the paper that reports the study, we will look for whether the researchers regularly calibrated and maintained their measurement instruments.

Diagnostic Question 7: During the Course of the Study, How Effectively Did the Researchers Control for Extraneous Variables?

A common theme runs through our process for evaluating research methods: Many of the diagnostic questions involve determining the extent to which a study's outcomes (the dependent variables) can be attributed to the conditions or treatments under investigation (the independent variables). Especially in research on humans, numerous factors can interfere with the effects of independent variables on dependent variables. The consequences are flawed results and invalid conclusions. In well-conducted studies, these extraneous variables are controlled, which means that they are eliminated completely or at least equalized across groups. The influences of extraneous variables on study outcomes are illustrated in Figure 3.11 (page 146). Ideally, an independent variable should have a direct, or untainted, effect on its associated dependent variable. In poorly controlled studies, however, extraneous variables cause confounding effects, either by interacting with an independent variable or by direct influences.

By screening potential participants for selected characteristics and assigning subjects to groups randomly, researchers can control for some extraneous variables before administering experimental treatments. Once a study is underway, however, other extraneous variables may emerge and cause confounding effects. So it's worthwhile to think ahead about what these potential confounders might

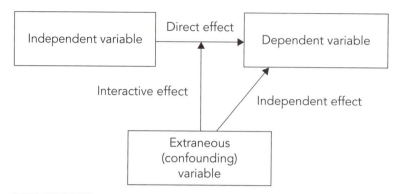

FIGURE 3.11 Relationships between independent, extraneous, and dependent variables.

be. The think-ahead routine for this diagnostic question entails brainstorming all factors other than the prescribed treatments or conditions that might influence outcomes during the study. It's also worth thinking ahead about the control measures that must be taken to avoid confounding effects.

A special type of control measure, called a *placebo control*, is essential in studies on treatments that influence subjects' expectations and beliefs. Placebo controls are used to avoid misinterpretations of results that are due to placebo effects, or outcomes that are not caused directly by a treatment's active properties. Entering a study on glucosamine sulfate for treating pain associated with osteoarthritis, for example, subjects will naturally hope for and expect improvement in their condition. Through psychological mechanisms, the subjects' positive beliefs might very well reduce their pain. But in this case the diminished symptoms of osteoarthritis might be due to a placebo effect rather than to the direct effect of glucosamine sulfate. To avoid misinterpreting the results, the researcher must include placebo controls. Both the experimental and control groups would need to take treatments on the same schedule. All of the subjects would be *blinded* to, or unaware of, their specific treatment. Of course, the experimental group would take glucosamine sulfate while the control group takes an inactive substance. This method balances out possible placebo effects across the two groups so that the treatment effect can be isolated.

It's difficult to adequately control for all extraneous variables in life science research. This is obviously a problematic concern for scientists. However, for students writing scientific papers, the problem can be advantageous. By critically diagnosing the extent to which studies control for extraneous variables, student writers can generate especially insightful and original arguments.

Diagnostic Question 8: How Appropriate and Accurate Were the Study's Statistical Analyses?

The routines for diagnosing statistical analyses in research go beyond the scope of this book. If, however, you are pursing graduate studies or a career in science, you will certainly need to raise the diagnostic question at hand in your future work. Skillful evaluations of the statistical analyses reported in published studies are essential for constructing strong scientific arguments. And the reality is that published studies vary widely in the appropriateness and quality of their statistical procedures.

SUMMING UP AND STEPPING AHEAD

This chapter addressed the most fundamental challenge in all of scientific writing—it's the challenge of generating content, or figuring out what to say in our papers. We approached the process through five activities: independent and collaborative brainstorming, reading for relevance, interpreting study data, synthesizing study outcomes, and constructing convincing scientific arguments. The activities demand specialized critical thinking and reading skills; in addition, they ideally rely on applying rhetorical goals to identify and develop effective ideas and arguments.

Our approach to the overall writing process separates generating content from organizing content and writing a draft. A major reason for separating these stages is to encourage the development of prolific ideas and substantial arguments without getting bogged down by concerns of a paper's global organization, paragraph coherence, and sentence phrasing and structure. At some point, of course, we must translate the rough content we generate into structured patterns and prose. This is our focus in the next chapter, which covers the stages of organizing content and writing a first draft.

CHAPTER 4

Organizing Content and Writing a Draft

The Science of Animal and Human Movement

Life science naturally encompasses the study of movement in living organisms. For animals and humans, survival fundamentally depends on the ability to move the whole body and its specialized parts. Animals and humans must move to obtain food, to tune their sensory organs to life-saving information in the environment, to avoid or fight against physical threats, and to procreate. And, of course, movement is at the heart of so many artistic and recreational activities that enrich human life. A number of subdisciplines in the life sciences are devoted to the study of animal and human movement. Under the general heading *movement science*, these research fields include biomechanics, motor control, motor learning, and ergonomics. Among other areas of study, movement scientists seek to understand the neural control of muscle activity; the mechanics of efficient posture, limb function, and whole-body locomotion; and the causes and cures of diseases that impair normal motor behavior.

One of the primary objectives of movement science is to explain how purposeful intentions, or plans, to move are translated into forms and actions that serve the overall functions of living organisms. In a way, this objective runs parallel to our concerns in the present chapter, which covers the stages of organizing content and writing a draft. Here we will investigate how intentions—specifically, the rhetorical goals and rough ideas developed in early stages of the writing process—are translated into organized forms such as linear outlines and the structured sentences and paragraphs that ultimately serve the overall functions of scientific papers.

INTRODUCTION

Imagine for a moment that you are a world-renowned biologist whose research focuses on understanding how prehistoric animals walked, ran, swam, and flew. You have a special interest in the locomotor behavior of Triceratops, a three-horned plant-eating dinosaur that lived during the late Cretaceous period, approximately 70 million years ago. You've been commissioned by a major museum of natural history to build a functional model of Triceratops. The museum is relying on your expertise to create an exhibition that accurately captures how Triceratops moved its body and limbs as it traversed the landscape of what is now North America. A team of paleontologists has excavated many hundreds of Triceratops bones and packed them in boxes that are en route to your laboratory. Your research team will take stock of the boxes, organize the bones by the different parts of the dinosaur's body, construct the model by integrating its parts, and add devices (perhaps some artificial muscle-tendon units, driven by a computer) to enable Triceratops to locomote.

Why have I asked you to imagine taking on this extraordinary challenge? It happens to be an analogy for this chapter's pursuits. We will begin with activities for organizing the content of scientific papers, which is a bit like coordinating the numerous and intricate parts of our Triceratops model. Both endeavors demand a principled approach to developing plans for structuring coherent and functional entities. The chapter then focuses on the process of writing a draft. Like the challenge of enabling Triceratops to move, drafting requires translating structural plans, as well as ideas and mental images, into functional elements. For our purposes in drafting scientific papers, these elements include titles, abstracts, paragraphs, sentences, and graphics.

ABOUT THE PROCESS: ORGANIZING CONTENT

Our approach to organizing content involves ordering and integrating relatively large parts of scientific papers—specifically, their major sections and subsections. At this stage of the writing process, the main objective is to produce what we will call an *organizing plan*. This might be a formal linear outline or a nonlinear, map-like diagram. If you have already developed a goal-based plan through the previous two chapters' activities, organizing content is a fairly straightforward process. To appreciate this point, let's modify the details of our Triceratops-building project. Suppose that the paleontologists who excavated the bones had prearranged them according to the dinosaur's major sections: its head, body, tail, and extremities. The bones were sent to your laboratory in boxes labeled by these sections. Now your task of constructing Triceratops will be quite manageable—that is, compared to building the model from totally disorganized parts. Of course, your task will be infinitely easier than having to dig up the dinosaur bones, organize them, and build the model all at the same time.

While the analogy isn't exactly perfect, you can gain similar sorts of advantages from prearranging the elements of your papers in the form of goal-based plans. In constructing a linear outline, for example, you won't be overwhelmed by simultaneously having to dig up all its contents. Indeed, you will have *already* developed your outline's main structural elements, which correspond to your rhetorical goals and strategies. Moreover, following the structure of goal-based plans as presented in Chapter 2, these elements will be prearranged by your paper's major sections. This preparation will allow you to concentrate fully on the task at hand: deciding on the best ways to order and integrate your rhetorical goals and strategies, along with any associated content that you have generated in note form. Ultimately, this process results in papers that are distinguished by two superior characteristics of organization: global unity and global coherence. Global unity is the quality of thematic *oneness*, or overall consistency, in the numerous sections that compose complex documents. From the reader's perspective, global unity is the very favorable experience of recognizing that every section of a paper relates logically to its overall topic, message, and goal. Global coherence refers to the *flowing* quality of consecutive sections of well-organized papers.

The chapter's activities for organizing content are diagrammed in Figure 4.1 (page 152) and summarized as follows.

Activity 1: Choose a design for your organizing plan. One logical design for an organizing plan is a traditional outline that lists topics to cover, or rhetorical goals to accomplish, in linear order. However, writers can choose from other formats, including bubble diagrams and concept maps that organize content spatially rather than linearly. This first activity is simply for selecting a design for an organizing plan that works best for you.

Activity 2: Take a principled approach to organizing content. For the most part, organizing content is a process that is ideally guided by a principled, problem-solving approach. This activity presents a set of powerful organizing principles and demonstrates their applications.

Activity 3: Organize your paper's major sections. The major sections of scientific papers are usually predesignated and ordered by convention; as described in Chapter 1, an example is the IMRAD format of research papers. This activity involves evaluating organizing plans for whether they contain the required major sections, whether the sections are ordered correctly, and whether the content planned for each section is indeed fitting.

Activity 4: Organize your paper's subsections. The greatest challenges in structuring the content of scientific papers involve deciding how to order and integrate their subsections, or the content that comes within major sections. In this activity, you'll learn how to

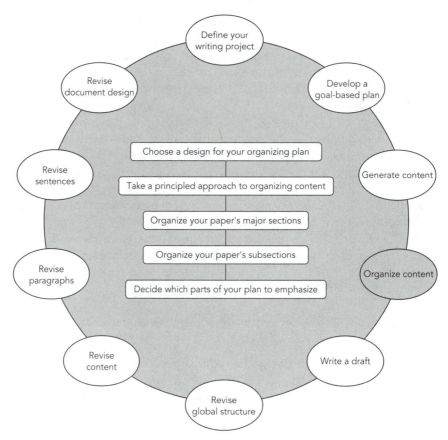

FIGURE 4.1 Process map for *organizing content.*

meet these challenges through problem solving and by applying key organizing principles.

Activity 5: Decide which parts of your plan to emphasize. As a consequence of extensive planning, you may reach this stage of the writing process with more goals, strategies, ideas, and arguments than you have space to write about. You will thus be faced with some tough organizational decisions: Which parts of your plan should you emphasize? Will you need to eliminate content completely? How much space should you devote to carrying out given parts of your plan? These questions are central to this activity.

The time and effort that you must spend organizing content will depend on a number of factors, including the complexity of your writing project, your experience with the paper type, your topic knowledge, and the extent of your preliminary planning. Imagine that you are writing a short research proposal with explicit instructions for how to structure its sections. If you have previously written research proposals in the same format and you have developed a full set of rhetorical goals, you might be able to create an effective outline in no time at all. In contrast, consider a complex thesis project that scales to 50 pages or more, with an assigned format that you are not familiar with. To do your best, you will obviously need to spend substantial time organizing the paper's content, using various tools and techniques to guide the process. The upcoming sections of the chapter present a full set of such organizing tools and techniques. Considering their applications, you can choose the ones that meet your specific needs.

Choosing a Design for Your Organizing Plan

What will your organizing plan look like? Will it be an informal list of your paper's topics or rhetorical goals scratched out on the back of an envelope? Or would you prefer a formal outline that systematically labels and indents its items? As an alternative design, how about constructing your plan as a nonlinear graphic, such as a bubble diagram? These are important questions because your organizing plan's design can influence the nuts-and-bolts decisions you must make about how to structure your paper.

Back in grade school, you may have learned to construct formal linear outlines with indented sections labeled by Roman numerals, capitalized and lowercase letters, and Arabic numbers, as in the following format.

I. Major section
  A. Level 1 subsection
    1. Level 2 subsection
      a. Level 3 subsection
        (1) Level 4 subsection

An excerpt from a formal linear outline is presented in Figure 4.2 (page 154). It's part of an organizing plan for an informative review paper on movement-related disorders and diseases, including cerebral palsy, myasthenia gravis, Parkinson's disease, and Huntington's disease. Following conventions for the structure of review papers, this one will have three major sections: an introduction, a body, and a conclusion. The excerpt in Figure 4.2 outlines one subsection of the paper's body, specifically a subsection devoted to informing readers about Parkinson's disease.

There are a number of advantages to designing organizing plans as formal linear outlines. The layout makes it easy to establish recognizable structural patterns and to check for matters of consistency, completeness, and order before drafting. Consider, for example, the subsections corresponding to the Arabic numbers in Figure 4.2. The structural pattern is clearly evident and, as a consequence,

II. Body of the Paper

A. Parkinson's disease
  1. Physiological explanations of the disease
    a. Loss of dopaminergic neurons in the substantia nigra
    b. Abnormal neural activity in the basal ganglia
    c. Abnormal neural activity in the motor cortex
  2. Etiology
    a. Genetic causes
    b. Environmental causes
    c. Trauma to the head
    d. Age-related causes
  3. Effects on everyday movement capacities
    a. Muscle rigidity
    b. Tremor
    c. Bradykinesia (slowness of movement, especially in walking)
    d. Poor postural control (balance)
  4. Treatments
    a. Drug treatments
      (1) Levodopa
      (2) Monoamine oxidase inhibitors
      (3) Dopamine agonists
      (4) Surgical implants
    b. Physical therapies
      (1) Electrical muscle stimulation
      (2) Aerobic exercise
      (3) Stretching
  5. Future directions for research and treatment
    a. Gene therapy
    b. Stem cell implants
    c. Transcranial magnetic stimulation

FIGURE 4.2 An excerpt from a formal linear outline for a review paper on movement-related disorders and diseases.

effective. Subsections 1 and 2 will begin with well-established background knowledge about Parkinson's disease (*Physiological explanations of the disease* and *Etiology*). Subsections 3 and 4 will address everyday concerns for Parkinson's disease patients (*Effects on everyday movement capacities* and *Treatments*). Finally, subsection 5 will pose new methods for studying and treating the disease (*Future directions for research and treatment*). This logical pattern obviously reflects past, present, and future concerns regarding the subtopic of Parkinson's disease. In the outline of the review paper's other subsections, which cover additional movement-related disorders and diseases, the author will follow this sound pattern.

Figure 4.2 reflects a standard topic-based outline, formally arranged in linear order. Its elements are subtopics of the overall topic, *movement-related disorders and diseases*. Formal linear outlines are also ideal for papers in which authors seek to accomplish diverse rhetorical goals. As a case in point, consider a position paper that a student named Marvette is writing on acupuncture as a treatment for movement-related disorders and diseases. Acupuncture, the healing technique that involves inserting fine needles into specific points of the body, is a form of traditional Chinese medicine. In Western societies, its popularity as a treatment for various illnesses is growing rapidly. However, research on acupuncture for patients with movement-related disorders and diseases has yielded contradictory results. The bulk of the scientific evidence actually supports arguments *against* the treatment. Acupuncture has thus been generally rejected, or at least ignored, by many mainstream health care providers in Western societies. Nonetheless, after critically analyzing studies on this issue, Marvette believes that the strongest research indeed supports acupuncture. In her position paper, Marvette aims to convince readers that the treatment is effective for patients with movement-related disorders and diseases, specifically when it is combined with conventional Western medical approaches. She is directing her paper to an audience of Western physicians and physical therapists. A complete outline of Marvette's paper is presented as follows (Figure 4.3).

I. Introduction

1. Present the contrasting sides of the research issue to help readers understand the debate
   1.1. Summarize studies that support acupuncture for patients with movement-related disorders and diseases
   1.2. Summarize studies that reveal no benefits of acupuncture for patients with movement-related disorders and diseases
   1.3. Introduce theories for how acupuncture might effectively treat movement-related disorders and diseases
       1.3.1. Introduce the Gate Theory (acupuncture blocks pain signals to the brain)
       1.3.2. Introduce the Endorphin Theory (acupuncture works by releasing pain-killing chemicals in the brain)

FIGURE 4.3 A formal linear outline for a position paper on the use of acupuncture in treating movement-related disorders and diseases.

1.3.3. Introduce the Trigger Point Theory (acupuncture relaxes concentrated points of muscle tension)

1.3.4. Introduce the Electrical Conduction Theory (acupuncture enhances the conduction of neural impulses to the muscles)

1.4. Introduce arguments *against* theories that explain positive effects of acupuncture

1.4.1. Introduce the argument that direct evidence to support the theories is lacking

1.4.2. Introduce the argument that any "positive" effects of acupuncture are due to psychological placebo effects

2. Discuss the importance of the research issue, to convince readers that my paper is relevant to their interests and concerns

2.1. Document the growing number of people in our society who seek alternative medical treatments like acupuncture for movement-related disorders and diseases

2.2. Explain why it is important for Western clinicians to be concerned about whether acupuncture is an effective complementary treatment

2.2.1. Discuss the importance of giving well-informed advice to patients who express interest in trying acupuncture

2.2.2. Discuss the implications for Western health care providers if research actually supports the effectiveness of acupuncture

2.2.3. Address the implications for Western health care providers if research does *not* support the effectiveness of acupuncture

3. Present my claim and provide an overview of how I will support it

3.1. State my claim: Acupuncture, when combined with conventional Western medicine and physical therapy, is an effective treatment for movement-related disorders and diseases

3.2. Briefly outline the evidence and reasoning that I will present in support of my claim

II. Body

1. Present evidence and reasoning from published research that supports my claim, to convince readers to accept my argument

1.1. Present research showing positive effects of acupuncture on balance and mobility in Parkinson's disease patients

1.2. Present research showing positive effects of acupuncture on walking biomechanics in children with cerebral palsy

1.3. Present research showing positive effects of acupuncture on cognitive deficits associated with movement disorders

1.4. Present research showing positive effects of acupuncture on the performance of daily living activities in older adults with movement disorders

FIGURE 4.3 *continued.*

2. Present the physiological rationale to support my argument
   2.1. Elaborate on the Gate Theory of pain reduction
   2.2. Elaborate on the Endorphin Theory of pain reduction
   2.3. Elaborate on the Trigger Point Theory
   2.4. Elaborate on the Electrical Conductance Theory
   * For all of the theories, explain how acupuncture ultimately enhances movement-related capacities in patients
3. Acknowledge and refute opposing arguments, to convince readers that my argument is the strongest
   3.1. Acknowledge and refute the argument that positive results of acupuncture are only due to placebo effects
       3.1.1. Explain the concept of placebo
       3.1.2. Summarize studies that support the argument that acupuncture works through placebo effects
       3.1.3. Refute the arguments by presenting studies in which researchers used strong controls against placebo effects
   3.2. Acknowledge and refute the opposing argument by pointing out methodological weaknesses in studies that support it
       3.2.1. Argue that many studies revealing no effects of acupuncture were flawed because acupuncture needles were not inserted properly
       3.2.2. Argue that many studies revealing no effects of acupuncture were flawed because subjects did not receive a sufficient number of treatments
4. Acknowledge limitations to my argument, convincing readers that they are noteworthy concerns although they do not weaken my overall claim
   4.1. Acknowledge that there are relatively few published studies that back my argument; respond that all of the supporting studies have strong methods and therefore that their results are convincing
   4.2. Acknowledge that acupuncture treatments are expensive and that they are usually not covered by health insurance companies; respond that the cost-benefit relationship supports using the treatment
   4.3. Acknowledge that theories explaining how acupuncture works have not been confirmed through research; respond that the lack of support for the theories does not discount positive effects of acupuncture observed in controlled research

III. Conclusion

1. Summarize the key points of my argument
2. Suggest future studies to advance the research field and its clinical applications

FIGURE 4.3 *continued.*

Marvette's linear outline looks a lot like the goal-based plans that were presented in Chapters 2 and 3. The items preceded by whole and decimal numbers are rhetorical goals and strategies, respectively. Goal-based plans can be handily converted into linear outlines by rearranging their constituent rhetorical goals and strategies to promote global unity and coherence. The process ideally applies sound organizing principles, one of which is to address concerns of readers. For a quick demonstration of this principled approach to organizing the elements of a goal-based outline, let's examine Marvette's plan from the perspective of her primary readers. These are Western doctors and physical therapists, many of whom do *not* use acupuncture as a primary treatment for movement-related disorders and diseases. They likely know that, at least in Western medical journals, much of the research does not support acupuncture as a viable treatment.

In Marvette's outline, notice that the first subsection of the paper's body is devoted to the rhetorical goal of presenting published research that directly supports her claim for the beneficial effects of acupuncture. The third subsection of the body is devoted to the rhetorical goal of acknowledging and refuting opposing arguments. Considering that many of Marvette's readers might be biased against acupuncture, think about whether the subsections are ordered most effectively for making the strongest argument. Assume the mindset of Marvette's audience as they begin to read the body of her paper, immediately after she presents her claim for acupuncture's positive effects. How will readers respond to the first line of supporting evidence, which runs *counter* to their beliefs and practices? With every study that Marvette presents at the start of the paper's body, skeptical readers may respond negatively: "So you have some data from a few studies that back acupuncture," they will say. "What about all the studies that do *not* back it?" By the time Marvette presents those opposing studies and attempts to argue against them, these skeptical readers may have dismissed her claim entirely.

Now think about how Marvette's readers will respond to a different structure for the paper's body. Picture an outline that reverses the order of the first and third subsections. In this arrangement, the body begins by acknowledging and refuting opposing arguments. The new order gives Marvette a greater advantage in her argument because she will address her readers' most immediate concerns *immediately*. This sort of reflection, as well as the application of a key organizing principle for considering concerns of readers, is facilitated by the linear goal-based design of Marvette's outline.

A formal linear outline, however, is not the only design for an organizing plan. You may be familiar with graphic alternatives such as tree diagrams, concept maps, and bubble diagrams. A sample bubble diagram, laying out the parts of Marvette's position paper in nonlinear fashion, is presented in Figure 4.4 (page 160). Projecting from the diagram's largest bubble, which is labeled by the paper's topic and type, are smaller bubbles that correspond to its three major sections: the introduction, body, and conclusion. Lines from each major section go to clusters of bubbles representing its rhetorical goals and associated strategies. Nonlinear organizing plans can be constructed with various tools, including paper and pencil, computer-based drawing programs, white boards or black boards, and sticky notes pasted on a wall.

Nonlinear organizing plans are ideal for writers who naturally take a spatial approach to structuring their paper topics, goals, and ideas. The graphic design affords a bird's-eye view of a document's structure, which helps the writer establish whether the plan is sufficiently developed and whether its parts are logically related to promote strong global unity. In addition, it's usually easier to experiment with different structural patterns, to figure out what is optimal, when a plan's parts are first arranged spatially instead of linearly.

For an example of the advantages to nonlinear organizing plans, have a look at the cluster of bubbles in the top-right portion of Marvette's plan in Figure 4.4. These bubbles represent the writer's rhetorical goal to discuss the importance of her research issue. Marvette's strategies involve explaining why the debate about acupuncture is relevant to her audience. It's an essential rhetorical goal for the introduction section of Marvette's position paper because its audience may not naturally appreciate the issue's relevance to their specific interests and medical practices. As Marvette scanned her entire bubble diagram, however, she realized a potential problem: The goal to convince readers of the research issue's importance is not reinforced in her plan for the paper's body and conclusion sections. Marvette reasoned that if readers lose sight of how the issue relates to them, her overall argument would be compromised. This insightful analysis was prompted and aided by the nonlinear design of Marvette's organizing plan. In revising her plan, Marvette decided to add a rhetorical goal to the conclusion section, a goal to reinforce ideas about why her paper's motivating issue is indeed meaningful for her specific audience.

In the lower-left portion of Figure 4.4 (page 160), the cluster of bubbles represents Marvette's rhetorical goal to acknowledge limitations to her argument, which include shortcomings in the supporting research and theory. Notice the line drawn from this goal to the goal for suggesting new studies in Marvette's conclusion section. The connection represents Marvette's plan to demonstrate explicit relationships between the two sections, specifically by suggesting new studies that have the potential to overcome limitations to her argument. The nonlinear design of the organizing plan helps to establish logical and unified relationships across the different sections of a paper.

If you prefer to work with nonlinear plans, you must still ultimately organize their elements in a linear format, corresponding to how they will be ordered in your paper. This might simply involve numbering the bubbles in a bubble diagram or converting your nonlinear plan into a formal linear outline.

Taking a Principled Approach to Organizing Content

As mentioned earlier, organizing content is largely a problem-solving process. Experienced scientific authors make optimal decisions about coordinating the structural elements of their papers by applying sound principles. Our approach

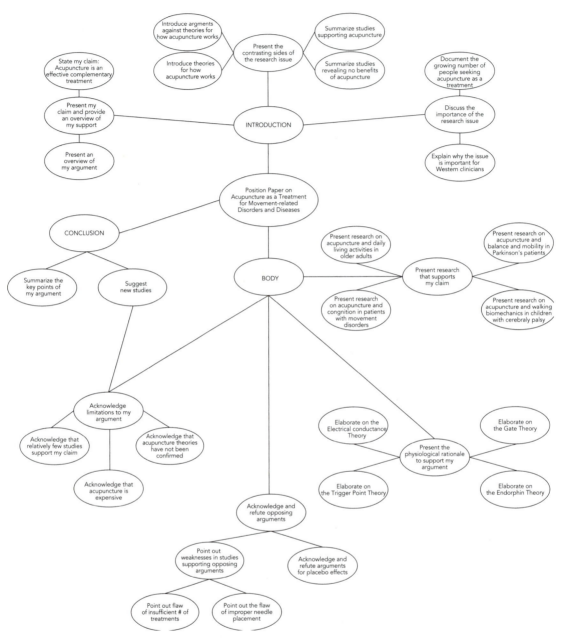

FIGURE 4.4 A sample nonlinear organizing plan—in this case, a bubble diagram.

will be guided by the seven organizing principles that are outlined in Figure 4.5 and introduced next.

Organizing Principle 1: Follow Assignment Instructions That Directly Address Matters of Structure

This is the most straightforward way to decide how to organize content in scientific papers. If your assignment comes with directions concerning document structure, you should follow them to the letter. Picture an assignment for a research proposal in which your professor's instructions are to "begin by directly stating three specific aims for your proposed project, and then present the background information to elucidate the aims." Given these explicit directions, you should *not* take the liberty of reversing the order of the introductory content by beginning with background information and then stating your three specific aims. Your professor will be looking for the aims to appear right away, so your paper's structure should meet this expectation.

Organizing Principle 2: Follow Discipline-Specific Conventions for Organizing Scientific Papers

Chapter 2 demonstrated a modeling approach to developing goal-based plans; applying the approach, you derive rhetorical goals and strategies for your papers by studying what scientific authors do and say, by convention, in published articles. Modeling also works well for solving problems in organizing content. Suppose that you are writing a lab report based on the IMRAD format of published research papers. You are well aware of the overall structure, but perhaps you are uncertain about how to organize details of the results section. Should the section be divided into subsections with separate headings? If so, how should

---

1. Follow assignment instructions that directly address matters of structure.
2. Follow discipline-specific conventions for organizing scientific papers.
3. Rely on familiar and consistent organizing themes.
4. Arrange your paper's parts to serve its overall functions.
5. Apply audience analysis to determine the optimal order and integration of content.
6. Aim for the exemplary quality of global unity.
7. Use good old-fashioned common sense to ensure logical organization.

FIGURE 4.5 Seven key principles for organizing the content of scientific papers.

the subsections be titled and ordered? And what sort of information should the subsections contain? Taking the modeling approach to answering these questions, you would get a handful of published research papers on your topic and then study the structure of their results sections. You'll notice that most are effectively organized by subsections that correspond to key outcome variables that relate directly to their studies' motivating questions, problems, and hypotheses. You can also learn about conventions for organizing content through educational resources on scientific writing. In this book, for instance, detailed conventional approaches to organizing results sections are presented in Chapter 8.

Organizing Principle 3: Rely on Familiar and Consistent Organizing Themes

Familiar patterns enable us to negotiate our way through the world successfully. Take the simple example of walking through a furniture-filled lecture hall. Suppose that, upon entering the doorway, you decide to sit in a chair on the room's farthest end. Immediately, you recognize a familiar pattern: The chairs are organized in perfect rows, one after the next. Moving toward your destination, you don't give a second thought to any tricky maneuvers. And you have no worry about stumbling along the way. Just the simple recognition of a pattern ensures that you will move confidently through the room. But what if the chairs were in such complete disarray that you could not identify any pattern at all? Walking across the lecture hall, you might have to slow down or stop to avoid obstacles, to figure out your next turn, and perhaps even to assess whether the journey to the other side is worthwhile after all. It's a similar situation for audiences of scientific papers. When readers clearly recognize structural patterns in a document, they move confidently from section to section; in addition, this familiarity enhances reading comprehension and memory. We'll refer to easy-to-recognize and consistent structural patterns as *organizing themes*. As follows, I describe 12 conventional organizing themes that apply to scientific writing.

1. Goal-based: This global organizing theme was introduced in Chapter 2 and is demonstrated in the present chapter by several examples, including the outline in Figure 4.3 (pages 155–157). The main structural elements of a goal-based organizing theme are rhetorical goals and their strategies. A goal-based approach to organizing content is ideal for all types of scientific papers.

2. Topic-based: Like the previous organizing theme, this one is *global* in the sense that it can apply to entire documents. Its main structural elements are the subtopics of a paper's overall topic. Topic-based themes are especially useful for organizing straightforward informative review papers.

3. Chronological: Some events and processes are ideally described chronologically, or on a detailed timeline. As an example, consider a paper on the control of rapid movements. In one subsection, the writer is planning to describe the brain and muscular processes involved in abruptly stopping a car in response to a red traffic light. From the light's appearance until the foot's breaking movement is completed several hundred milliseconds later, the underlying mechanisms occur at discrete points in time. In this case, a chronological approach to organizing the description makes perfect sense.

4. **Sequential:** Like the chronological organizing theme, this one applies to describing processes and events that are defined primarily by temporal relationships. The difference is that the sequential theme is not strictly tied to *specific* points in time. It's a logical choice for organizing the methods sections of research papers and proposals because they describe study procedures and analyses that are executed in a defined sequence.

5. **Spatial:** A spatial organizing theme is fitting for describing objects and systems that are distinguished primarily by their physical layout. Take the example of describing the anatomy of the neuromuscular system. A logical scheme would begin literally at the *top* by describing how brain cells form nerve fibers that connect with cells in the spinal cord, the *middle* level of the spatial layout. Then, moving to the *bottom* level of the system, the author would describe how spinal nerves innervate muscles in the body's extremities. Depending on the object or system, the ideal spatial pattern might be top-to-bottom, left-to-right, inside-to-outside, or some other arrangement.

6. **Most-to-least important:** Some rhetorical goals for scientific papers may call for ranking information and ideas in order of importance. To support an original conclusion, for example, an author might have numerous lines of evidence and reasoning that vary in convincibility. By ordering content from most-to-least important, or in this case most-to-least convincing, the author will make a strong first impression and reinforce direct messages about the relative emphasis of each line of support.

7. **Least-to-most important:** Suppose that, in a twist on the preceding scenario, the writer wanted to quickly dismiss a few relatively minor lines of support in order to set up arguments for bigger and better ones. In this case, the least-to-most-important theme is warranted. It's also the best pattern for leaving readers with strong lasting impressions.

8. **Whole to part:** For describing complex systems and clarifying elaborate concepts, a whole-to-part organizing theme makes good common sense. Suppose, for example, that you are faced with describing the intricate anatomy of skeletal muscle, which is a truly complex system. Whole skeletal muscles comprise membrane-enclosed bundles of elongated muscle cells, which may number in the thousands. Each cell is composed of tiny fibers called myofibrils, which contain

even smaller filaments made of highly organized protein molecules. A whole-to-part description of this architecture would be ideal.

9. Part to whole: Complementing the previous theme, this one applies to describing complex systems and concepts by focusing first on their smallest elements and ideas. These are progressively integrated into larger parts, one after the next, until the whole is revealed.

10. General to specific: This is a common theme in scientific writing because many rhetorical goals require backing general statements, such as conclusions and claims, with specific and detailed lines of support. For any rhetorical goal that entails constructing an argument, the general-to-specific organizing theme is effective.

11. Specific to general: Reversing the previous theme, this one begins with specific supporting details and works its way to more general statements. It's a useful pattern for setting up claims and conclusions that can be backed with relatively small amounts of straightforward data.

12. Compare and contrast: This is the most logical organizing theme for any rhetorical goal that engages writers in synthesizing research. One such goal, which applies to the discussion section of IMRAD-structured papers, is to relate your study's findings and conclusions to those from previous studies on your research issue. Wielding the skills of synthesis (as presented in the previous chapter), you would naturally compare and contrast the studies' outcomes.

Organizing Principle 4: Arrange Your Paper's Parts to Serve Its Overall Functions

One of the fundamental principles in movement science is that *form serves function*. Consider the webbed feet of a duck, which are structured to enable the most efficient and powerful paddling movements in the water. Another example is the architecture of skeletal muscles, which determines their force-generating capabilities. Form also serves function in well-written scientific papers. Take the IMRAD format for research papers. It is the ideal arrangement to serve the overall function of a research paper, which is to report a study to readers who did not witness it firsthand. The IMRAD format is set by convention, so authors naturally use it to organize the major sections of research papers.

Other more challenging structural problems in scientific writing are ideally solved by purposefully structuring papers to serve their overall functions. Recall, for example, the position paper that a student named Marvette was planning. The overall function of any position paper is to convince readers to accept a claim in a debatable argument. This end is logically achieved partly by the paper's form, as illustrated in Marvette's outline in Figure 4.3 (pages 155–157). For the introduction section, Marvette's first rhetorical goal is to present the contrasting sides of her debatable issue on acupuncture as a treatment for movement-related disorders and diseases. This goal is the structural foundation for Marvette's claim (that acupuncture is a viable treatment), which comes at the end of the introduction. What's the best form for the position paper's body section? It's the structure revealed in Marvette's outline: The body is organized by rhetorical goals and strategies for presenting separate lines of support for the writer's claim. Each subsection of the body provides a different answer to the reader's logical question, "What evidence and reasoning back your claim that acupuncture can effectively treat disorders and diseases that affect movement?"

The principle to arrange form to serve a paper's overall function is so logical that it may seem trivial. But, in reality, many cases of poorly organized scientific papers could have been avoided if their authors had purposefully applied this principle.

Organizing Principle 5: Apply Audience Analysis to Determine the Optimal Order and Integration of Content

In an audience-centered approach to writing, this principle is extremely powerful for guiding sound decisions about how to order and integrate a paper's ideas, topics, and rhetorical goals. In applying the principle at hand, skilled writers reflect on their readers' needs, expectations, and values concerning matters of structure. The use of audience analysis in organizing content was demonstrated in our discussion of a potential problem in Marvette's outline. Recall that the first rhetorical goal for the body of Marvette's paper was slated to present research that directly supports her claim for the treatment efficacy of acupuncture. As we considered, however, Marvette's audience is likely to have a strong bias against this claim. So readers might have responded negatively to Marvette's beginning her argument without considering their views. The solution, derived through audience analysis, was to begin the position paper's body with the rhetorical goal to acknowledge the audience's views and then to present scientific evidence and reasoning to refute them.

Organizing Principle 6: Aim for the Exemplary Quality of Global Unity

As defined earlier, global unity refers to the overall consistency of well-organized papers, which readers experience as the sense that every bit of content has its logical place. The key to achieving strong global unity is to plan purposefully for this desirable outcome. Let's say that you are planning a hypothesis-driven research paper structured by the IMRAD format. Following convention, you will present your hypothesis in the paper's introduction section. To promote strong global unity, your plan for the methods, results, and discussion sections should include elements that logically relate to your hypothesis and that maintain the hypothesis-testing theme of your paper. For the methods section, for example, selected rhetorical goals and strategies should reflect your plan to explain how certain procedures and analyses were aimed specifically at testing the hypothesis. Moreover, the plan for your discussion section should include goals and strategies for presenting and supporting conclusions that resolve the hypothesis.

Organizing Principle 7: Use Good Old-Fashioned Common Sense to Ensure Logical Organization

Which section of a lab report should present the purposes of its associated study? Naturally, the answer is *the introduction section*. In the discussion section of a research paper, where is the best place to offer suggestions for future studies—at the beginning or toward the end? This is a no-brainer—the best place is *toward the end*. When you are writing about study methods, should your descriptions of statistical analyses come before or after details about procedures for collecting data? Of course, the most logical position is *after*. The point of these questions and their simple answers is that some organizational problems in scientific writing are best solved through using good old-fashioned common sense. It's another principle that may seem obvious; however, writers sometimes need a little self-prompting to apply it.

Organizing Your Paper's Major Sections

For the primary types of scientific papers, major sections and their linear order are usually prescribed by convention. An example is the IMRAD format for standard research papers and their related forms, such as lab reports and theses. Or consider that most review papers have three major sections—an introduction, a body, and a conclusion—that are ordered accordingly. And many research proposals and grant applications have major sections devoted, in linear order, to outlining the specific aims and hypotheses of proposed studies, providing background information and discussing matters of significance, describing preliminary studies and pilot work, detailing proposed research methods, and presenting and justifying planned budgets. When the order of a paper's major sections is prescribed by convention, the process of organizing them is straightforward. There are, however, a few details concerning the organization of major sections that warrant attention in planning. You should double-check your organizing plan for whether it follows assignment instructions and conventional guidelines for including all required major sections. In addition, you should make sure that your organizing plan's major sections are ordered appropriately. Instructions for scientific papers occasionally call for their major sections to be ordered in nonstandard forms. In some journals, for example, methods sections come at the end of research papers rather than immediately after the introduction.

From Chapter 2, you may recall the advice to check your goal-based plan for whether its rhetorical goals are placed in the appropriate major sections. In the present stage of the writing process, you might double-check this feature of organization. As explained earlier, the evaluation requires knowing the conventions for what to do and say in the major sections of scientific papers. Details about these conventions are presented in Chapter 8.

Organizing Your Paper's Subsections

Within the major sections of scientific papers, subsections are composed of paragraphs that cover related topics or that serve common rhetorical goals and strategies. Subsections are often designated by formal headings. For example, in the methods sections of most IMRAD-structured papers, consecutive subsections have headings such as *Subjects*, *General Procedures*, *Measurements*, and *Statistical Analyses*. In contrast to major sections, the order and integration of subsections are less likely to be predetermined by assignment instructions and conventions. So to effectively organize the content of subsections, skilled writers rely heavily on problem-solving approaches.

To demonstrate the activity at hand, we'll suppose that you are writing a lab report for a class in biomechanics, the field that applies principles of physics and mechanical engineering to the study of animal and human movement. Your class conducted an experiment to analyze the biomechanics of a phenomenon called the walk–run transition. This term refers to the change in gait patterns that occurs when a progressive increase in walking speed naturally induces the action of running. Your class studied this event as subjects walked and then ran on a treadmill set to increase its speed at gradual intervals. The key study outcome was the movement rate (measured in meters per second) at which the walk–run transition occurred.

For the major sections of your lab report, the assignment instructions call for using the IMRAD format in its conventional linear order. Our demonstration of how to organize a paper's subsections will be based on your lab report's discussion section, which you have wisely decided to structure according to a goal-based theme. In other words, the subsections will be organized by the key rhetorical goals and strategies for discussion sections of IMRAD-structured papers. Briefly outlined in no particular order, here are the rhetorical goals that you have planned:

1. Compare my results and conclusions to those from previous studies on my research issue.
2. Suggest future research.
3. Discuss methodological limitations that may have compromised my study's outcomes.
4. Reintroduce the research issue that motivated my study.
5. State and support my conclusions.
6. Explain my study's results in terms of underlying mechanisms.
7. Call attention to my study's methodological strengths.

To order the goals and their corresponding subsections, we'll rely on our organizing principles (Figure 4.5, page 161). Let's begin by recalling the straightforward principle to *follow assignment instructions that directly address matters of structure.* In the assignment for your lab report, maybe your professor designated a specific order for topics to cover, goals to accomplish, or questions to answer in the discussion section. These instructions would take precedence over other approaches to ordering the content. Suppose, however, that your professor gave no explicit directions for structure. In this case, you should turn next to the organizing principle to *follow discipline-specific conventions for organizing scientific papers.* For example, in published research papers it's conventional, as well as commonsensical, to begin discussion sections by reorienting readers to the questions, problems, hypotheses, and purposes on which reported studies were based. Following this convention, you would simply move Rhetorical Goal 4 in the preceding unordered list (*Reintroduce the research issue that motivated my study*) to the plan's first position. Another convention that also makes good common sense is to end discussion sections with suggestions for future studies. Accordingly, Rhetorical Goal 2 (*Suggest future research*) would move to the last position. So far, so good.

To determine the best order for the other rhetorical goals in our hypothetical plan, we'll rely on the powerful organizing principle that involves applying audience analysis. What will your readers expect you to do and say first, second, third, and so on in your discussion section? What will be on their minds, for instance, at the section's beginning? At this point, they will be most curious to know your conclusions, based on interpretations of data presented in the paper's results section. Where, then, in the discussion section should you state and support your conclusions (Rhetorical Goal 5 in the randomly ordered list)? Some authors save their conclusions for last—that is, to end the discussion section. This creates a *bottom-up* structure, something like that of a murder mystery novel. The authors may aim to build suspense by leaving the most important message, the conclusion, until the very end. However, from the reader's perspective, a bottom-up structure of the discussion section is problematic for two main reasons. First, it fails to address the most pressing question on readers' minds: *What are the author's conclusions?* Second, the bottom-up structure often undermines the organizing principle to arrange a paper's parts to serve its overall functions. The overall function of a well-written discussion section is to argue for conclusions that resolve the research issue that motivated the associated study. When authors delay their conclusions until the end of the discussion section, readers cannot immediately grasp how the preceding information and ideas serve the section's essential argument. Through audience analysis, then, we have figured out that the rhetorical goal to state and support conclusions (the fifth goal in our list) should be positioned *early* in the discussion section's plan. A fitting position is immediately after the goal to reintroduce readers to your study's motivating research issue.

Presented as follows are two more examples of how to solve problems in organizing the subsections of a scientific paper, specifically subsections in the discussion section of our hypothetical lab report on the walk–run transition.

Example 1

Rhetorical goal: Discuss methodological limitations that may have compromised your study's outcomes.

Scenario: Suppose that the methodological shortcomings to your study were minor and unlikely to have had a significant negative influence on its outcomes.

Organizational solution: Wait until relatively late in the discussion section to discuss the limitations.

Alternative scenario: Now imagine that your readers are likely to strongly criticize certain aspects of your study's methods, which might indeed have influenced the outcomes.

Organizational solution: To address your readers' concerns and criticisms, you should acknowledge and respond to them soon after stating and supporting your conclusions.

Example 2

Rhetorical goal: Compare your results and conclusions to those from previous studies on your research issue.

Scenario: Suppose that your results and conclusions contrast sharply with those from previous studies on the walk–run transition. Let's say that the values you obtained for the rate at which the transition occurs are much higher than those reported in the literature. After you present your conclusions, readers who know the research field will certainly ask, "What about the research that contrasts your findings? How do you explain the contradictions?"

Organizational solution: Audience analysis and common sense bring about the solution to place this rhetorical goal directly following the goal to state and support your conclusions.

The preceding demonstrations involved deciding how to order the subsections in one major section of a scientific paper. Greater challenges come with organizing content at more refined levels, such as within one subsection devoted to accomplishing one rhetorical goal. As an example, consider the following rhetorical goal for the discussion section of a research paper: **Discuss the underlying mechanisms that may offer plausible explanations for your study's results**. In the plan for your lab report on the biomechanics of the walk–run transition, suppose that you have positioned this goal in the middle of the discussion section. It's a fitting goal because experienced readers will respond to your results by asking, "What actually causes the transition from walking to running as the speed of locomotion increases?" One explanation is that the transition is hard-wired in the brain—in other words, at some critical speed of locomotion, the brain commands the response. Another explanation is that the walk–run transition is controlled by networks of nerve cells in the spinal cord. Other mechanisms involve the influence of sensory feedback from the contracting muscles and moving joints. To simplify the matter, we'll say that there are four plausible mechanisms, which we'll refer to as A, B, C, and D. How will you determine the best order and integration of these mechanisms?

At this refined level—that is, organizing content within subsections—we usually do not have explicit instructions and established conventions to follow. So other organizing principles must come into play. As always, the principle to apply audience analysis is extremely useful. Suppose, for example, that in order to understand mechanism B, your readers first need to know about mechanisms D and C. In this scenario, the organizational problem is solved through considering the knowledge needs of your audience: Your discussion of mechanisms D and C should come before your discussion of mechanism B.

In another scenario, imagine that the four mechanisms are highly debatable, calling for a scientific argument about which ones offer the most plausible explanations for the walk–run transition. To solve this structural problem, you might apply the organizing theme that ranks information, ideas, and arguments in most-to-least-important order—that is, you would argue for the mechanisms in the order of their explanatory power. Or maybe the mechanisms are best described by their spatial relationships in the body. Some of the mechanisms are based in the brain, which is the *highest* level of the motor control system. Other mechanisms operate in the spinal cord and the muscles, which are the *middle* and *lowest* levels, respectively. In this analysis, the spatial organizing theme would be ideal.

Deciding Which Parts of Your Plan to Emphasize

All of the planning activities presented in earlier chapters are powerful antidotes for the dreaded experience of writer's block. If you have done the groundwork—by defining your writing project, developing goal-based plans, and generating content—you will reach this stage of organizing content with plenty to write about. Indeed, your plan might even overstretch the boundaries of your assignment, such as the length restriction for your paper. This situation calls for solving the problem of which parts of your organizing plan to build on as well as which parts to pare down or perhaps even delete. It's actually quite a nice problem to have, especially if you are accustomed to having writer's block!

Skilled writers base their decisions about content to emphasize versus de-emphasize on assignment instructions and considerations of audience. Take the hypothetical case of Professor Joseph Augustine, who is writing a research paper to report a study on the biomechanics of walking in children with cerebral palsy (CP). Professor Augustine will submit the paper to *The Journal of Applied and Clinical Biomechanics*, which has a primary audience of clinical researchers and health professionals. The journal gives explicit instructions to highlight clinical applications of research. In addition to setting rhetorical goals that draw attention to the clinical focus of his study, Professor Augustine has other goals to achieve in the paper. They include describing the study's methods, presenting and interpreting the results, and placing the research in the larger context of clinical biomechanics. However, given the journal's instructions and readership, Professor Augustine makes the sound decision to emphasize the clinically relevant rhetorical goals in their respective sections. In the introduction section, for example, he will devote a relatively large amount of content to explaining how current knowledge gaps regarding the biomechanics of walking in children with CP have hampered clinicians' efforts to develop effective therapeutic modalities. Then, in the discussion section, Professor Augustine plans to write considerably more about the clinical applications of his study's results than about how the study fits into the larger research field.

This activity—assigning different levels of emphasis to parts of an organizing plan—is not just for scientists writing for publication. It's equally important for school writing projects. The principles and practices that guide career scientists in the activity also apply to student writers. For instance, students should attend closely to assignment instructions that explicitly indicate rhetorical goals and topics to emphasize in their papers. In addition, sound decisions result from audience analysis. If your professor is your primary audience, it's worth asking that all-important question, "What is he or she looking for students to emphasize in their papers?" The flip side is equally powerful: "What is he or she looking for students to *de*-emphasize in their papers?" Another key factor in the decision at hand involves identifying the hottest aspects of the research issue you are writing about. Which concepts in the field are most interesting and deserving of elaborate explanations? If, for example, you are writing about original research, which of your hypotheses and findings are most noteworthy? In making debatable arguments, which claims and conclusions require the most extensive supporting evidence and reasoning? And, generally speaking, what information, ideas, and arguments do you want to impress most of all on your readers?

By thinking carefully about which parts of your organizing plan to emphasize, you will set helpful boundaries on your paper before setting out to draft. As an example, suppose that you deem a particular strategy for accomplishing a rhetorical goal to be extremely important. To successfully achieve the strategy, you realize that you will need to write much more than you originally planned. This evaluation might lead you to convert the strategy into a full-blown rhetorical goal and then to flesh out a new set of strategies for it. In a contrasting scenario, you might find that several strategies for a goal are relatively unimportant, so you will combine them into one strategy. You may even have to delete parts of your plan because you simply don't have space in your paper for them. My point is that these decisions and actions are essential for setting up an efficient and productive drafting process, which we turn to next.

ABOUT THE PROCESS: WRITING A DRAFT

Like organizing content, the process of writing a draft involves giving form to relatively abstract and unordered elements of plans. However, drafting poses specialized challenges, the most demanding of which entail translating rough plans, ideas, and mental images into grammatical sentences and coherent paragraphs. The remainder of the chapter presents guidelines and techniques for drafting the following elements of scientific papers: titles, abstracts, section headings, paragraphs, sentences, graphics, and references. We'll also cover strategies for avoiding plagiarism. The overall process is diagrammed in Figure 4.6.

Drafting Titles

Titles are examples of small things that can make a big difference in scientific papers. Using databases to find literature on their topics, for example, readers obviously rely heavily on titles to decide which articles are relevant to their needs and interests. Only a handful of words may thus determine whether a published article

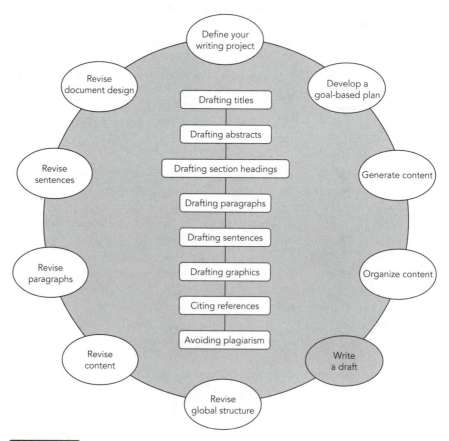

FIGURE 4.6 Process map for *writing a draft*.

ultimately reaches its intended audience. Of course, professors cannot decide to pass over student papers because their titles are not personally relevant or especially appealing. Nonetheless, well-titled student papers instill positive first impressions that can influence professors' evaluations. The following guidelines for drafting effective titles are aimed at promoting favorable initial responses from readers.

Title Your Paper to Convey the Defining Features of Your Research Issue

The best titles contain specific words and phrases that directly convey their papers' motivating research issues. Judged according to this guideline, the following sample title has room for improvement.

> Effects of Drug Treatments on Movement Control in Parkinson's Patients

It's the title for a research paper that reports a study to determine how levodopa, a drug for treating Parkinson's disease, affects a property of balance called postural sway. The study focused exclusively on patients in the first stage of the disease. The title is weak because it does not convey the specific, defining aspects of the research issue. The study did not address the effects of drug treatments in general. Instead, it focused on a specific drug, *levodopa*. Moreover, the study did not address the effects of levodopa on general properties of movement control. It focused specifically on *postural sway*. The title also omits key information about the specific population of patients studied. Here's a revision that better reflects the defining features of the research issue:

> Effects of Levodopa on Postural Sway in Stage I Parkinson's Patients

The improvement is the result of adding the necessary specific words and phrases.

Restrict Your Titles to Approximately 8 to 15 Words

Whereas specificity is a hallmark of effective titles, there is a limit to how much information titles can convey before overwhelming readers with details. The following exaggerated example makes the point.

> Effects of a 2.6-gram Daily Dose of Levodopa on Postural Sway as Measured by Center-of-Foot-Pressure Displacement in 60- to 75-year-old Stage-1 Parkinson's Disease Patients as Evaluated by the Hoehn and Yahr Scales

The guideline to restrict titles to 8–15 words is adaptable to a degree. Often, however, titles that contain fewer than six or seven words lack specificity. Most titles that exceed 15 words are just too cumbersome.

Introductory phrases such as *A review paper on* and *A study concerning* are usually unnecessary in titles, as demonstrated in the following strikeout revision.

> ~~A Review Paper on the Analysis of the~~ Biomechanics of Wing Movements of the Desert Locust *Schistocerca gregaria*

Readers will know whether they are reading a review paper or a study (i.e., a research paper) from the document's design; in addition, most scientific databases label papers by their types. In the preceding example, the phrase *on the analysis of the* is deleted because its meaning is implied.

The following 16-word title is awkwardly phrased and contains unnecessary details.

> Temperature and Viscosity and Their Effects on Underwater Swimming Movements of the Antarctic Pteropod, *Clione antarctica*

Here's an effective 11-word revision:

> Effects of Temperature and Viscosity on Swimming Movements of *Clione antarctica*

The revision's phrasing is ideal for titles of papers that report experiments on the influence of treatments and conditions on targeted outcomes.

Highlight the Novel and Unique Features of Your Research Issue and Paper

A well-written title satisfactorily answers the reader's question, "Why should I read this article instead of, or in addition to, others on the topic?" For an example of this guideline, consider a new study on the development of eye–hand coordination in infants and children. Suppose that all previous studies on this issue have used cross-sectional designs, comparing eye–hand coordination skills across separate age groups. The new study was the first with a longitudinal design. The researchers followed the *same* subjects over time, from infancy through childhood, enabling a more accurate assessment of skill development. The following titles effectively draw attention to the study's novel and unique design.

> Longitudinal Development of Eye–Hand Coordination in Infants and Children

> Development of Eye–Hand Coordination in Infants and Children: A Longitudinal Study

For emphasis, word groups that reflect novel and unique features of a paper should be placed at the start of the title or at its end, as demonstrated in the preceding examples.

Include Specific Words and Phrases That Reflect
Your Paper's Overall Goal

Compare the following four titles for the same research paper.

> Example 1: Aerobic training and movement efficiency in adults with cerebral palsy

> Example 2: Effects of aerobic training on movement efficiency in adults with cerebral palsy

> Example 3: Does aerobic training improve movement efficiency in adults with cerebral palsy?

> Example 4: Aerobic training improves movement efficiency in adults with cerebral palsy

Example 1 implies that the author will discuss the paper's topic in a general way. This is problematic because the overall goal of a research paper is *not* to generally discuss its topic. Instead, the goal is to report the study's aims, methods, and outcomes. This is effectively conveyed in Example 2. The beginning phrase, *effects of*, reflects the overall goal of determining how a specific independent variable (aerobic exercise) affects a specific dependent variable (movement efficiency).

A nontraditional but often effective way to express a paper's overall goal is to phrase the title as a question, as in Example 3. This approach intrigues readers and can even help them navigate challenging content. In the middle of a paper, if readers are confused about the relevance of complex information and ideas, they may be enlightened simply by reflecting back on the title. Another nontraditional type of title directly presents the author's conclusion, as in Example 4. A title phrased as a conclusion reflects sound audience analysis because readers appreciate getting the take-home messages of scientific papers in a nutshell. Titles conveying especially interesting conclusions engage readers from the outset.

For Directions on Formatting Titles, Refer to Author-Instruction
Documents and Discipline-Specific Style Guides

Which letters in titles should be capitalized—the first letter only or the first letter of every word? Should your title have a particular font style and size? Should it be centered on a cover page? There are no all-encompassing rules for capitalization, font design, and positioning of titles in scientific manuscripts. Instead, directions are specific to individual assignments, journals, and grant agencies. So you will need to consult your assignment instructions and, if necessary, a *style guide* for the scientific discipline in which you are writing. If you are unfamiliar with style guides, make sure to read about them in Figure 4.7 (page 176).

Drafting Abstracts

In a typical published journal article, including research papers and review papers, the title is followed by a list of the authors and their affiliations. Next comes the

How should numbers be written in scientific papers—spelled out or as Arabic numerals? Is it acceptable to abbreviate certain words? If so, what are the correct abbreviations? For which terms can acronyms be used? How should mathematical equations be integrated into texts? What are the rules for capitalization in titles and section headings? What are the correct symbols to use for representing human and animal genes? Should drug treatments be designated by their commercial or chemical names? To answer literally hundreds of technical questions that concern matters of scientific nomenclature, format, and style, this chapter would need to be transformed into a book. In fact, numerous books would be necessary, one for each of the major subdisciplines of the life sciences. The reason is that guidelines for nomenclature, format, and style vary across disciplines of biochemistry, biology, physiology, medicine, psychology, and other life sciences. Books that contain the directions for specific disciplines are called style guides. When general writing guidelines and author-instruction documents do not answer your specific questions about technical matters for manuscript preparation, consult the style guide for the scientific field in which you're writing. Citations for commonly used style guides in the life sciences are presented below. Most university libraries hold these books in their reference sections.

American Chemical Society (2006). *The ACS style guide: Effective communication of scientific information (3rd edition)*. Washington, DC: American Chemical Society.

American Medical Association (2007). *American Medical Association manual of style: A guide for authors and editors (10th edition)*. New York: Oxford University Press.

American Psychological Association (2001). *Publication manual of the American Psychological Association (5th Edition)*. Washington, DC: American Psychological Association.

American Sociological Association (2007). *American Sociological Association style guide (3rd edition)*. Washington, DC: American Sociological Association.

American Society for Microbiology (1991). *The ASM style manual for journals and books*. Washington, DC: American Society for Microbiology.

Style Manual Committee, Council of Science Editors (2006). *Scientific style and format: The CSE manual for authors, editors, and publishers (7th edition)*. Reston, VA: Council of Science Editors.

FIGURE 4.7 Information about style guides for writing in various life science disciplines.

abstract, which is a summary of a paper's main contents and messages. In conducting literature searches, writers use abstracts to get more information about articles than titles can obviously provide. Abstracts thus inform final decisions about whether to obtain and read articles of interest. Many professors include abstracts as required parts of school writing assignments because they challenge students to organize scientific content clearly and concisely. By convention, abstracts for scientific papers are usually restricted to a maximum length of 200 to 300 words.

The upcoming guidelines for drafting abstracts are effectively demonstrated by the sample abstract in Figure 4.8. The model comes from a published research paper on the effects of exercise on the carotid arteries in young and older men (Tanaka et al., 2002). The scientists, Hirofumi Tanaka and coworkers, focused on a layer of tissue called the intima-media, which lines the inner walls of the carotid arteries. As a normal consequence of aging, the intima-media becomes thicker. The result is a narrowing of the vessel's lumen, or the hollow tube through which blood flows; this condition increases the risk for stroke because the reduced lumen diameter restricts blood flow from the heart to the brain. Studies have shown

---

(1) Carotid artery intima-media thickness (IMT), an independent risk factor for stroke, increases with age. **(2)** Habitual exercise is associated with a lower prevalence of stroke, but it is unclear whether this protective effect could be mediated through a favorable influence on carotid IMT. **(3)** We examined this possibility using both cross-sectional and intervention approaches. **(4)** First, 137 healthy men (age 18-77 yr) who were either sedentary or endurance trained were studied. **(5)** In both groups, carotid IMT and IMT-to-lumen ratios were progressively higher with age ($P < 0.05$). **(6)** There were no significant differences in measures of carotid IMT between sedentary and endurance-trained men at any age. **(7)** Carotid systolic blood pressure increased progressively with age and was related to carotid IMT ($r = 0.63$, $P < 0.01$). **(8)** Second, 18 healthy sedentary subjects (54 ± 2 yr) were studied before and after 3 mo of endurance training. **(9)** Carotid IMT, IMT/lumen ratio, and carotid systolic blood pressure did not change with exercise intervention. **(10)** Our results do not support the hypothesis that regular aerobic exercise exerts its protective effect against stroke by attenuating the age-related increase in carotid IMT. **(11)** This lack of effect on carotid IMT may be due to the apparent inability of habitual exercise to prevent or reduce the age-associated elevation in carotid distending pressure.

Key words: atherosclerosis; lifestyle; ultrasound

FIGURE 4.8 A well-written abstract from a published research paper authored by Tanaka et al. (2002). Reproduced with permission from the American Physiological Society.

that people who exercise regularly have a relatively low risk of stroke. However, researchers have not determined exactly *how* exercise exerts this protective effect. Tanaka et al. tested the hypothesis that exercise reduces the risk of stroke by slowing the age-related increase in intima-media thickness, or IMT.

Organize Your Abstract to Correspond to the Major Sections of Your Paper

In Tanaka et al.'s abstract in Figure 4.8, consecutive groups of sentences effectively mirror the overall IMRAD structure of a research paper (which is the type of paper that Tanaka et al. wrote). Sentences 1–3 summarize the introduction section. Sentence 4 summarizes the key methods of a cross-sectional study in which Tanaka et al. compared measures of carotid IMT in sedentary subjects versus subjects who normally participated in endurance training. Sentences 5–7 present the cross-sectional study's results. Sentence 8 summarizes the key methods of an intervention study in which measures of carotid IMT were compared in sedentary subjects versus subjects who participated in a 3-month exercise program. The results are summarized in sentence 9. Corresponding to content in the paper's discussion section, sentences 10 and 11 reflect the researchers' overall conclusion and their rationale for it. The same tactic of organizing consecutive groups of sentences in abstracts to correspond to a document's major sections also works well for review papers.

Use Your Abstract's First Sentence (or First Set of Sentences) to Summarize Your Paper's Motivating Research Issue and Any Related Hypotheses

This guideline is effectively demonstrated in sentences 1–3 of Tanaka et al.'s abstract.

- Sentence 1 presents background information that readers need in order to understand the study's research issue.
- Sentence 2 directly conveys the issue, with the idea that "it is unclear whether this protective effect [of exercise] could be mediated through a favorable influence on carotid IMT."
- Sentence 3 outlines the study's purpose and design.

To Summarize Study Methods, Include Essential Details to Help Readers Understand How the Primary Results Were Obtained

These methodological details should include relevant characteristics of the subjects, a brief overview of the study procedures, and the essential independent and dependent variables. Sentences 4 and 8 of Tanaka et al.'s abstract effectively demonstrate this guideline.

Summarize Your Study's Most Important Results

This guideline applies to abstracts of papers that report original studies, including research papers, lab reports, and theses. The key is to summarize the essential results that you used to reach your conclusions. The summary may include group means, measures of variability, and values from statistical tests.

To Summarize the Discussion Section of an IMRAD-Structured Paper, Directly Present Your Overall Conclusions

Conclusions may be answers to research questions, solutions to problems, or statements indicating whether hypotheses were supported. In Tanaka et al.'s abstract, sentence 10 exemplifies this guideline.

Drafting Section Headings

Section headings play a number of important roles in scientific papers. They break up long stretches of text, inviting readers to rest for a moment, to reflect on the take-home messages of the previous section, and to gather mental energy for the upcoming section. In addition, section headings provide useful advance information that can enhance the reader's focus and comprehension. Presented as follows, guidelines for drafting effective section headings are similar to those for drafting titles.

Include Specific Words and Phrases That Convey the Upcoming Section's Main Topic and Rhetorical Goal

Consider a hypothetical NIH grant application proposing a study that involves a new technology for measuring brain activity. The authors call it *spectral frequency imaging,* or SFI. NIH grant applications have a predesignated major section that is headed *Background and Significance.* Within this section, authors create their own subsections with corresponding headings. Suppose that the authors of our hypothetical grant application are devoting a subsection to discussing the clinical significance of their proposed study. They have a great idea, which is to use SFI to predict and diagnose movement-related diseases. Here's how the subsection's heading turned out in the authors' first draft:

| Prospective Study Outcomes

It's not a very useful heading at all, because it lacks the necessary specificity to cue readers into the topic and rhetorical goal of the subsection. The heading contains *no* words that reflect the authors' intention to discuss their proposed study's clinical significance. The second try is better but still not specific enough:

| Clinical Significance

The third version makes all the difference:

| Clinical Applications of SFI for Predicting and Diagnosing Movement-related Diseases

Restrict Your Section Headings to Approximately 5–10 Words

By convention, section headings are shorter than paper titles. This makes sense because headings provide advance information about more narrowly defined content. Headings shorter than four or five words tend to be too vague, and those that exceed 10 words or so are often overwhelming.

For Directions on Formatting Section Headings, Refer to Author-Instruction Documents and Discipline-Specific Style Guides

As is true for paper titles, the directions for formatting section headings can vary across assignments, journals, and entire scientific subdisciplines. So writers must consult author-instruction documents and style guides for appropriate formatting guidelines. One common design convention is worth noting: In most published scientific papers, section headings are displayed in a typeface that lacks *serifs*, which are short lines and hook-like markings that are added to the tops and bottoms of letters, mostly for decoration.

- The text that you are reading now has a serif typeface (*Times New Roman*). In the word *typeface*, for example, note the extra lines that look like tails and feet on the letters *t*, *p*, and *a*.
- The text that you are reading now has a sans-serif typeface (*Arial*), which means that it lacks serifs.

Scientific publishers regularly use serif typefaces for the text body of printed documents. One (debatable) theory is that serifs enhance the readability of print documents by smoothing the progression of the eyes from letter to letter in long lines of text. In addition to adding a nice touch of variety to the design of printed documents, the sans-serif typeface of section headings reinforces their role in breaking up long strings of text. Scientific publishers also commonly use sans-serif typefaces for article titles, the axis titles and legends of graphs, the titles and notes of tables, and the captions of figures. These same conventions for the typefaces used in published scientific articles are ideal for designing student papers (unless specific instructions to authors indicate otherwise).

Drafting Paragraphs

As any experienced scientific author will agree, it's a real challenge to produce perfect paragraphs in first drafts. A major reason is that the paragraphs of scientific papers are generally composed of sentences that have complex relationships in logic, meaning, and structure. From sentence to sentence, flawless paragraphs shine in their coherence, or logical flow. The problem posed in drafting paragraphs is a Catch-22 of sorts: When writers stop to concentrate on evaluating local (sentence-to-sentence) coherence, they can lose the vital sense of a paragraph's overall flow, which is actually necessary to create strong coherence in the first place. The solution is to draft without constantly stopping to analyze and correct matters of coherence. In other words, the solution is to delay the process of revising paragraphs until a first draft is complete. We will cover paragraph-level revision in Chapter 6. In the meantime, a few general pointers for drafting paragraphs efficiently are in order.

Aim for Paragraph Unity

A scientific paper has strong global unity if its successive sections serve its overall topic, message, and goal. In a similar way, a paragraph has strong unity if its successive sentences (a) adhere to and develop a central topic, (b) convey and support a main message, and (c) contribute to achieving a focused rhetorical goal. The following pointers take aim on achieving paragraph unity.

1. Narrow the focus of your paragraphs. Paragraphs that lack unity give readers the sense of being led in too many disparate directions without purpose. In contrast, strong unity is characterized by the quality of *oneness* in a paragraph's topic, message, and goal. Such a narrow focus depends on a sound organizing plan. A plan composed of rhetorical goals and strategies is especially useful because its elements can be logically scaled to individual paragraphs. For example, a simple rhetorical goal that requires relatively little content to accomplish might correspond to just one paragraph in a draft. Then again, consider a more extensive rhetorical goal that has many related strategies. In this case, each focused strategy might scale to its own paragraph.

2. Reinforce unity by repeating key words, phrases, and grammatical structures. It's easy to identify the main topics, messages, and goals of paragraphs with strong unity. The reason is that their authors remind readers of these elements quite regularly. Take the sample paragraph that follows (Figure 4.9), which comes from the introduction section to a review paper on movement-related functional capacities in older adults. The paragraph's central topic concerns approaches to assessing risks of losing functional independence among the elderly.

(1) For many older adults, the most pressing health concern is the potential loss of functional independence and the consequent reliance on family members and professional caregivers for assistance in carrying out daily activities. **(2)** Risks of losing functional independence have traditionally been assessed with self-report questionnaires that primarily address the older individual's social, cultural, and economic needs. **(3)** A promising new approach focuses on movement-related capacities, which are assessed through clinical tests of motor skill performance. **(4)** For example, manual dexterity is assessed through timed performance on tasks that include opening and closing latches, writing short sentences, and picking up small objects. **(5)** Postural control is measured with

FIGURE 4.9 Demonstrating strong unity, a sample paragraph from the introduction to a review paper on methods for assessing physical capabilities in older adults.

balancing tasks such as standing on one foot or in a heel-to-toe position. **(6)** Locomotor skill and mobility are evaluated by performance on simple obstacle courses. **(7)** In prospective studies, researchers are using elderly subjects' scores on these motor performance tests to determine long-term risks of losing functional independence.

FIGURE 4.9 *continued.*

The sample paragraph's strong unity is partly a product of repeated words and phrases that keep readers focused on the central topic.

- *the potential loss of functional independence* (sentence 1)
- *Risks of losing functional independence* (sentence 2)
- *A promising new approach* [to assessing risks of functional loss] (sentence 3)
- *long-term risks of losing functional independence* (sentence 7)

Paragraph unity can also be reinforced by repeating grammatical structures and their associated ideas. As a case in point, consider the main subject–verb units in sentences 4, 5, and 6 of Figure 4.9:

- *manual dexterity is assessed* (sentence 4)
- *Postural control is measured* (sentence 5)
- *Locomotor skill and mobility are evaluated* (sentence 6)

Each sentence begins with a grammatical subject that names a characteristic of motor skill. Each subject is immediately followed by a verb phrase that expresses the idea of measuring the characteristic. Then, all of the sentences end with a prepositional phrase that conveys how the characteristic of motor skill is measured:

- *through timed performance on tasks that...* (sentence 4)
- *with balancing tasks such as...* (sentence 5)
- *by performance on...* (sentence 6)

The consistent grammatical structures reveal and reinforce the paragraph's unified rhetorical goal to introduce the new assessment approach, which entails administering clinical tests of motor skill performance, by presenting examples of its applications.

3. Assign every sentence a purposeful role. In paragraphs with strong unity, every sentence serves a well-defined purpose. The sample paragraph in Figure 4.9 demonstrates this point. The paragraph's main message is that researchers are using a promising new approach to assess the older adult's risks of losing functional independence. The first two sentences of the paragraph deliberately set the scene for introducing the new assessment approach; they do so by providing background information about the motivating problem and traditional assessment approaches.

Sentence 3 expresses the main message directly, and sentences 4 through 6 present supporting examples of the new assessment approach. Finally, sentence 7 plays the role of reinforcing the overall message, which strengthens the paragraph's unity.

How can you develop purposeful sentences while drafting paragraphs? Before starting a new sentence, try pausing briefly to contemplate its function. If the sentence does not clearly serve the paragraph's overall topic, message, or goal, either omit the sentence or shape its idea to better reflect its useful purpose. If this method interrupts the flow of your drafting, you can always apply it in the revision process. Another technique for promoting strong unity of purpose is to outline the goals and rough ideas of each sentence (or set of related sentences) in a paragraph before drafting it. (In Chapter 6, you will see examples of this paragraph outlining technique applied to revision.)

Drafting Strong Topic Sentences

Way back in grade school, in your first experiences with writing school papers, your teachers likely encouraged the use of topic sentences. You learned that a topic sentence captures a paragraph's main subject, which the remaining sentences serve to develop. Because you first learned about topic sentences so long ago, and because they may appear to be simple elements of writing, you might be inclined to dismiss them as unsophisticated and trivial. But doing so would be a major mistake. Well-written topic sentences are extremely useful and advanced rhetorical devices.

Take the topic sentence of the sample paragraph in Figure 4.9:

> A promising new approach [to assessing risks of losing functional independence in the elderly] focuses on movement-related capacities, which are assessed through clinical tests of motor skill performance.

Simple as it may seem, the sentence is pivotal. In a nutshell, it answers important questions that thoughtful readers will naturally raise about its paragraph.

- **What is the paragraph about?** It's about a promising new approach to assessing risks of losing functional independence in the elderly.
- **What is the paragraph's take-home message?** The approach entails administering clinical tests of motor skill performance.
- **What is the paragraph's goal?** The topic sentence indirectly reflects the goal to introduce examples of the promising new assessment approach.
- **How are the paragraph's key ideas structured?** From the topic sentence, we might surmise that the paragraph will list examples of the clinical tests of motor skill performance.

As demonstrated by these questions and answers, strong topic sentences do much more than capture the main subjects of paragraphs. Considering their elaborate functions, perhaps we should call them *topic-message-goal-structure sentences*. While I am tempted to use this term, I will spare you its clumsiness!

In the drafting process, it's worth attending to your paragraphs' topic sentences. Without worrying about perfection, take aim on drafting topic sentences that address the four questions in the preceding bulleted list. You will likely find that the effort to draft a sound topic sentence for a paragraph makes it easier to draft the remaining sentences. As we will discuss in Chapter 6, perfection usually demands revising to evaluate whether all paragraphs in a draft actually need topic sentences, whether they are strategically placed, whether they are sufficiently specific, and whether they make promises that their paragraphs actually keep.

Go with the Flow: Drafting Paragraphs with Strong Coherence

Well-written paragraphs have a characteristic flowing quality, every sentence being seamlessly connected to its neighboring sentences in logic and meaning. Readers thus progress smoothly, without having to struggle to understand the relationships between successive ideas. As defined earlier, this desirable quality is called paragraph coherence. A few helpful techniques for fostering coherence in draft paragraphs are presented as follows.

1. Begin sentences with word groups that relate to the key topics or messages of previous sentences. This advice is adapted from a classic article about scientific communication, titled "The Science of Scientific Writing," authored by George Gopen and Judith Swan (Gopen & Swan, 1990). They describe two types of information that sentences convey in paragraphs: *old information*, which comprises familiar topics and messages from previous sentences, and *new information*, which introduces topics and messages for the first time. In their must-read article, Gopen and Swan explain how the placement of old and new information in sentences can influence the reader's sense of paragraph coherence.

> In reading, as in most experiences, we appreciate the opportunity to become familiar with a new environment before having to function in it. Writing that continually begins sentences with new information and ends with old information forbids both the sense of comfort and orientation at the start and the sense of fulfilling arrival at the end. It misleads the reader as to whose story is being told; it burdens the reader with new information that must be carried further into the sentence before it can be connected to the discussion; and it creates ambiguity as to which material the writer intended the reader to emphasize. All of these distractions require that readers expend a disproportionate amount of energy to unravel the structure of the prose, leaving less energy available for perceiving content. (p. 555; used with permission)

An excellent model of paragraph coherence is presented in Figure 4.10. Authored by biologists Thomas Roberts and Jeffrey Scales, the paragraph comes from a research paper on the biomechanics of running in wild turkeys (Roberts & Scales 2002). The sample paragraph nicely demonstrates the technique of beginning sentences with word groups that relate to the key topics or messages of previous sentences.

(1) During running, muscle–tendon units operate like springs, storing and recovering mechanical energy as the limbs flex and extend with each step. **(2)** Some of this cyclical work is done by muscle contractile elements that absorb work as they are actively stretched and produce work as they shorten. **(3)** However, most of the spring-like function of the limb could be performed by the passive stretch and recoil of tendons because steady-speed running on level ground involves no net change in the average mechanical energy of the body. **(4)** Elastic mechanisms can allow muscle contractile elements to operate as near-isometric struts, developing force without shortening or producing significant power. **(5)** The spring-like function of muscle–tendon units allows for economic force development by minimizing muscular work.

FIGURE 4.10 A coherent paragraph from a published paper authored by Roberts and Scales (2002). Reproduced with permission from The Company of Biologists Limited.

Sentence 1 conveys the message that muscles and tendons generate power for the running stride through a spring-like action. In the first sentence's closing phrase, the action is described as a cycle in which mechanical energy is stored as the limbs bend (flex) and then is released as the limbs straighten (extend). Sentence 2 begins with a noun phrase, *Some of this cyclical work*, that clearly relates back to the main message of sentence 1. This obvious reference—that is, to the cyclical and spring-like activity of the limbs—engenders a sense of familiarity as we begin to read sentence 2. In addition, the recognizable topic of sentence 2 prepares us for the new information that is appropriately placed at the sentence's end. There we learn that some of the cyclical work is produced by the active stretching-and-shortening contractions of muscles. (In this case, *active* refers to the role that the nervous system plays in causing muscle contractions.) The coherence is also strong across sentences 2 and 3. We can easily recognize the topic of sentence 3, *most of the spring-like function of the limb*, as a familiar one. Again, the old information enhances the flow across the two sentences, and it establishes the context for the new information that comes toward the end of sentence 3. This new information conveys the message that *most* of the spring-like action of muscles can be carried out by passive stretch-and-recoil mechanisms (as opposed to active, neutrally controlled muscle contractions). Sentences 4 and 5 also begin with old information that relates to key ideas in the previous sentences.

To appreciate the positive effect of this coherence-cultivating technique, compare Roberts and Scales's original paragraph (Figure 4.10) with a version that I deliberately tinkered with as follows (Figure 4.11) (page 186). In the altered version, sentences 2–5 begin with new information and end with old information, which is the structure that Gopen and Swan described as so problematic for readers.

(1) During running, muscle–tendon units operate like springs, storing and recovering mechanical energy as the limbs flex and extend with each step. (2) Muscle contractile elements that absorb work as they are actively stretched and produce work as they shorten do some of this cyclical work. (3) Because steady-speed running on level ground involves no net change in the average mechanical energy of the body, however, the passive stretch and recoil of tendons could perform most of the spring-like function of the limb. (4) Muscle contractile elements can be allowed to operate as near-isometric struts, developing force without shortening or producing significant power, by elastic mechanisms. (5) Economic force development by minimizing muscular work is allowed for by the spring-like function of muscle–tendon units.

FIGURE 4.11 An altered version of Roberts and Scales's effective paragraph in Figure 4.10. As explained in the text, this altered paragraph demonstrates the importance of beginning sentences with word groups that relate to the key topics and messages of previous sentences.

Consider, for example, the coherence across sentences 2–3 in Figure 4.11. After reading sentence 2, we get the point that some of the cyclical work in running is produced by the active contractions of muscles. At the start of sentence 3, we are naturally looking for familiar cues—perhaps a phrase relating to *some of this cyclical work* or to *the active contractions of muscles*. But sentence 3 begins with completely new ideas. Take the introductory dependent clause: *Because steady-speed running on level ground involves no net change in the average mechanical energy of the body*. As you read the start of sentence 3, listen closely to your mind's voice. You might hear self-talk that goes something like this:

> "Steady-speed running on level ground"? What does that have to do with the cyclical work and contractile activity of muscles? And why is the sentence talking about the lack of a "net change in the average mechanical energy of the body"? That idea doesn't clearly relate to what the first two sentences are talking about.

This sort of mental noise underlies the confusion we experience when reading incoherent paragraphs.

It's a simple enough technique to begin sentences with old information that relates back to the key topic and message of the previous sentence(s). Be careful, however, to avoid taking this advice to extremes. Imagine, for example, a paragraph in which every sentence begins with a subject that literally repeats a noun phrase from the end of the previous sentence. This mechanical application of the technique may undermine effective paragraph coherence by creating a distracting monotonous tone.

2. Begin sentences with transitional words and phrases. This advice applies when readers might need extra help understanding relationships in meaning and logic across consecutive sentences. The appropriate transitional words and phrases to use depend on the nature of the relationships. For example, as demonstrated in this very sentence, a phrase like *for example* signals readers that the present sentence explains, clarifies, or exemplifies an idea in the previous sentence. In Roberts and Scales's paragraph (Figure 4.10), notice the transitional word *however* at the start of sentence 3. It plays an important role in promoting paragraph coherence. The word *however* alerts us to a noteworthy upcoming idea, one that we can easily predict will contrast a key idea in sentence 2. The transition informs us that we are going to learn something *importantly different* about the cyclical, spring-like work of the muscles in running. The message of sentence 2 is that *some* of the work can be performed by active mechanisms. In contrast, the message of sentence 3 is that *most* of the work could be performed by passive mechanisms. The word *however* provides a helpful heads-up. We will talk more about the uses and potential abuses of transitional words and phrases in Chapter 6.

3. Consider your readers' expectations and needs. Here it is once again: the advice to solve problems in the writing process through audience analysis. This time, it's a matter of reflecting on what readers will expect and need as they progress from one sentence to the next in a paragraph. In drafting, you might pause for a brief moment before each sentence to take the perspective of your audience. Will the idea in the upcoming sentence meet readers' expectations and answer their questions about the idea(s) in the previous sentence(s)? If this focused audience-centered analysis bogs down your drafting flow, you can always delay the analysis until the revision process.

Drafting Sentences

Earlier, in describing the difficulty of producing perfect paragraphs in first drafts, I mentioned the challenges of establishing clear relationships in logic and meaning across successive sentences. Another major reason for the rarity of perfect draft paragraphs is that it's extremely challenging to craft flawless sentences on first attempts. Conscientious writers can easily get caught up in evaluating and refining features of sentence logic, structure, grammar, and punctuation. As a result, drafting momentum can grind to a halt and coherence falls apart. So, as is true for paragraphs, it's usually best to delay concentrated revisions at the sentence level until drafts are complete. Chapter 7 covers the process of revising sentences. Here, I offer a few pointers for promoting efficiency and momentum in drafting sentences.

Build Complex Sentence Constructions on Simple Foundations
Aiming for variety and sophistication in their prose, some writers may try too hard to build elaborate sentence structures in first drafts. The results are usually counterproductive. Most often, the best approach to drafting complex ideas is to

start with relatively simple sentence structures. In the revision process, the skeleton sentences can be fleshed out with words and phrases that add elaboration and clarity. In addition, the revision process is often the ideal time to use sentence-combining techniques to enhance variety and sophistication (these techniques are covered in Chapter 6).

When You're Struggling with Choices of Individual Words and Phrases, Quickly Note the Alternatives and Then Move on

One of the most frustrating drafting experiences is getting stuck on a single word or phrase and spending an inordinate amount of time mulling over alternative versions. Here's how the self-talk goes:

> *Should I say **scientists**, or is it better to use the word **researchers**? Then again, maybe I should use **investigators**. Or how about **experimentalists**? Yea, **experimentalists**—that might be a good one to add some variety. After all, it seems like I've been using the words **researchers**, **scientists**, and **investigators** too much. Maybe I better go back and read early parts of my draft to make sure that I'm not overusing these words. But, wait a second: I'm not even sure that **experimentalist** means the same thing as **scientist**, and **experimentalist** sure sounds weird. Maybe I better look it up in the dictionary.*

Before you know it, 15 minutes have passed, you haven't made a decision, and your drafting process has stalled. To avoid this maddening experience, you first have to be mindful of its arising. When you catch yourself in futile indecision, try quickly noting all of the alternative words or phrases that come to mind. Then, simply move on to the next sentence. You will likely find that the decisions are much easier to make *after* you have completed the entire draft and are working on revisions.

Talk to Your Imagined Audience and to Yourself

Scientific writing is, of course, essentially an act of communication. But during the drafting process, when authors are putting pen to paper or striking their computer keyboards, they are not really communicating with anyone directly. This partly explains why drafting can be difficult and, at times, quite frustrating. To remedy the situation, experienced writers conjure mental images of their audiences and, quite literally, talk to these imagined readers. The imagination of an intended audience changes the dynamics of drafting sentences, making it easier to get points across clearly and efficiently. The same advantage can be gained by talking to yourself as you draft. Try it sometime: When an idea refuses to shape up as you intend on paper, try speaking it out loud as quickly as you can. You may find that doing so naturally yields a well-structured and grammatically correct sentence.

Drafting Graphics

One of the most important skills in all of scientific writing is not what we normally think of as writing *per se*. It's the skill of conveying information, ideas, and arguments in the form of graphics. All scientific graphics are either tables or figures. Tables present numerical values, verbal information, or both in organized columns and rows. Figures include line drawings, photographs, flow charts, and various types of Cartesian-coordinate graphs such as line graphs, column graphs, and scatter plots. This section of the chapter covers the basics of designing and constructing graphics for drafts of scientific papers.

Graphics serve many vital functions in scientific communication. As a case in point, consider the task of reporting data from a study on age-related differences in the ability to control the rate of repetitive movements. Participants in four age groups (6–10, 21–25, 48–52, and 71–75 years) performed 15 trials of a finger-tapping task on an electrical switch. Over the first 10 seconds of each 30-second trial, the subjects were paced by a metronome that beeped every 550 milliseconds (ms). Over the last 20 seconds of each trial, the metronome was turned off and the subjects attempted to maintain the same movement rate, with a 550 ms inter-tap interval (ITI). The electrical switch was connected to a computer that ana-lyzed the mean ITIs and the total ITI variability for each trial. The latter measure, calculated as the standard deviation of ITIs for each trial, reflects consistency in timing the repetitive movements. Lower values for ITI variability indicate greater consistency and timing control. Using a complex theoretical model (the details of which we will conveniently overlook), the researchers estimated how much of the total ITI variability was due to variability of control mechanisms in the brain compared to the tapping finger. The researchers referred to these two compo-nents of total ITI variability as clock variability (for the brain mechanisms) and motor delay variability (for the finger mechanisms). Here's one way of reporting the study's data:

> For children in the 6-10 year-old age group, the mean ITI value was 560.66 ms and the total ITI variability was 48.03 ms. Of this total ITI variability, clock variability was 42.33 ms and motor delay variability was 5.70 ms. For young adults in the 21-25 year-old age group, the mean ITI value was 545.18 ms and the total ITI variability was 25.22 ms. Of this total ITI vari-ability, clock variability was 19.35 ms and motor delay variability was 5.87 ms. The values for middle-aged subjects, who ranged between 48 and 52 years old, were 544.19 ms for the mean ITI and 26.87 ms, 20.75 ms, and 6.12 ms respectively, for total ITI variability, clock ITI variability, and motor delay ITI variability. In the oldest age group, 71-75 years, motor delay ITI variability (6.43 ms) and clock ITI variability (40.70 ms) accounted for a total ITI variability of 47.13 ms, and the mean ITI was 558.39 ms.

The problems are obvious: The text is overloaded with numbers, which sorely impairs its readability; in addition, the cumbersome repetition of phrases convey-ing contextual information (phrases such as *the mean ITI value was* and *clock vari-ability was*) obscures key information. Readers will have considerable difficulty comparing pertinent data values across age groups and interpreting the results to reach their own conclusions. These problems can be solved by replacing the text with a graphic—specifically Table 4.1 (page 190). As presented in the table's orga-nized columns and rows, the study's ITI data are easy to compare and interpret. We need not struggle to unpack data values from dense and disorganized sentences.

TABLE 4.1 Mean Inter-Tap Interval (ITI) and ITI Variability across Age Groups

Age group (yr)	Mean ITI (ms)[*]	ITI variability (ms)		
		Total[*]	Clock*	Motor delay[†]
6–10 (n = 24)	560.66	48.03	42.33	5.70
21–25 (n = 26)	545.18	25.22	19.35	5.87
48–52 (n = 22)	544.19	26.87	20.75	6.12
71–75 (n = 26)	558.39	47.13	40.70	6.43

Notes. *n* refers to the number of participants in each age group.
[*]Values across age groups were significantly different ($P < 0.05$) except for comparisons of (a) 6–10 yr and 71–75 yr and (b) 21–25 and 48–52 yr.
[†] There were no statistically significant differences across age groups.

In the upcoming sections, we will walk through the following five-step process for drafting effective graphics.

Step 1: Decide whether to use a graphic.

Step 2: Determine the optimal type of graphic.

Step 3: Apply essential principles for designing effective graphics.

Step 4: Produce and integrate the graphic into your paper.

Step 5: Write the graphic's accompanying text.

Step 1: Decide Whether to Use a Graphic

One factor, above all, informs good decisions about whether to use a graphic instead of text or to complement text: It is whether the graphic would be a superior means of achieving the rhetorical goal at hand. Many rhetorical goals for scientific papers, including the examples described as follows, are best accomplished with graphics.

1. Presenting large amounts of data to support claims and conclusions in scientific arguments. As addressed in the previous chapter, a major criterion of a strong scientific argument is a critical mass of data, or a sufficient amount of support for its claim or conclusion. When presented exclusively in the form of text, extensive supporting data can be so unwieldy, and therefore so difficult to read, that their associated arguments are compromised. Graphics are thus ideal for achieving all rhetorical goals that require developing data-driven arguments.

2. Summarizing and synthesizing numerous studies on a research issue. Chapter 3 presented a type of table called a summary chart, which served to demonstrate the critical thinking skill of synthesis (Table 3.1, pages 123–125). Commonly used in the body of review papers, summary charts are ideal for guiding readers through syntheses of numerous studies on a focused research issue, highlighting important similarities and differences in their methods, results, and conclusions.

3. Outlining complicated study methods. This goal-directed application of graphics applies to the methods sections of IMRAD-structured papers and research proposals. When a study's procedures and analyses are especially detailed and complex, readers may have trouble understanding descriptions and explanations that are presented solely in text. Ideal solutions to this problem are graphic outlines, flow charts, and timelines.

4. Describing data trends and patterns, as well as relationships between variables. Scientific graphics are most commonly used for this purpose. In any report of research involving a complex data set and more than one variable, readers want to see for themselves whether meaningful patterns and relationships exist. If two variables are related, for example, is the correlation linear or nonlinear? Does the data pattern suggest a cause-effect relationship or a loose association? Is there a recognizable trend in the data over time? These sorts of questions cannot be answered sufficiently through text alone.

5. Illustrating complex structures and processes. This goal is central to writing with graphics in the life sciences. Take the case of a scientist writing a comprehensive description of the anatomy and mechanisms underlying the neural control of human movement. The author would need to address questions like these:

- How do different structures of the brain communicate with each other in planning movements?
- How do electrical signals travel across neuromuscular junctions to trigger muscle contractions?
- What is the underlying architecture of muscle fibers?
- By what processes do muscles generate tension and contract?

As you might imagine, the most insightful answers to these questions would come in graphics, including line drawings and perhaps photographs that illustrate the intricate structures and processes. This is truly a case of a picture being worth a thousand words—or maybe even more!

Step 2: Determine the Optimal Type of Graphic

After deciding that a graphic is ideal for accomplishing a rhetorical goal, you will of course have to decide which type of graphic to use. Sometimes the choice is obvious. For instance, to illustrate the highly detailed architecture of muscle fibers, a line drawing would be most appropriate. In other cases, the best choice is not immediately clear. Take the general goal of presenting data to support a conclusion. Should you use a table or a figure? If a figure is best, should it be a line graph, a column graph, a scatter plot, or another type? The following guidelines apply to answering these sorts of questions.

1. Use tables to present large amounts of multivariate data with precision.
Tables are ideal for the following purposes.

- Summarizing key details about the methods, results, and conclusions from multiple studies.
- Providing a framework for synthesizing research—for example, with summary charts.
- Presenting multivariate data to support claims, conclusions, hypotheses, and concepts.
- Presenting precise data values—for example, values with numerous digits following a decimal point.
- Enabling readers to directly compare data values across numerous variables.
- Avoiding text that is overloaded with numbers and repetitive phrases.

Because tables present discrete data values, they do not effectively convey overall trends, patterns, and relationships between variables. For these purposes, graphs are superior.

2. Use line graphs to illustrate data patterns and relationships involving continuous variables. Line graphs represent relationships between two main variables: an independent variable, which is usually graphed on the horizontal axis (also called the X-axis); and a dependent variable, which is usually graphed on the vertical axis (also called the Y-axis). The axes are labeled by the variables they represent and marked at regular intervals by data points within the range of values obtained from a study. In general, line graphs are used for the following purposes.

- Depicting data for dependent variables, or study outcomes, that correspond to changes in continuous independent variables, such as time, distance, speed, or medication dose.
- Illustrating the degree to which relationships are linear or nonlinear.
- Enabling readers to identify critical data values, such as the highest or lowest points on a curve.
- Illustrating statistical interactions, or inconsistent (nonparallel) trends in the data representing different levels of an independent variable.

A sample line graph is shown in Figure 4.12. Based on a hypothetical data set, the graph reveals a linear association between running speed and oxygen consumption (a measure of the energy required to fuel muscle contraction) in subjects representing three age groups. The line graph is ideal for directly conveying the study's key findings: (a) oxygen consumption increases linearly with faster running speeds in all three age groups and (b) the oxygen cost of running is lower as age increases from 8 to 22 years.

3. Use column and bar graphs to illustrate data patterns and relationships involving discrete study groups or conditions. In column and bar graphs, data are represented by rectangular shapes oriented vertically (column graphs) or horizontally (bar graphs). In column graphs, study groups or conditions are

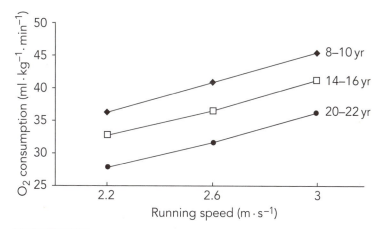

FIGURE 4.12 A sample line graph depicting hypothetical changes in oxygen consumption with increasing running speeds in subjects representing three age groups.

designated on the horizontal axis, and dependent variables are designated on the vertical axis. In bar graphs, the axes are reversed. Like line graphs, column and bar graphs illustrate data patterns and relationships in study outcomes. Column and bar graphs are especially suited for presenting *discrete* data, which represent study groups or conditions and do not vary on continuous scales.

A sample column graph is shown in Figure 4.13 (page 194). It presents the ITI variability data from the previously described study on age-related differences in controlling repetitive movements. These are the same results for total, clock, and motor delay variability that were presented in Table 4.1 (page 190) . Let's say that the study authors opted for the column graph instead of the table, reasoning that the graph would better support their main conclusions. For this purpose, the authors realized that readers did not need to know the precise values for ITIs, which are accurate to hundredths of a millisecond in the table. Instead, the authors wanted their audience to grasp several interesting overall patterns that correspond to their conclusions. One conclusion is that overall timing consistency, as reflected by lower total ITI variability scores, improves from childhood to middle age but then declines in older age back to levels observed in childhood. To support this conclusion, the authors would direct readers to view the *U-shape* formed by the four columns for total ITI variability in Figure 4.13. The authors would compare the relatively low values in the middle two age groups versus the youngest and oldest age group. A second key conclusion is that, compared to motor delay variability, clock variability contributes more to total variability. Figure 4.13 clearly supports this conclusion, given the greater heights of the columns for clock variability versus motor delay variability. The graphic patterns that support the authors' conclusions are much easier to grasp in the column graph versus the table.

4. Use scatter plots to illustrate correlations between two variables. Consider a hypothetical study to determine whether muscular strength increases as a

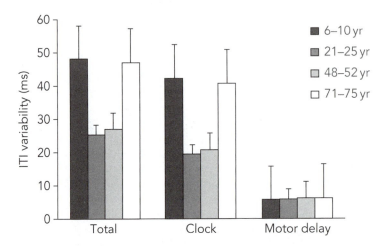

FIGURE 4.13 A sample column graph depicting hypothetical data from a study on age-related differences in the timing control of repetitive movements. The vertical lines above each column are error bars, which represent measures of variability such as standard deviations.

function of body weight. Twenty subjects participated. After being weighed, the subjects performed a knee extension task that measured maximum force production of the quadriceps muscles. Each subject thus had two scores: one for body weight and one for maximum force production. Both scores were expressed in kilograms. To illustrate the detailed relationship between these two variables, the data can be graphed in a two-dimensional scatter plot with body weight (the independent variable) on the X-axis and maximum muscular force production (the dependent variable) on the Y-axis. Figure 4.14 presents such a scatter plot. Each of the 20 data points corresponds to an individual subject's scores. For instance, moving from left to right, the first data point represents a subject who weighed 55 kg and produced 59 kg of muscular force.

Scatter plots are ideal for illustrating statistical correlations between two variables. In Figure 4.14, the statistical correlation between body weight and maximum force production ($r = 0.31$) is indicated in the graph's top-right portion. Following conventions for constructing scatter plots, Figure 4.14 includes a line of best fit, which was automatically generated by the software used to create the graph.

5. Use line and shape drawings or photographs to illustrate realistic features of research subjects and objects. For showing the subjects and objects of research in fully realistic conditions, photographs are the best choice of graphics. Take the example of a research paper reporting the study to determine the correlation between body weight and muscular strength. In the methods section, a photograph would be ideal for illustrating the machine used to measure muscle force

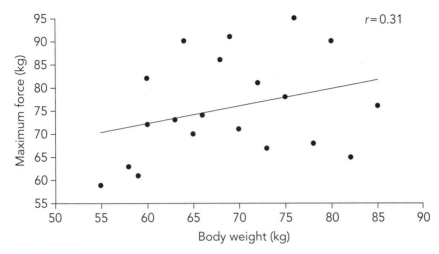

FIGURE 4.14 A sample scatter plot depicting hypothetical correlations between body weight and maximum force production of the knee extensor muscles in 20 subjects.

production. When especially fine details of subjects or objects call for focused attention, line and shape drawings are often superior to photographs. A model line drawing, reproduced from a published research paper introduced earlier in the chapter, is presented in Figure 4.15 (page 196). Recall that the authors, Thomas Roberts and Jeffrey Scales, studied the biomechanics of running in wild turkeys. The investigation focused on force production and leg movement patterns as the turkeys ran at a steady rate (part A of Figure 4.15) and then accelerated (part B of Figure 4.15). The line drawings highlight key biomechanical features that include the leg joint angles, force vectors, and center of mass in the running turkeys. Obviously, the drawings are not realistic representations. No feathers, eyes, or spurs are depicted. These realistic features, however, would obscure the most important information, on which our attention is focused in the line drawings.

6. Use flowcharts and timelines to outline complex sequential procedures and processes. Methods sections of scientific papers are sometimes difficult to read and understand because essential details are hidden in sentences with superfluous information. A solution is to outline study methods in flow charts and timelines. The same solution applies for presenting complex biological processes that are carried out in sequential steps.

Step 3: Apply Essential Principles for Designing Effective Graphics
Entire books have been written on the subject of writing with scientific graphics—and justly so. This vital craft demands extensive knowledge and skill. At their core, however, all effective scientific graphics manifest a fairly small set of essential design principles, which are summarized as follows.

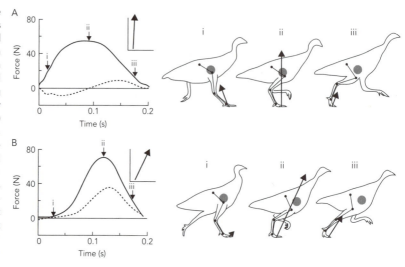

Fig. 2. Representative force traces and kinematic diagrams for a steady-speed run (A) and a high acceleration (B). High accelerations involved high peak propulsive horizontal forces (dotted lines), no braking horizontal forces and a delay in vertical force (solid line) development. The insets in the force graphs represent the mean orientation of the ground reaction force vector during stance. The turkey diagrams illustrate the changes in limb angle during acceleration that help to maintain the alignment between the forward-oriented ground reaction force (bold arrow) and the centre of mass. These diagrams were traced from video frames at the times indicated by arrows on the force graphs. The gray shaded circle represents the approximate location of the animal's center of mass.

FIGURE 4.15 A model figure and figure caption from a published research paper on the biomechanics of running turkeys (Roberts & Scales, 2002). Reproduced with permission from The Company of Biologists Limited.

1. Don't use graphics to decorate your documents; instead, use them to achieve your rhetorical goals. Many of the most problematic figures and tables in scientific papers can be traced to their authors' view of graphics as superficial decorations. In contrast, the ideal perspective holds that graphics serve the same ultimate function as text, which is to accomplish appropriate rhetorical goals for scientific communication. This key principle informs the best decisions about when to use graphics, which types to use, what information to include and exclude, and how to construct effective tables and figures.

2. Apply audience analysis to graphical design. The same questions that writers ask about readers' needs, expectations, and values concerning the content and structure of text should apply to graphics. Consider a research paper on movement-related capacities in patients categorized by different stages of Parkinson's disease. The subjects performed tests of maximum strength, reaction time, and coordination. The study's author must decide on the appropriate content and structure of the graphics for presenting the results. These decisions will ideally be informed by considerations of the audience's needs and reasons for reading the paper. An audience of movement scientists, for instance, might be looking for a synthesis of the findings. So an ideal graph would display the data in *crunched* form, or as mean scores for patients categorized by the disease stages. But suppose that the author is preparing the paper for clinicians, such as physicians and physical therapists. For their interests and purposes, the best graphics might present representative data for *individual* patients rather than group means.

3. Apply the Goldilocks Principle to designing graphics: Not too little, not too much, but just the right amount of information. Graphics that lack a critical mass of information are simply unnecessary. Picture, for instance, a column graph presenting mean reaction time scores for three subject groups. Suppose that the means are not statistically different and that the three columns are practically matched in height. The graph fails to reveal any pattern of interest. By the Goldilocks Principle, this is a case of *too little* data to warrant a graphic. In just one sentence in the text, the author could clearly convey the picture: *There were no statistical differences in mean reaction time, which ranged from 220 to 223 ms across the three groups.* On the other end of the continuum, graphics are problematic when they contain *too much* information and thereby overwhelm readers.

There are no general-purpose rules for determining the upper limits of information to include in scientific graphics. As is true for scaling content in text, writers must reach sound decisions by considering their rhetorical goals and audiences. Another useful guideline extends the concept of unity, which we have discussed for designing whole documents and paragraphs. An effective graphic has the quality of unity in its topic, message, and goal. In other words, every bit of information in the graphic serves these three functions without being redundant or overbearing.

4. Exclude all embellishments (or features that don't directly serve a graphic's rhetorical goal). For an example that grossly violates this principle, have a look at Figure 4.16. It's an embellished version of Figure 4.13, which we have already discussed. The problematic embellishments in Figure 4.16 include the shaded background, the heavy gridlines (one of which covers an item in the figure's legend), the decorated gradient column fills, and the 3-D effect. These added adornments are unnecessary because they do not convey essential information about the study's results; moreover, they undermine the rhetorical goal

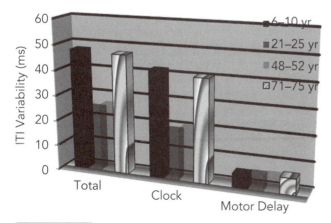

FIGURE 4.16 A problematic column graph that violates the design principle to avoid embellishments in graphics.

to support conclusions based on the results. The embellishments actually draw attention away from the essential information. Take the 3-D effect, which is especially problematic because, in reality, the ITI variability data do *not* extend into the third dimension (the Z-dimension in Cartesian coordinates).

5. Use the space of your graphics to highlight their key data and take-home messages. To demonstrate this principle, I'll pose a hypothetical study conducted to document physical activity levels of American teenagers from the 1960s to the 21st century's first decade. Imagine that the researchers compiled published reports on the extent to which teenagers participated in vigorous physical activity. The data, expressed in weekly minutes of vigorous exercise for girls and boys, are presented in a problematic line graph in Figure 4.17A.

The graph's design fails to highlight the most important study data and conclusions. The key problem is the white space above and below the plot lines. This expansive area obviously contains no information at all. The problem stems partly from the Y-axis scale, which ranges well beyond the study's maximum data values. The Y-axis ranges from 0 to 500 weekly minutes of exercise, while the actual data range from approximately 100 to 210 weekly minutes. The scaling error restricts the presentation of data to a tiny slice of the graph. The error also flattens the plot lines, giving readers the false impression that weekly exercise participation has remained fairly constant over the decades. This problem is compounded by a design flaw in a feature called the *aspect ratio*, which is defined as a graph's height divided by its width. The aspect ratio in Figure 4.17A is inordinately low—in other words, the X-axis is stretched too far. The hypothetical study results indicate that exercise participation has declined significantly, specifically by 33% and 50% for boys and girls, respectively, over the five decades. But the graph is not designed to emphasize this important take-home message. In addition, the design obscures a very interesting and important observation: Between 1980 and 1990 weekly exercise participation dropped sharply among

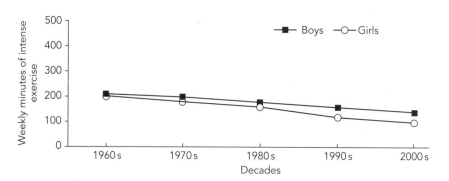

FIGURE 4.17A A problematic line graph that fails to accentuate its most important information and ideas.

teenage girls. A revised version of the graph is presented in Figure 4.17B. Scaled appropriately, the revision accentuates the study's most important data and conclusions.

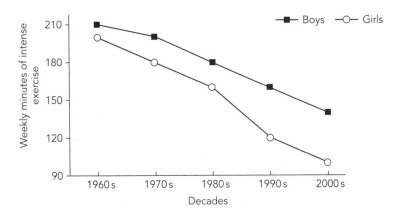

FIGURE 4.17B An effective revision of Figure 4.17A.

6. Aim for ethics and honesty in graphical design. In some fields of communication, graphics are deliberately designed for profit and ethically questionable outcomes. Picture a politician who opposes government spending for after-school exercise programs for overweight teenagers. The politician might actually use the line graph in Figure 4.17A to argue that the programs are unnecessary, claiming that a lack of exercise is not the problem at all. "The data clearly show," the politician would contend, "that since 1960 the amount of vigorous physical activity performed by teenagers has not changed markedly. It's almost a flat line."

The scientific community naturally views deliberate distortions and fabrications of data as thoroughly reprehensible. However, many cases of misleading scientific graphics are inadvertent, resulting because authors fail to critically question whether they are representing study findings accurately, validly, and honestly. Sometimes authors carelessly accept problematic default settings in their graphics software, such as distorted aspect ratios and poorly fitting ranges of data on graph axes. This is especially a concern for student writers who use general-purpose graphics programs.

The design of ethical and honest scientific graphics begins with accurate interpretations of the statistical and practical significance of study data, following guidelines that we covered in Chapter 3. If, for example, you misinterpret a study result by overestimating its practical significance, your graphical representation of the result will likely be inaccurately exaggerated. To prevent such distortions, you must design graphic elements—such as slopes of lines, heights of columns, and dispersions of data points—to agree with the truest representations of study outcomes and the most valid conclusions that they support.

Step 4: Produce and Integrate the Graphic into Your Paper

After deciding to use a graphic, selecting the appropriate type, and applying sound principles for graphical design, you will carry out the practical tasks of producing the graphic and integrating it in your paper. Before discussing the logistics of constructing original graphics, let's raise the option of using tables and figures from published journal articles, books, and Web site material. That's what I did, for example, to include Figure 4.15 (page 196) in this book. The line drawing was created by biologists Thomas Roberts and Jeffrey Scales for their published research paper on the biomechanics of running turkeys. Because I don't draw very well, and Roberts and Scales's figure was perfect for achieving my goal to demonstrate an effective line drawing, I wanted to use the figure here. If I had copied it without permission and without giving appropriate credit to the authors, I would have been guilty of violating copyright laws and plagiarizing. To request approval to use the published figure in this book, I sent an e-mail letter to the permissions manager for *The Company of Biologists*, which is the organization that publishes the journal in which the figure originally appeared. In the letter, I provided information about my project, explained my specific intentions for using the figure, and asked for the publisher's consent to reproduce it. A few days later, I received e-mail from the permissions manager, who granted approval and gave me instructions on how to cite the original source to give appropriate credit to the authors and the publisher. This is the same essential approach that scientists, as well as science students, must take to acquire permission to reproduce published graphics, as well as excerpts of published text, in their papers.

These days, computer technologies make it simple to scan printed graphics and to copy digital graphics from Web sites and downloaded PDFs. In no time at all, we can copy and paste high-quality, published tables and figures into our papers. Note, however, that all scientific information on the Web, including text and images, is published and automatically copyrighted. This means that writers who use graphics obtained from the Internet without first getting permission from the publishers, and in some cases the authors, may be violating copyright laws. If you would like to use published graphics in your papers, heed the following guidelines.

1. Ask your professors whether they encourage students to use published graphics in class papers. Some professors may discourage the practice of using published graphics, reasoning that students should take on the challenges and gain the experience of producing their own graphics.

2. Carefully follow publishers' instructions for requesting permission to reproduce materials. On their Web sites, scientific journals and professional organizations provide copyright information that specifies whether text and graphics can be freely used, how to request permission to use their materials, how to credit sources, and whether requesters must pay a fee. These days, many publishers and professional organizations are providing Web forms for authors to submit their permission requests.

3. Always credit the original author and source by following appropriate citation guidelines. Right around the corner we will discuss how to cite sources of scientific information.

If you are taking on the challenge of creating original graphics, you will need to answer many practical questions, such as the following.

- Where in your paper should you place graphics—immediately after referring to them in the text or on separate pages at the paper's end?
- How should you number your graphics?
- Should your tables include horizontal and vertical lines?
- Should titles and captions be placed above or below graphics?
- What are appropriate dimensions for graph axes?
- Is it okay to use colors for plot lines and to fill columns and bars?
- What are the rules for capitalizing text in axis labels, figure captions, and table titles and notes?
- Is there a limit on the number of graphics that you can include?

The answers vary by school assignments, publication organizations, journals, and grant agencies. So, once again, the best advice is to consult author-instruction documents and discipline-specific style guides. Another suggestion is to study graphics in published articles from the scientific field in which you are writing. Let's say that you are writing a psychology paper, with instructions to follow the publication guidelines of the American Psychological Association (APA). To learn the design specifications for your paper's tables, you should read the section on tables in the APA style guide (American Psychological Association, 2001). In addition, you might examine how tables are prepared in published articles that follow APA format. Among other design features, you will observe that table titles are italicized and that they follow the rules of capitalization for sentences. In addition, informative table notes are placed below the graphic and are preceded by the word *Note*, which is italicized and followed by a period.

To construct original tables and figures, you will need to use graphics software. Examples of programs that scientists and science students commonly use are listed in Table 4.2 (page 202). If you lack experience working with a graphics program, make sure to allow enough time for learning and practicing its applications. In addition to the *help* features and tutorials that come with the software, you might find educational resources on the Internet.

Step 5: Write the Graphic's Accompanying Text

For the most part, effectively designed graphics are easy to navigate and understand all on their own. All graphics, however, require at least a little accompanying text. For example, sentences or phrases are necessary to refer readers to graphics and to summarize their take-home messages. The following examples demonstrate these roles of text that accompanies graphics.

The baseline physical characteristics of the study participants are presented in Table 3.

TABLE 4.2 Software for Constructing Original Graphics

Type of graphic	Software programs
Tables	Word processing programs such as OpenOffice Writer and Microsoft Word Spreadsheet programs such as Microsoft Excel
Graphs	General-purpose graphics programs such as Microsoft Excel Specialized scientific graphics programs such as SigmaPlot, DeltaGraph, MATLAB, and GraphPad
Line and shape drawings	Drawing programs such as OpenOffice Draw, Adobe Illustrator, Paint (which comes with the Windows Operating System), and Freeverse Lineform (for Macintosh computers)
Flow charts and concept maps	Word processing programs such as Microsoft Word Drawing programs (examples provided above)

Note. To learn more about the different software programs, enter their names into an Internet search program and check out their Web sites.

> Our main finding was that maximal strength and the cross-sectional area of muscles were linearly related (see Figure 2).

At the top of every table in a scientific paper, a title should introduce the graphic. Without overwhelming readers with details, the titles should include specific words and phrases that convey the main contents of their tables. An effective example is the title presented earlier for Table 4.1 (page 190):

> Mean Inter-tap Interval (ITI) and ITI Variability across Age Groups

To appreciate the effect of specificity in table titles, compare the preceding model to the following vague title for Table 4.1:

> Timing Control Data

Tables may also be accompanied by explanatory notes, which are placed below the graphic. As demonstrated in Table 4.1, table notes are used to indicate the meaning of abbreviations and to explain symbols that indicate outcomes of statistical analyses. More elaborate explanations of table contents, especially explanations of how a table's specific data support the author's conclusions and arguments, should be included in the text body.

All figures in scientific papers should include captions, which are explanations that help readers navigate and understand key graphical elements. For complex figures that are not self-explanatory, well-developed explanations are necessary. Placed in the paper's text body or in a caption below or adjacent to a figure, the explanation should do the following.

• Summarize the figure's overall layout.

- Give the reader directions and strategies for navigating and understanding the figure.
- Highlight key patterns and relationships among variables that the figure is intended to convey.
- Focus the reader's attention on specific features or data points that are relevant to the figure's rhetorical goal.
- Explain *how* the figure accomplishes its rhetorical goal—for example, how the data in a graph actually support an author's conclusions.

A superior example of a figure caption for a complex graphic is presented in Figure 4.18 (page 204). The model comes from a published research paper on the neuromuscular mechanisms that control hindlimb posture and locomotion in the American alligator (Reilly & Blob, 2003). In the paper's introduction section, biologists Stephen Reilly and Richard Blob used the line drawing reproduced in Figure 4.18 to present two models for explaining the neuromuscular and biomechanical mechanisms underlying different patterns of the hindlimb movements of alligators. To fully understand the graphic and its caption, you would need to be a member of Reilly and Blob's primary audience of peer scientists who study crocodilian locomotion. Nonetheless, reflecting on the complexity of the line drawing's features, you will appreciate how effectively the figure caption achieves the strategies in the preceding bulleted list. The extensive development of Reilly and Blob's figure caption is somewhat uncommon in life science papers; however, given the situation, the lengthy and detailed caption is completely appropriate and extremely helpful in guiding readers' comprehension of the complex graphic.

Citing References

Experienced scientists are exceptionally conscientious about recognizing and praising predecessors and peers who have influenced their work; moreover, they are keenly aware that failing to appropriately credit original sources constitutes plagiarism, one of the most serious instances of academic dishonesty. Scientific authors give credit by citing references, which means documenting the journal articles, books, Web sites, and personal communications through which they derive information and ideas. Citations come in two places in a typical scientific paper: in the text and in a closing reference list. As follows, guidelines are presented for citing references in life science papers.

For Directions on Formatting References, Consult Assignment Instructions and Discipline-Specific Style Guides

There is no universal format for citing references in scientific papers. Instead, formats vary by research fields, journals, publishing organizations, and grant agencies. Most published life science articles, however, are based on one of the following three general citation systems.

- Citation-name system: In a paper's text body, citations are designated by numbers that are offset in parentheses, superscripted, or subscripted. The

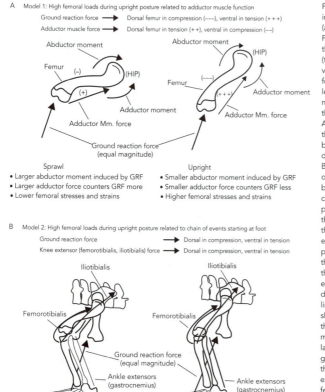

Fig 1. Diagrammatic illustrations of alternative models to explain increases in femoral loading during upright locomotion in alligators (adapted from Blob, 1998, 2001; Blob and Biewener, 1999, 2001). For visual clarity of the forces and moments bearing on the model, the views are presented from an oblique posterolateral perspective (this causes the femur to appear not to project perpendicular to the vertebral axis as it does at mid-stance). For each arrow depicting a force or moment, a difference in thickness (not length) between the left and right sides of a panel indicates a difference in force or moment magnitude between the postures illustrated in those panels(with thicker arrows indicating larger forces or moments). Note that panels A and B are drawn to different scales (A is magnified for clarity), and that comparisons of force and moment magnitudes are not intended between A and B. Because the ground reaction force (GRF) does not differ in magnitude between sprawling and upright steps (Blob and Biewener, 2001), both models are based on changes in the action of hindlimb muscles between these postures. (A) Bending induced by the ground reaction force (red arrow) places the dorsal femur in compression (-), and the ventral femur in tension (+). In sprawling posture (left), the GRF might have a longer moment arm about the hip than in upright posture (right), resulting in a larger abductor moment that would tend to rotate the femur dorsally. To keep the hip joint in equilibrium, the hip adductors might exert a larger force in sprawling posture (left) and a smaller force in upright posture (right). Because the hip adductors bend the femur in the opposite direction from the GRF, larger adductor forces during sprawling steps could more effectively mitigate strains induced by the GRF, resulting in lower dorsal and ventral stresses and strains during sprawling steps. (B) As limb posture becomes more upright, the centre of pressure of the GRF shifts away from the ankle, increasing the moment arm of the GRF at the ankle (RGRF). Consequently, ankle extensors (e.g. gastrocnemius) must exert higher forces during upright steps in order to counter the larger ankle flexor moment and maintain joint equilibrium. Because gastrocnemius also spans the knee, it makes a greater contribution to the flexor moment at the knee during more upright steps, and knee extensors (femorotibialis and iliotibialis, on the dorsal aspect of the femur) must exert greater force to counter this moment and maintain equilibrium at the knee. Increases in knee extensor forces could then raise dorsal and ventral femoral strains and stresses as alligators use more upright posture. Data from previous force platform studies (Blob and Biewener, 2001) are consistent with the model proposed in B, but changes in muscle activity patterns have not been tested prior to this study. Iankext, moment arm of the ankle extensor muscles at ankle (no change between sprawling and upright stance).

FIGURE 4.18 A figure caption that, in exceptional detail and development, explains a complex graphic. From Reilly and Blob (2003). Reproduced with permission from The Company of Biologists Limited.

numbers correspond to complete citations that are ordered alphabetically in the paper's reference list.

- Citation-sequence system: In a paper's text body, citations are designated by numbers that are offset in parentheses, superscripted, or subscripted. The numbers correspond to complete citations in the reference list that are ordered by the sequence in which they appeared in the text.

- Name-year system: In a paper's text body, citations are designated by authors' last names and the year that the source literature was published. Complete corresponding citations are presented in alphabetical order in the paper's reference list.

Within each general system, there are numerous subsystems for formatting citation information—that is, for ordering the elements of citations, using punctuation, capitalizing letters, and choosing font styles. This point is illustrated in Figure 4.19, which presents sample citations from the reference lists of four life science journals.

Some of the formatting differences are quite subtle. In Examples 3 and 4, for instance, notice the difference in how the authors' names are joined. In Example 3, following instructions for *The Journal of Experimental Biology* (JEB), the names are joined by the word *and*. In Example 4, which follows APA guidelines, the last two authors' names are joined by an ampersand (the symbol &). The two formats also differ in font styles. In the JEB citation, the volume number is in bold font; in the APA citation, the same element is italicized. Another difference is that in APA format, reference list citations have hanging indents, which means that for each citation, the first line is on the left margin, and all other lines are indented 0.5 inch.

Scientists who submit publishable papers with citations in the wrong format are reprimanded by editors to go back and follow the instructions. For student

---

Example 1: Journal of Applied Physiology (Publisher: American Physiological Society)
Ivanenko YP, Poppele RE, Lacquaniti F. Five basic muscle activation patterns account for muscle activity during human locomotion. *J Physiol* 556: 267–282, 2004.

Example 2: The Journal of Biochemistry (Publisher: Oxford University Press)
Frock, R.L., Kudlow, B.A., Evans, A.M., Jameson, S.A., Hauschka, S.D., and Kennedy, B.K. (2006) Lamin A/C and emerin are critical for skeletal muscle satellite cell differentiation. *Genes. Dev.* **20**, 486–500

Example 3: The Journal of Experimental Biology (Publisher: The Company of Biologists Limited)
Pelletier, Y. and McLeod, C. D. (1994). Obstacle perception by insect antennae during terrestrial locomotion. *Physiol. Entomol.* **19**, 360–362.

Example 4: Journal of Family Psychology (Publisher: American Psychological Association)
Sheeran, P., Conner, M., & Norman, P. (2001). Can the theory of planned behavior explain patterns of health behavior change? *Health Psychology, 20,* 12–19.

FIGURE 4.19 Citation formats in different life science journals.

writers, incorrectly cited references may result in point and grade deductions. So it's important to carefully examine assignment instructions and discipline-specific style guides for details about how to format references. The process is greatly simplified by citation-managing computer programs such as EndNote (http://www.endnote.com/) and Reference Manager (http://www.refman.com/). Your campus library's Web site may provide access to a citation-managing program at no charge.

Cite Sources of Information and Ideas That Are Neither Your Own nor Common Knowledge

If you are new to writing in a research field, figuring out whether information and ideas are common knowledge is no trivial matter. The following examples would *not* be considered common knowledge and would therefore require citations.

- Original data and conclusions from published research that you did not conduct.
- Recently developed concepts and theories that are not widely accepted in the scientific community.
- Any elements of debatable arguments in the published literature, including claims, supporting data, and conceptual reasoning.
- Concepts, theories, and arguments that have clearly been attributed to one scientist or scientific team.

The following examples *would* be considered common knowledge and would therefore not require citations.

- Facts in nature and science, such as the fact that muscular strength declines in old age or that a motor unit consists of a spinal neuron and all the muscle fibers it innervates.
- Consistently defined and uncontested terms, such as the definition of *reaction time*, which is the duration between the presentation of a stimulus and the initiation of a movement response.
- Widely accepted and uncontested concepts, theories, and arguments that cannot be attributed to any one scientist or scientific team.

When you are uncertain about whether to cite a source, ask your professor for guidance. Or you can go ahead and cite the source just to be safe.

Use and Cite Primary Sources

This guideline reiterates advice offered in Chapter 1, where we discussed how to judge the credibility of various sources of scientific knowledge. Recall that *primary* sources are peer-reviewed research papers that report original studies. Secondary and tertiary sources include review papers, textbooks, and material from Web sites and class lectures. Knowledge derived from second-hand sources often lacks essential details and sometimes fails to represent corresponding primary sources accurately. So unless you are completely confident that information from secondary and tertiary sources is complete and credible, the best practice is to use and cite primary sources in your papers.

Avoid Excessive References to the Same Citation within a Paragraph

With the well-intentioned aim of avoiding plagiarism, student writers sometimes go overboard in citing references. A case in point is presented in Figure 4.20A. The writer's excessive reference to the same study clutters the text and thereby breaks the paragraph's flow. A more readable version, which still gives appropriate credit to the source, is presented in Figure 4.20B.

McLeod argued that a deep understanding of movement organization and control requires a conceptual approach that accounts for the dynamic interplay between neural and non-neural factors (McLeod, 2006). McLeod also proposed that the conceptual framework should encompass the role of sensory feedback in modifying ongoing movements (McLeod, 2006). The body is a mechanical system that is subject not only to neural commands but also to physical influences such as gravity, inertia, and reactive forces (McLeod, 2006). McLeod contends that these factors are not accounted for in theories of movement control that are strictly based on neural control mechanisms (McLeod, 2006).

FIGURE 4.20A A case of excessive and unnecessary citing of references.

McLeod (2006) argued that a deep understanding of movement organization and control requires a conceptual approach that accounts for the dynamic interplay between neural and non-neural factors. The argument holds that the conceptual framework should encompass the role of sensory feedback in modifying ongoing movements. Accordingly, the body is a mechanical system that is subject not only to neural commands but also to physical influences such as gravity, inertia, and reactive forces. McLeod contends that these factors are not accounted for in theories of movement control that are strictly based on neural control mechanisms.

FIGURE 4.20B An effective revision of the problematic use of citations in Figure 4.20A.

Avoid Stacking Long Strings of Citations in the Middle of Sentences

This guideline addresses an annoying problem that can crop up in papers that use the name-year system for citing references. Figure 4.21 illustrates the problem. The message is difficult to grasp because the sentence is continually interrupted by long strings of citations. One way to avoid this problem is to construct sentences so that all citations come at the end. This might require breaking long sentences into shorter ones. Another strategy for reducing the clutter is to delete citations that are not absolutely essential.

Refer Your Readers to Sources of Additional Background Information

Apply this guideline when the length restrictions for your papers prohibit presenting background information to help readers understand advanced concepts and arguments. Suppose that your audience might lack the foundational knowledge to grasp a complex argument in your paper, but you lack sufficient space to present the basics. If an appropriate presentation of the information is included in a published article or book, you can refer readers to the source by citing it.

Avoiding Plagiarism

Plagiarism is the highly unethical act of using someone else's information and ideas, when they are not common knowledge, without giving appropriate credit. A blatant case is lifting verbatim strings of unquoted text from published articles and books. More subtle but equally egregious cases transpire when writers paraphrase the information and ideas of others without documenting the sources. Scientists who are caught plagiarizing often lose their jobs; in addition, plagiarism may result in legal action against the offender. Students who plagiarize also face very serious penalties, which include failing grades and suspension or dismissal from educational institutions. The same dire consequences usually apply

> The risk of falling among older adults has been linked with cognitive impairments (Adams, Klein & Trujillo, 2007; Beller & Meyer, 1998; Camby, Elway, & Kendrick, 2005), which are associated with age-related neural degeneration in the hippocampus (Darnell, 2008; Edwards, Lundin, & Hurd, 2001; Franklin, 2004) and the prefrontal cortex (Gurule, Beem, & Perry, 2006; Lewis & Harrell, 1997), declines in vision (Harris, 1995; Murtaugh and Ford, 2000) and proprioception (Isaac, Scholz, Hinton, & Fitch, 2008), and the loss of muscular strength (Jacobs, Clarkson, & Cole, 1997; Kelso & Brody, 2005).

FIGURE 4.21 A case of excessive and disruptive citing of references within sentences.

whether the plagiarism is intentional or not. Student writers who plagiarize may report genuine feelings of overwhelming pressure and stress to meet deadlines and get good grades. The bottom line, however, is that the choice to cheat is never justified. In these situations, students are well advised to speak with their professors and counselors about how to manage time and reduce stress. The following advice applies to avoiding unintentional plagiarism.

Take Notes in Your Own Words

In a common instance of inadvertent plagiarism, writers use their notes as draft material but forget having copied their notes verbatim from published literature. Even an unquoted string of four or five words could be plagiarism. So from the earliest stages of the writing process, make sure to take notes in your own words. If a noteworthy idea is difficult to paraphrase, put the source aside for a moment or two, letting the author's phrasing dissipate from memory. Then, talk yourself through the idea. Try speaking it out loud. After recording the idea in your notes, check it against the original article to ensure that the phrasing is not identical. If it is, rearrange word groups and use synonyms to express the idea in your own words.

Use Direct Quotations When They Are Necessary

Unlike writing about genres of classical literature such as fiction and poetry, scientific writing rarely requires reproducing an author's words verbatim. Indeed, you can read hundreds of published scientific articles without ever coming across a single set of quotation marks. Direct quotations are generally discouraged in scientific writing. Many professors respond negatively to long strings of directly quoted text in student papers, concluding that their authors deliberately avoided the effort to summarize and paraphrase published literature. There are, however, exceptions to the rule to refrain from using direct quotations in scientific writing. Quotations are acceptable for calling special attention to precisely stated ideas and arguments in the literature. Suppose that you are making an original argument in response to a claim in a published journal article. You deem that the exact wording of the author's claim is crucial to setting up your argument. In this case, a direct quotation would be appropriate. Here's an example:

> In Stamford's (2005) argument against using dopamine agonists to treat movement disorders, the author concludes that the drugs "play no role whatsoever in protecting against neurodegeneration." In response, I will argue that dopamine agonists are in fact effective in protecting against neurodegeneration when they are targeted at specific receptors.

Avoid Copying Patterns of Ideas from Published Articles

This guideline addresses a form of plagiarism in which authors copy patterns of ideas from published works. Take the case of two research papers involving the same disease. In both papers, the introduction section begins with a paragraph reporting the same statistics on the number of people who suffer and die from the disease. Next, both introductions have a paragraph that summarizes the same previous studies in matching order. Then, the third paragraph of both introductions describes common treatments for the disease—the same treatments presented in identical order. For any *single* paper on the topic, this pattern of idea development might indeed be ideal. But when two papers are practically indistinguishable in their pattern of ideas, one author likely copied the other. Even if no strings of text were reproduced verbatim, this is still a serious case of plagiarism. One way to prevent it is to deliberately avoid relying on the patterns by which authors develop their ideas in published articles on your topic. In addition, you can prevent this form of plagiarism by purposefully adapting the articles' contents in order to achieve your specific rhetorical goals for your intended audience.

SUMMING UP AND STEPPING AHEAD

We certainly have covered a lot of ground in this chapter on organizing content and writing a draft. In closing, let's recall a few of the chapter's high points. For organizing content, a key take-home message is that the process can be straightforward and productive for writers who have done the essential groundwork in earlier stages of planning. Specifically, this means reaching the organizing stage with a well-developed plan of rhetorical goals, strategies, and rough notes. The most important lesson for organizing content is to take a principled approach. Most decisions about how to order and integrate information in scientific papers *cannot* be resolved by following hard-and-fast rules. As emphasized throughout the chapter, the best decisions emerge through considering discipline-specific instructions and conventions, applying audience analysis, developing recognizable structural patterns, and arranging parts of papers to serve their overall functions.

A strong organizing plan greatly enhances the process and products of drafting. To help you write strong first drafts, this chapter presented guidelines for composing their titles, abstracts, paragraphs, sentences, citations, and more. As emphasized, the essential challenge of drafting is to produce well-structured sentences and coherent paragraphs without getting bogged down by applying rules and overanalyzing what you have written. Think about it this way: The main reason for writing a draft is to have good material to *revise*. As all experienced authors know, many of the best opportunities to excel at writing come about through revision, which is the process we turn to next.

CHAPTER 5

Revising Document Design, Global Structure, and Content

The Science of Aging and Longevity

In old-fashioned Hollywood movies, scientists are often portrayed as mad masterminds, holed up in dank underground laboratories, stirring steamy solutions in glass beakers, and concocting mysterious elixirs to revive the dead and bestow eternity to the living. In reality, there is actually a grain of truth to these Hollywood caricatures. In laboratories around the world, scientists are currently experimenting with methods that may extend the normal human lifespan. Unlike their movie counterparts, however, mainstream life scientists are not driven by the crazed desire to discover the Fountain of Youth. Instead, they seek to understand the causes of aging and age-related declines in health and functional capacity. It is truly noble work because many older adults in our society are living well beyond the average life expectancy only to suffer debilitating physical and mental diseases.

A paradoxical lesson learned from research on aging and longevity is that the body's life-sustaining processes, such as those for producing energy, ultimately yield molecular byproducts that slowly but surely take life away from us. Another lesson is that our chances of thriving in old age are greatly enhanced if we adopt healthy behaviors in youth. But perhaps the most important lesson about aging and longevity comes from common experience rather than scientific research: As we age, we naturally gain wisdom and improve the quality of our lives through reflection and problem solving. Coincidentally, reflection and problem solving are central to our concerns in this chapter, which introduces the process of revising scientific papers.

INTRODUCTION

In my scientific writing classes, students turn in first drafts for comments and suggestions for revision. As advertised in the course syllabus, as long as first drafts are revised, they do not count toward final grades. On days that the drafts are due, to convey an important lesson about expertise in writing and to encourage some lively conversation, I make my students an offer: "If you're completely satisfied with your draft," I announce, "you don't have to revise it. I'll just grade your draft and we'll count it as your final grade for the assignment." Then, I half-jokingly suggest that instead of devoting upcoming class meetings to the revision process we go out for coffee or lunch. To my great delight, and often to the students' own surprise, the class invariably rejects the offer. They do so quite passionately. Before I can even finish extending the proposal, heads start to shake back and forth vigorously, and the cry of "NO WAY!" echoes throughout the classroom. When the excitement finally settles, I encourage everyone to think about why they protested so strongly. By and large, the students realize that they are not satisfied with their drafts. They all know that they can do better, and most are truly motivated to improve their papers. Here's the important lesson: The strong motivation to improve one's writing through revision is a defining characteristic of expert authors. Indeed, it figures heavily in the ultimate success of any scientific paper because skillful revision can make a world of difference in writing quality.

Writing instructors define revision literally as looking (*vision*) again (*re-*). Specifically, the task entails identifying strengths to reinforce and problems to solve. In all forms of academic writing, and especially in science, revision is challenging because so many features of drafts may vie for the writer's attention. These features range from the correctness of the tiniest punctuation marks to the convincingness of extensively developed arguments. It makes good sense, then, that an expert approach to revision must be systematic, comprehensive, and manageable. This is how we will take on revision here and in Chapters 6 and 7. Our approach divides the overall process into five units, or *levels of discourse*. They are outlined in Table 5.1. This chapter focuses on revising at the levels of document design, global structure, and content. Then, Chapters 6 and 7 cover revision at the paragraph and sentence levels, respectively. At each level of discourse, our approach will follow the general guidelines for successful revision that are presented next.

Revision Guideline 1: Start Revising by Ignoring Your Draft for a While

Often, the best first step to take in the revision process is a step backward—that is, away from your draft. Immersed in diligent work on a paper for a long time, you can easily lose an objective view of it. You may, for example, have difficulty separating ideas in your mind from those on paper. Or you may assume that your audience will naturally accept arguments that are deeply engrained in your belief system. Time and distance away from first drafts will help you regain a fresh,

TABLE 5.1 **Key Features for Revision at Five Levels of Discourse**

Level of discourse	Key features
Document design	Font type and size, line spacing, margin width, page numbering, presentation of graphics, and citation format
Global structure	Ordering of sections, conceptual unity across sections, parallel structure across sections, and matters of repetition and redundancy
Content	Clarity, accuracy, development, consideration of audience, success in accomplishing rhetorical goals, and strength of arguments
Paragraphs	Unity, coherence, use of topic sentences, sentence variety, and matters of paragraph design
Sentences	Punctuation marks, grammar, word choice, logic in expressing ideas, and matters of style

objective, and critical view. So make it a practice to turn your attention to other projects for at least a few days after completing a draft. But make sure to give yourself sufficient time to devote to revision before your final paper is due. Any advantage gained from taking a mental break from a draft will not help if you are completely stressed about a fast-approaching deadline!

Revision Guideline 2: Revise Proactively

In the revision process, if no obvious errors jump up from the surface of a text, the novice writer may call it a day. Of course, this would be a mistake if problems exist at deeper levels. In contrast, expert writers take a proactive approach to revision. They know *what* to look for at all levels of discourse, and they know *how* to go looking. This rich knowledge, which is central to our concerns, distinguishes novices and experts at work in revision.

Revision Guideline 3: Approach the Process as Revision by Division

As explained in the book's introduction, a process-based approach to scientific writing helps us avoid the frustration and the poor outcomes that result when we try to do too many activities in the writing process at once. A process-based approach to revision is especially important because its activities are inherently complex. Our approach, which we will call *revision by division*, works systematically through the five levels of discourse outlined in Table 5.1. It is an organized, comprehensive, and manageable system; however, it is not meant to be followed mechanically. Throughout this chapter and the next two chapters, I encourage you to reflect on and experiment with adaptations of our divide-and-conquer approach to meet your needs. Along the way, I will offer general suggestions for adapting the approach to account for individual circumstances.

Revision Guideline 4: Don't Forget the Re- In Revision

By definition, revision is an iterative process. In practice, this means making repeated passes through a draft to evaluate single features or small sets of related features at each level of discourse. For example, draft paragraphs should be evaluated for their unity, topic sentences, coherence (flow), and sentence variety. To avoid becoming overwhelmed and overlooking important features of paragraphs, expert writers concentrate their revision. One pass through a draft might be focused on the strength of its topic sentences. Another pass might be devoted to evaluating sentence-to-sentence flow. Still another might target qualities of paragraph unity. It's definitely a lot of work, but it comes with the territory of successful scientific writing.

An iterative approach also entails checking to make sure that changes to improve drafts do not inadvertently create new problems. Consider, for example, a draft paragraph composed of many short, choppy sentences that create a monotonous tone. As we will discuss in Chapter 6, the solution is to apply sentence-combining techniques in order to form longer, more complex and varying structures. However, these techniques can introduce new problems in paragraphs, such as awkwardly structured sentences, grammatical run-ons, and gaps in coherence. Considering the potential for solutions to create new problems, the best writers take the *re-* in revision to heart and practice.

Revision Guideline 5: Approach Revision as Problem Solving

If revision means *looking again*, you might ask, "What, *exactly*, should I be looking again for?" This is a fitting question because the answer essentially guides the revision process for ideal outcomes. Our approach to revision takes dead aim on *problems* to solve. At the content level, for example, we will look for problems such as inaccurate information, underdeveloped ideas, and insufficient evidence and reasoning in arguments. At the paragraph level, we will look for disunity, weak topic sentences, poor coherence, and a lack of variety in sentence structure. At the sentence level, we will look for various types of grammatical errors, inappropriately used words, incorrect punctuation marks, and stylistic flaws. While we certainly will not overlook identifying strengths, or excellent features to reinforce in drafts, we will concentrate on problem solving for several reasons. Problems are normally things to avoid because they cause us nothing but trouble. However, in the context of revising scientific papers, problems are extremely useful. Just think about it: If your drafts contain problems (and, really, whose drafts don't?), by successfully solving them, you will of course improve the quality of your final papers. In addition, through learning and applying advanced skills for solving specific writing problems, you will naturally avoid them in future drafts.

Figure 5.1 presents a procedural map of our problem-solving approach, highlighting the three levels of discourse covered in this chapter: document design, global structure, and content. (In Chapters 6 and 7, similar maps represent the processes of revising paragraphs and sentences, respectively.) The graphic is quite busy, so let's walk through it step by step. First, focus on the three boxes labeled

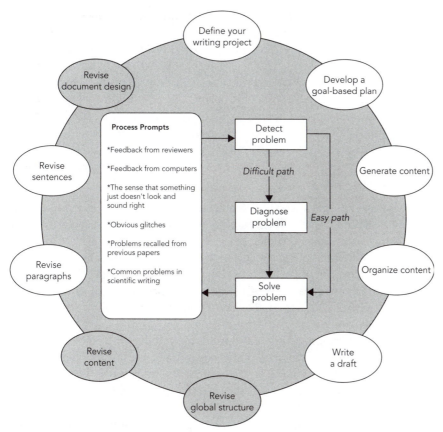

FIGURE 5.1 Process map for *revising document design, revising global structure,* and *revising content.*

Detect Problem, Diagnose Problem, and *Solve Problem.* These labels are adapted from seminal studies on the cognitive strategies of expert writers working at revision. These studies were conducted in the 1970s and 1980s by two separate teams of cognitive scientists—Linda Flowers and John Hayes, in the United States; and Carl Bereiter and Marlene Scardamalia, in Canada (Bereiter & Scardamalia 1987; Flower & Hayes 1981; Flower et al., 1986). Among other interesting findings, the research revealed that when experts revised their drafts, they focused on identifying mismatches between their intentions and what they actually wrote. They treated the mismatches as problems to solve. To devise fitting solutions, the experts formally diagnosed the problems. In this context, diagnosis refers to applying specialized knowledge about writing in order to determine the underlying causes of detected problems in drafts.

In keeping with cognitive models of the revision process, we will define *detecting* as becoming aware of a problem, either vaguely, absolutely, or somewhere between these extremes. Upon detecting problems in our drafts, we naturally want to solve them immediately and move on. For some problems, such as blatant typographical and spelling errors, solutions do indeed come to mind instantly. These revisions take the *easy solution* path illustrated in Figure 5.1. Other problems, however, do not automatically lead us to solutions. Examples are unconvincing arguments that lack relevant supporting evidence, incoherent paragraphs composed of disjointed ideas, and incomprehensible sentences with deep structural flaws. These sorts of problems take the *difficult solution* path in our process map. They require writers to apply diagnostic routines in order to determine underlying causes and thereby identify strong solutions.

Feeding into the detect-diagnose-solve pathway in Figure 5.1 is a set of six *process prompts*. Described as follows, these are common factors that cue writers to begin the revision process.

- **Feedback from reviewers**: To complement their autonomous revision, experienced writers solicit comments from reviewers who may be coauthors, peers, classmates, professors, editors, the staff of a campus writing center, or even friends and family members. Reviewers may offer direct solutions to problems, or they may support authors in carrying out independent revision.
- **Feedback from computers**: This prompt applies mostly to revising at the sentence level, where computer-based grammar and spelling checkers may initiate the process.
- **The sense that something just doesn't look and sound right**: All writers have this experience in reading their own drafts. Maybe a sentence sounds awkward when read aloud, a paragraph does not seem to flow, or an idea feels incomplete. The sense that something just isn't right prompts deeper diagnoses of problems.
- **Obvious glitches**: Including typos and glaring grammatical errors, these sorts of problems may be inadvertently overlooked in drafting; however, they often pop up during revision. Their solutions usually take the easy path in our process map in Figure 5.1.
- **Problems recalled from previous papers**: Conscientious writers know their strengths as well as their weaknesses. This awareness stems from honest self-assessments of their work and from feedback offered by reviewers on previous writing projects. Take the case of a student writer whose professors have regularly noted room for improvement in constructing strong, original arguments. The student's recollection of this problem from previous papers is a powerful prompt for revising current and future drafts.
- **Common problems in scientific writing**: Some flaws in drafts come with the territory of communicating in science. Examples include misuses of word groups such as *compose/comprise* and *affect/effect*, a lack of variety in sentence structure, gaps in paragraph flow caused by missing information, and the use of unnecessary jargon. We can initiate revision by taking stock of such common problems in scientific writing.

Even though we are emphasizing a problem-solving approach to revision, you should not overlook strengths in your drafts. Imagine, for example, that you have identified an exceptionally strong argument in a section of your draft. The argument is especially powerful because you have backed it with well-developed and highly detailed evidence and reasoning from research. You can capitalize on this diagnosis and reinforce the strength by evaluating arguments in your draft's *other* sections, checking for whether they are backed with similarly effective support.

Revision Guideline 6: Approach Revision as Critical Thinking

In many respects, revision is nothing other than critical thinking. When we identify strengths to reinforce and problems to solve in drafts, we think in ways that are analytical, evaluative, and well reasoned. At its best, revision engages a special critical thinking method, which we will call *criterion-based* thinking. This method entails judging the quality of something by applying specific criteria that distinguish its excellence. Many demonstrations of criterion-based thinking are presented in this chapter and the next two chapters.

Revision Guideline 7: Get a Little Help from Your Friends

Few activities in the writing process are more challenging than independent revision—that is, detecting, diagnosing, and solving problems in our own papers without external feedback. One reason that independent revision is difficult is that we do not intentionally create problems in our drafts. In drafting an explanation of a complex concept, for example, you would never purposefully scheme to confuse your readers. Instead, you try to elucidate the concept as clearly and completely as possible. This is exactly how the explanation may come across to you when revising independently. Every now and again, however, what is crystal clear to us as authors may be mostly muddy or beyond baffling to our intended audiences. This is just a reality of the challenges of successful communication. Indeed, it's a reality that underscores the great importance of getting feedback from external reviewers. Later in the chapter, I will talk about the specialized knowledge and skills for giving useful feedback (as a peer reviewer) and for evaluating and applying it (as a writer). For now, heed the following general suggestions for soliciting and using feedback from reviewers.

- Seek feedback from people who know a lot about your paper's topic and understand scientific discourse conventions.
- Seek feedback from people who are members of your intended audience or who can easily take their perspective.
- Guide your reviewers by asking for feedback on *specific* strengths and weaknesses in your draft.
- If possible, seek feedback from several reviewers, so that you can compare and contrast their critiques and suggestions for revision.
- Don't automatically accept and apply reviewers' critiques and suggestions for revision—instead, first judge the validity and usefulness of the feedback for yourself.

Revision Guideline 8: Master Independent Revision

While feedback from reviewers can be extremely valuable, our approach to revision emphasizes autonomous problem solving. Through learning and mastering the skills of independent revision, you will gain great confidence in your writing ability. In turn, you will work more efficiently and productively in all stages of the writing process. Your drafting, for example, will go more smoothly because you will not be slowed by excessive worries about making mistakes. Proficiency at independent revision will help you critically evaluate the feedback that you receive from reviewers on your drafts. In addition, the skill will enhance the feedback that you give to peers and coworkers on their drafts.

ABOUT THE PROCESS: REVISING FOR MATTERS OF DOCUMENT DESIGN

The term *document design* has different meanings in various fields of written communication. The definition that we will adopt focuses on matters of page layout, format, and style in manuscript preparation. Revising at this level of discourse, we evaluate our drafts for features such as margin width, line spacing, font type and size, placement of section headings, and citation format. The process is fairly straightforward, especially when writing assignments come with clear directions. Even so, you might expect that your first drafts will possibly overlook important instructions or guidelines. Maybe a few reference-list citations are not formatted correctly, or page numbers are not positioned as instructed. Perhaps you have used the typeface *Calibri* when the instructions call for *Times New Roman*. To avoid being penalized for failing to follow directions, take a pass through your draft to double-check its document design features. From Chapter 1, you may recall the suggestion to create task-analysis checklists and note forms that include such features (see Figure 1.2 on page 6). I advised listing and noting features that you do not normally and automatically attend to when preparing manuscripts. Revising for matters of document design simply involves applying task-analysis checklists and notes to make sure that final papers follow instructions, guidelines, and evaluation criteria.

In the overall writing process, it's usually most productive to revise document design last of all, after any problems involving global structure, content, paragraphs, and sentences have been resolved. This order is reflected in the process map in Figure 5.1 (page 215). However, for convenience and balance in organizing this book's chapters, I decided to quickly cover revising for document design here. We can now move on to bigger and better things.

ABOUT THE PROCESS: REVISING FOR MATTERS OF GLOBAL STRUCTURE

As defined in Chapter 4, global structure refers to the organization of major sections and subsections of written documents. Revising at this level of discourse, we focus on matters of order, integration, and coordination of these large structural units. As illustrated in Figure 5.1, revision of global structure should precede revision at the levels of content, paragraphs, sentences, and document design. This order makes sense because revisions at the more local and refined levels may be counterproductive if problems exist in a draft's overall organization. Revision of global structure is a problem-solving process that is ideally guided by the same seven organizing principles that were presented back in Figure 4.5 (page 161) and demonstrated throughout the previous chapter. As follows, we will revisit these principles and apply them to detecting, diagnosing, and solving key problems that weaken global structure.

Disordered Sections

The previous chapter demonstrated how to apply selected organizing principles to determine the most effective linear order for the major sections and subsections of drafts. In the revision process, we reapply the principles to check for the flaw of disordered content, or sections that are not lined up most fittingly. Take the principle to follow discipline-specific conventions for organizing content. It can be handily converted into a powerful diagnostic question for guiding revision: Are the major sections and subsections of my draft ordered according to established conventions for the paper type? Other principle-based diagnostic questions, adapted for evaluating the order of a draft's sections, are as follows.

- Are my draft's major sections and subsections ordered according to instructions for my assignment?
- Does the order of my draft's major sections and subsections account for my readers' needs, expectations, and values?
- Are my draft's major sections and subsections ordered in ways that make good common sense?

If your answers to these questions reveal any instances of disordered content, the solutions will obviously entail re-outlining the associated sections of your draft.

Weak Global Unity

As defined in Chapter 4, global unity refers to the quality of oneness in the overall structure and function of a written document. Each section of a paper with strong global unity clearly relates to the overall topic, supports the overall message, and contributes to achieving the overall goal. In papers that lack this desirable quality, major sections and subsections fall short of supporting these three global functions. To better understand the problem, consider a conceptual review paper authored by a student named Ben. His overall goal is to argue for a theory that attributes human aging to damage occurring in cellular structures called *mitochondria*. Nearly all of the body's trillions of cells contain hundreds to thousands of mitochondria, which are known as the cells' "energy powerhouses." In these structures, oxygen and food-derived fuels are used to synthesize adenosine triphosphate (ATP), a molecule that releases energy for supporting nearly all of the body's functions. The mitochondria are unique because they contain their own DNA, which enables them to maintain their structure and function somewhat independently of other cellular components. However, mitochondrial DNA can be damaged in the process of ATP formation. As a natural byproduct of this metabolic process, unstable molecules called free radicals are produced. A particular type of free radical, called reactive oxygen species (ROS), can damage mitochondrial DNA and thereby impair the energy-producing functions of cells. According to the theory on which Ben is basing his review paper, the harmful effects of ROS can directly cause the deterioration of bodily structure and function that occurs with aging.

In his introduction section, Ben presented the following three overall purposes for his paper:

1. To present the detailed processes by which ROS are formed and damage the mitochondria.
2. To explain how the damage caused by ROS directly contributes to aging.
3. To present evidence from research that supports the "mitochondrial damage theory" of aging.

The global unity of Ben's draft would be strong if its body were structured by subsections that clearly relate to these overall purposes and follow through on achieving them. As it turned out, however, the draft had room for improvement in this area. For instance, in the first subsection of the body, Ben extensively discussed the influence of various types of free radicals on many different cellular structures. This subsection weakened the draft's global unity by veering off course from Ben's declared aim, which was to present the detailed processes by which a specific type of free radical (ROS) damages a specific cellular structure (the mitochondria). In another extensively developed subsection, Ben explained the mechanisms by which ROS cause various types of cancer. This subsection also compromised the draft's global unity because the subtopic did not directly relate to the mitochondrial damage theory of aging. If uncorrected, these deviations from Ben's introductory purposes will elicit a sense of disorientation in readers as they work their way through the paper's body.

To evaluate the global unity of a draft, first focus attention on the overall topic, message, and goal of your paper. Ideally, these elements will be conveyed in the draft's introduction section. For example, in your introduction you may have listed the specific purposes for your paper, reflecting its overall goal. The next steps are to evaluate all of the draft's sections, following the introduction, for how effectively their topics, messages, and goals relate to their overall counterparts. For each section, raise the following questions.

- Does it logically fit into the paper's global scheme?
- How well do its subtopics relate to and develop the paper's overall topic?
- How closely do its main messages support the paper's overall message?
- How effectively do its rhetorical goals contribute to accomplishing the paper's overall goal?

If the answer to any of these questions indicates a problem, a deeper diagnosis may be necessary to formulate a strong solution. Take the example of Ben's draft, in which the global unity was weakened by a subsection that discussed the effects of ROS on cancer rather than on the intended topic of aging. One solution is to delete the subsection entirely. If, however, Ben deems that it is actually relevant, his best solution is to add content that explains the connections between mechanisms of cancer and aging, specifically in the context of mitochondrial damage caused by ROS.

An especially effective technique for strengthening global unity is to continually refocus the reader's attention on a paper's overall topic, message, and goal. In a research paper, for example, this might involve reintroducing the study's motivating questions at pivotal junctures, such as in the methods section where key procedures for answering the questions are presented. Or, in a position paper, strengthening global unity might involve restating essential claims at strategic points.

Mismatched Organizing Themes

As defined in Chapter 4, organizing themes are recognizable structural patterns in papers. Recall, for example, that scientific papers may be organized by goal-based, chronological, sequential, most-to-least important, or part-to-whole themes. Here we consider the problem of mismatched organizing themes, or inconsistent patterns across a paper's sections (when consistency would actually be helpful). A review paper authored by a student named Valeria will serve as a case in point. Her project involves various medical and behavioral methods for extending the lifespan. Among other methods, Valeria is writing about growth hormone therapy, gene therapy, and calorie restriction. Each of these methods corresponds to a major section in Valeria's draft. We can say that the major sections are thematically consistent, because each represents an approach to life extension. We might thus expect that the subsections of Valeria's draft would also be consistent in structure. In other words, the subsections might logically be structured by the same organizing theme. As it turned out, however, Valeria's draft lacked this consistency in structure; the organizing themes for its subsections were mismatched. Rather than reproduce the entire draft, I will explain the problem by referring to an outline created from three major sections of Valeria's draft. The outline is displayed Figure 5.2 (page 222).

In the outline's first major section, which is devoted to growth hormone therapy, notice that the subsections are organized by key rhetorical goals for research papers. Valeria begins the section by introducing the topic of growth hormone as an antiaging therapy. Then she devotes subsections to presenting common methods for studying the therapy's effects, summarizing findings of previous research, and discussing conclusions from the research. On its own, this organizing theme seems logical enough. However, Valeria completely abandons the theme in the next major section, which focuses on gene therapy. This section happens to be organized by a part-to-whole theme. That is, Valeria addresses the topic of gene

I. Life-extension Method #1: Growth Hormone Therapy
   A. Introduce the use of growth hormone as an anti-aging therapy
   B. Present commonly used methods for studying the effects of growth hormone on the aging process
   C. Summarize the results of key studies on growth hormone and aging
   D. Discuss the conclusions from the research on growth hormone and aging

II. Life-extension Method #2: Gene Therapy
   A. Explain how gene therapy affects cellular processes that regulate aging
   B. Explain how gene therapy affects the structure and function of the body's tissues
   C. Explain how gene therapy affects the structure and function of the body's organ systems
   D. Explain how gene therapy affects survival and longevity in living organisms

III. Life-extension Method #3: Calorie Restriction
   A. Present the most convincing argument for how calorie restriction may extend longevity: Calorie restriction reduces the body's metabolic rate and thereby inhibits free radical damage
   B. Present secondary arguments for how calorie restriction may extend longevity: Calorie restriction may reduce exposure to toxins in foods and prevent diseases that influence the rate of aging
   C. Present a less convincing argument for how calorie restriction influences longevity: Calorie restriction has a minor influence on longevity by influencing immune function

FIGURE 5.2 An outline demonstrating the problem of mismatched organizing themes.

therapy by explaining its effects at progressively more inclusive levels of biological systems. She begins at the cellular level and works up to explaining how gene therapy influences survival and longevity in whole living organisms. The pattern is logical on its own; however, it contrasts the organizing theme of the previous section. The organization of Valeria's third major section, devoted to calorie restriction, adds to the structural inconsistency. Its organizing theme involves presenting the most-to-least convincing arguments for how calorie restriction may extend longevity.

Inconsistency in the structural patterns of a paper's successive sections is not automatically problematic. In Valeria's case, however, the mismatched themes weaken the draft's global unity. Readers will not have a clear sense of the paper's overall purpose regarding the various methods for life extension. Is the paper intended to provide information about the procedures, results, and conclusions of previous studies? Is it intended to explain how the methods influence biological systems, from the cellular to organismic levels? Or is it intended to make arguments about the most and least important theories for explaining how the methods affect survival and longevity?

To check for consistency in your draft's organizing themes—that is, when consistency is indeed warranted—begin by creating an outline of the draft, as demonstrated by the outline of Valeria's draft in Figure 5.2. The next step is to reflect on which organizing themes would be ideal for your draft's sections and subsections. If the major sections naturally serve a similar purpose, the organizing themes for their subsections should be consistent. Upon diagnosing any mismatches between your reflections on the ideal structure for your draft and its actual structure, you must decide which organizing themes take precedence. The decision should be based on the overall purpose and function of your paper or selected parts of it. Suppose that Valeria decides that, above all, she wants her paper to inform readers about previous studies on various life-extension methods. She will then restructure all of the major sections to match the first one (for growth hormone therapy) in the outline in Figure 5.2 because its organization is based on a theme for reporting research studies.

Redundancy of Content Across Sections

In some circumstances, the repetition of information, ideas, and arguments across the sections of a scientific paper can be very helpful. As noted earlier, the strategic repetition of content reflecting overall topics, messages, and goals can effectively reinforce global unity. However, a fine line often exists between helpful repetition and redundancy, the latter of which is defined as needless and problematic repetition. Imagine, for example, a grant application in which the author devotes a subsection of her conclusion to arguing for the proposed study's significance to society. The author presented the very same argument, in fairly similar words, in the introduction section. The repetition will be helpful for readers, as well as for the writer in her endeavor to obtain funding, if the following conditions are met: (a) the content that comes between the introduction and conclusion is substantial and diverse in its topics and goals; and (b) the author's ideas about the study's significance to society are especially pivotal for the project's success. The first condition must be met because readers might respond negatively to the author's raising the same ideas in such close proximity. The second condition is noteworthy because winning ideas are often worth repeating.

If you identify repeated content across sections of your drafts, check for whether it is helpful or redundant by raising questions about proximity (*Am I repeating information, ideas, and arguments too closely?*) and importance (*Does the repeated content serve an essential rhetorical goal that will strongly influence the project's success?*). In addition, seek specific feedback from reviewers on this matter of revision. Ask them whether they find cases of repetition across sections to be helpful and necessary or distracting and therefore needless.

ABOUT THE PROCESS: REVISING FOR MATTERS OF CONTENT

When revising drafts for global structure, we intentionally overlook details of content—in other words, we are not concerned with evaluating specific information, ideas, and arguments. But, of course, a paper's content ultimately determines its success. So, after ensuring that matters of global structure are in good shape, we turn to revising content. At this level of discourse, key features include the accuracy, clarity, and development of our ideas, as well as the strength of our arguments. At its best, content-level revision focuses on how successfully drafts accomplish appropriate rhetorical goals and meet the needs, expectations, and values of intended readers.

It's essential to complete revisions of content before revising for matters at the paragraph and sentence levels. Taking this approach, we ensure that *what* we are saying is sound before refining *how* we are saying it. This revision sequence also helps us avoid the frustrating and counterproductive experience of spending lots of time correcting superficial flaws—such as awkward phrasing, inappropriate word choices, and grammatical errors—only to find that they are contained in problematic content that actually needs to be deleted from our drafts or completely reworked. Another reason for revising content first is that doing so often naturally engenders well-crafted prose. When we crystallize our thoughts, they are relatively easy to express in coherent paragraphs and pointed sentences.

This section of the chapter presents procedures for revising numerous content-level problems that commonly occur in drafts of scientific papers. The problems are outlined in Figure 5.3 and explained as follows.

Missing Content

For the most part, we revise content by evaluating what *is* in our drafts. This first problem, however, concerns content that *is not* in our drafts but should be. Sometimes writers inadvertently omit important content because they overlook assignment instructions, guidelines, and evaluation criteria. Detecting this problem requires referring back to assignments to double-check directions for content to include. Suppose, for example, that your professor has given instructions to discuss the relevance of your paper's topic to society. Or maybe your professor is reserving high grades for papers that propose original ideas for future studies. In addition to evaluating whether your draft contains content that follows assignment instructions, check for whether you have included everything from your organizing plan. You might think that all this double-checking is unnecessary—after all, if you planned to include certain content, there is no reason to think that it's missing. But under the heavy cognitive load of the drafting process, even

Revising for General Matters of Content
    Missing content (224)
    Ambiguous content (225)
    Inaccurate content (226)
    Content that misses the target on key rhetorical goals (228)
    Content that fails to adequately address concerns of audience (234)
    Saying too little or too much (236)

Revising Logical Fallacies
    Circular reasoning (238)
    Red herring (239)
    Hasty generalizations (240)
    Subjectivism and personal experience (243)
    False cause (243)
    False dichotomy (244)
    Appeal to ignorance (245)
    Appeal to authority (246)
    Bandwagon (247)
    Straw man (247)

FIGURE 5.3 Common content-level problems in scientific writing. The pages on which the problems are introduced are shown in parentheses.

the best writers sometimes omit important content despite having well-conceived plans for it.

In another scenario, writers may overlook important topics to discuss and key rhetorical goals to achieve because they never conceived them in planning. To detect this problem, read your drafts from the perspective of your intended audiences, imagining how they will respond to what you have written and *have not* written. Of course, feedback from skillful reviewers can be invaluable for detecting the problem of missing content.

Ambiguous Content

At the root of ambiguous content, we may find a single vaguely defined word, an incoherent paragraph, or an inadequately explained concept spanning an entire section. No matter how or where the problem begins, ambiguous content is an especially difficult flaw to detect independently because our ideas usually seem clear enough to us. But, of course, they might not be clear to our readers. This is why it is so important to seek external feedback aimed specifically at matters of clarity. Ask your reviewers to be completely honest in identifying draft content that makes them say, "I read it again and again, but I just don't understand what you're trying to say here." Most important, ask your reviewers to give specific reasons for their misunderstanding.

- Do they lack essential background knowledge that the draft fails to provide?
- Are certain words or concepts expressed too vaguely?
- Is the problem due to awkwardly phrased sentences?

The input on the underlying reasons for the ambiguity is essential for figuring out the best strategies for achieving clarity through revision.

While it's difficult to detect and diagnose unclear content independently, the task is certainly not impossible. First, put your draft aside for a while—at least a few days would be ideal—to gain a more objective view. Then, read it with a focus on identifying information, ideas, and arguments that you initially found difficult to convey. Along the way, isolate any content that is not *crystal* clear to you. Then assume the mindset of your readers, asking whether they will understand complex concepts in your draft. If you detect a lack of clarity, diagnose the problem by determining its underlying causes. Is the content unclear because you do not fully understand the science about which you are writing? This, of course, is a common problem for undergraduate students writing about new topics that are based on advanced scientific knowledge. In their ambitious efforts, these students rely on what they have learned through reading journal articles that were written for career scientists who are topic experts. If you find yourself in this situation—that is, perplexed about what you are writing—your path to a strong revision must backtrack to stages of planning that involve searching for understandable literature and reading to learn the science. To clarify and deepen your understanding of complex content in the published literature, try explaining it to yourself as well as to peers who can take the perspective of your intended readers. Get their feedback on the clarity of your ideas before rewriting your draft.

Through independently diagnosing the causes of unclear draft content, you might find that you understand your topic perfectly. In this case, the problem comes with the naturally challenging territory of expressing complex scientific information clearly to others. A simple but powerful revision tool for fostering clarity is the question, "What, *exactly*, am I trying to get across to my readers?" Through repeated cycles of asking yourself this question and then answering it, until you are satisfied with the clarity of your answer, you will figure out how to express difficult ideas most intelligibly. As every experienced writer knows, however, it can take more than just one or two cycles of this sort of questioning, answering, and rewriting to produce the clearest expressions of complex ideas.

Inaccurate Content

The accuracy of scientific information is commonly judged by whether terms are defined factually, statistics are reproduced exactly as they appeared in original sources, and concepts are explained precisely according to uncontested views in the scientific community. When we evaluate drafts for accuracy, we thus look for whether the content aligns with well-established scientific knowledge. If we detect anything that is obviously incorrect, the solution is straightforward: Just correct it. As authors, however, we generally do not include information in our drafts unless we believe that it is correct. But what if an inaccurate definition, statistic, or concept makes its way into your draft inadvertently? How in the world are you supposed to detect the flaw independently? Like revising for ambiguous content, this is another aspect of revision that ideally calls on external reviewers, preferably ones who are very knowledgeable about your topic. If your professors offer reviews on first drafts, ask them for feedback on matters of accuracy. Direct them to specific information and ideas that have raised questions and doubts in your mind. In addition, you might seek feedback from teaching assistants and classmates who know a lot about your topic.

Revising for accuracy is also challenging because so much scientific knowledge is dynamic and debatable rather than concrete and purely factual. This is true, for example, in fields of research on aging and longevity. Take the fundamental question, "What causes human aging?" New hypotheses and theories proposing answers to this question are being developed and adapted all the time. This dynamic situation, which is extremely common in all life science disciplines, naturally engenders debates about what *is* accurate and what *is not*. So when you are evaluating the accuracy of a paper on an evolving research issue, common definitions of right and wrong do not apply. Instead, accuracy should be defined by the extent to which the content does the following:

- includes all of the relevant information and ideas in the evolving field;
- acknowledges all sides of debatable issues; and
- represents the most reputable consensus views, or the current state of understanding shared by the field's leading scientists.

Consider, for example, a paper on theories of aging in which the author addresses only a few of the many theories that the scientific community recognizes as valid. A knowledgeable reviewer will respond to the author as follows: "You are saying that the main theories of aging are theories A and B, but according to Professor So-and-So's recent paper, viable theories include C, D, and E. As I do not have to remind you, Professor So-and-So is recognized as a world's leading expert on aging and longevity. You have also completely ignored theories F and G." Based on the qualities that define accuracy for dynamic and debatable research issues, the reviewer's feedback goes beyond criticizing the paper for missing content. It also conveys the problem that the content is not completely accurate.

To diagnose the accuracy of content in your drafts independently, ask the following questions.

- Have I defined specialized terms consistently with their definitions in the scientific community?
- Have I accurately reproduced statistics obtained from the scientific literature?
- Are my explanations of complex concepts and theories aligned with the most reputable and widely accepted explanations in the scientific community?

- Have I comprehensively represented all sides of debatable issues?
- Is my summary of consensus views up to date?

Answering these questions requires double-checking original sources of the knowledge, evidence, and reasoning that you have included in your draft. In addition, several of the questions may lead you to reevaluate sources for their validity and credibility (following guidelines presented in Chapter 1). If the process raises doubt about the accuracy of selected draft content, you may need to search for additional literature for confirmation.

Content That Misses the Target on Key Rhetorical Goals

Writing a draft of a scientific paper is a bit like throwing darts. With every toss of an idea or argument, we aim for the perfect result: hitting the bull's-eye dead center. On first attempts, however, a fair number of tosses will predictably land askew. Drawing the analogy to scientific writing, we can say that hitting the bull's-eye is akin to accomplishing the rhetorical goals that we set in planning. It follows that revision is a task of reckoning how far off-target our darts—I mean, our *drafts*—landed. In other words, it's a process of determining how successfully a draft's information, ideas, and arguments actually accomplished targeted rhetorical goals. Let's practice this goal-directed approach to revising content with a student named Derek, who is writing a position paper on the effects of exercise on aging and longevity. Based on his critical evaluation of published research on this issue, Derek is planning to argue that a lifetime of regular aerobic exercise (such as jogging, swimming, and bicycling) causes adaptations in the body that slow the normal rate of aging and can thereby extend longevity in humans.

In reality, whether regular aerobic exercise can slow the aging process and extend longevity in humans is quite debatable. While many aspects of health are undoubtedly enhanced by exercise, improved health *per se* does not necessarily influence the rate at which we age. Indeed, some scientists contend that people who exercise regularly will age *faster* than sedentary people. This argument is based on the free radical theory of aging. As explained earlier, free radicals are molecules formed in metabolic processes for producing ATP, the body's ultimate energy source. The theory holds that free radical accumulation can accelerate aging by damaging components of our cells, including their DNA. To meet the body's increased needs for energy during exercise, the rate of ATP production increases greatly compared to the rate in sedentary conditions. With increased rates of ATP production, greater numbers of free radicals are naturally produced. Theoretically, then, the increased exposure to free radicals in people who exercise might speed up the rate at which they age. There are, however, a number of contrasting theories as well as experimental findings that support the claim that aerobic exercise can increase longevity. So Derek's topic is perfect for making an argument in a position paper.

For the paper's introduction section, one of Derek's rhetorical goals is to convince readers that his research issue is unresolved and debatable. For achieving this goal, Derek developed the following two strategies.

1. Show readers that the research issue is unresolved by summarizing the conflicting findings and conclusions from previous studies that support its different sides.
2. Show readers that the research issue is debatable by explaining the contrasting theories that support its different sides.

For introduction sections to position papers, Derek's goal and strategies are highly appropriate—they are great targets to aim for. To understand why, consider how readers will respond if the introduction *does not* successfully convince them that the research issue is unsettled and open to debate. They will not see the point of Derek's paper at all. Why make a scientific argument about an issue that is not truly arguable? Also consider that if the introduction does not fairly represent all sides of the unresolved issue, readers may suspect that the writer is biased. They might thus dismiss his claim from the outset.

Figure 5.4 presents an excerpt from Derek's first draft, specifically the content that he intended to convince readers that the effects of regular aerobic exercise on aging and longevity are truly arguable. A goal-directed approach to revision begins with the question, "How successfully does the draft's content achieve the author's targeted rhetorical goals and strategies?" Let's apply this question to evaluating Derek's draft. Will readers be convinced that his research issue is indeed unresolved and debatable? (Before reading my upcoming critique of Derek's draft, take a few minutes to evaluate the content for yourself, assessing whether it hits the bull's-eye on its target.)

Many studies have convincingly shown that regular aerobic exercise significantly reduces the risk of premature death caused by the leading killers in our society, which include heart disease, cancer, and type 2 diabetes.[4, 7, 12, 15] For example, aerobic exercise has consistently been shown to lower cholesterol levels and blood pressure, to reduce body fat, and to improve insulin sensitivity. In addition, immune function is enhanced in individuals who perform aerobic exercise on a regular basis.[3,8,10] Based on these positive health outcomes, one might contend that aerobic exercise can slow the aging process and thereby extend the maximum lifespan potential. Support for this hypothesis comes from a study conducted by Goldman,[5] who found that mice subjected to lifelong wheel-running exercise lived beyond the maximum lifespan of normally

FIGURE 5.4 Draft content that misses the target on its rhetorical goal.

active mice. A possible casual factor could be the positive effect that exercise has on heat shock proteins, which are formed in response to physical exertion. These proteins play roles in maintaining cellular homeostasis and rebuilding cellular structures. Some have argued that Goldman's study was flawed. In addition, there is an accusation that exercise may accelerate the aging process by creating free radicals which damage cell membranes and mitochondrial DNA.[13] However, this has been countered by research showing that exercise actually leads to increased production of antioxidants, which neutralize the potentially harmful effects of free radicals.[6,9]

FIGURE 5.4 *continued.*

Derek's draft misses the target on its rhetorical goal, which is a fairly common occurrence in first drafts of scientific papers. Instead of fairly representing all sides of the research issue, most of the content backs the argument that exercise slows aging and prolongs the lifespan. Eight of the paragraph's 10 sentences convey views that support Derek's claim. Toward the end of the paragraph, Derek briefly acknowledges the other side of the argument. He does so by raising criticism that has been directed toward a study conducted by Goldman, which revealed an increase in the maximum lifespan of exercising mice. However, Derek's explanation of the criticism is vague and underdeveloped. Notice how he imprecisely alludes to "some" who have argued that Goldman's study was flawed. Where Derek recognizes the view that exercise might actually accelerate aging, he passes it off as an "accusation." Then he immediately discredits the view. The content reflects a lack of objectivity on the author's part. This is a problem because the rhetorical goal at hand demands absolute objectivity. To appreciate the consequences of the flawed content, think back to the rationale for Derek's rhetorical goal. It's all about convincing readers that the research issue is currently unresolved. Because the draft's content is so one-sided, readers who are not familiar with the published research on aerobic exercise and longevity will not be convinced that the issue is unresolved. Perhaps a bigger problem is that readers who *are* familiar with the research may accuse the author of being biased right off the bat. Of course, Derek will ultimately need to argue forcefully for his claim that exercise can extend longevity; however, this argument should come in the *body* of his position paper rather than in the introduction. In summary, we can say that Derek jumped the gun on supporting his claim.

We may miss the targets on our rhetorical goals as a natural consequence of the challenges of drafting. Focused intently on expressing our ideas in concrete terms, we can easily lose sight of our higher-level rhetorical goals, try as we might to keep them in view. Content-level revision thus demands helicopter thinking, the technique that we applied to developing rhetorical goals in Chapter 2 and to generating content in Chapter 3. As follows, I present a four-step process for goal-directed revision that relies heavily on helicopter thinking.

Step 1: Refocus on your rhetorical goals. Because it's such an important point, I will make it again: At its best, content-level revision is essentially a task of evaluating how successfully a draft's information, ideas, and arguments achieve the rhetorical goals that the writer set in planning and aimed for in drafting. So a logical first step in the revision process is to refocus attention on your rhetorical goals. This is simply a matter of scanning your draft and reflecting on what you originally intended to do and say in its successive sections and paragraphs. In the draft's margins, jot down short phrases that identify the targeted goals. For example, in the margin beside the draft material in Figure 5.4, Derek might write something like "convince unresolved and debatable."

Step 2: Focus on the criteria for successfully accomplishing your rhetorical goals. This step engages the think-ahead technique that I introduced in Chapter 3. Here we apply the technique to identifying the criteria for success in achieving selected rhetorical goals. Instead of reacting passively to what you have written in a draft, set your sights on very definite qualities of excellence. For each goal that you aimed for, raise the powerful question, "What does my draft need to do and say to succeed?" If, in stages of planning, you developed sets of strategies for achieving your rhetorical goals, use them to help you answer this question. As an example, consider again the two strategies that Derek developed for his goal to convince readers about the unresolved and debatable nature of his research issue:

1. Show readers that the research issue is unresolved by summarizing the conflicting findings and conclusions from previous studies that support its different sides.
2. Show readers that the research issue is debatable by explaining the contrasting theories that support its different sides.

As demonstrated in the next step, Derek will use these strategies as criteria for evaluating his draft's content.

Step 3: Use your think-ahead criteria to identify strengths and weaknesses in content. This is the core evaluative step in our process of goal-directed revision. The task is to judge whether selected content actually accomplishes its rhetorical goal. If it is indeed successful, you will note your strong goal-directed writing and reinforce it as you revise other sections of your draft. If, however, the content misses the target on the goal at hand, you will proceed to diagnose the problem. The process is ideally guided by the proactive thinking described in Step 2. Listen, for example, as Derek applies his think-ahead criteria to evaluating the draft material in Figure 5.4:

Okay, I know that I need to show readers the different sides of my research issue to convince them that it's unresolved and debatable. So I need to summarize results *from studies showing that, on the one hand, exercise increases longevity. But I also need to summarize studies showing no effects. And maybe I should present studies revealing that exercise speeds up the aging process and reduces longevity. So let's see how I did: I definitely summarized a lot of findings for positive effects of exercise. But I don't talk much about how exercise might **not** increase longevity. This part of my introduction doesn't present all the different sides of the issue equally. I don't even talk about the contrasting theories of how exercise might affect longevity. So readers might not be convinced that it's really an unresolved and debatable issue.*

Derek's self-assessment reveals a key problem to solve.

Step 4: Identify and implement appropriate solutions. When draft content falls short of accomplishing its rhetorical goals, the best solutions depend on the specific causes of the diagnosed problem. Suppose, for example, that you have identified a weak argument in your draft. You diagnosed the problem's specific cause: Several lines of evidence and reasoning presented in support of your claim are not fully relevant. The diagnosis obviously reveals a fitting solution, which is to go back to activities for generating effective content. This might entail brainstorming and rereading articles to extract more relevant support. Of course, you will redraft the argument accordingly.

Given the specific flaw in Derek's draft, his best solutions are to delete the content that reflects a biased view and to generate new content that more objectively represents each side of the issue. The actual changes that Derek made are presented as follows (Figure 5.5). Derek's added content is underlined.

Many studies have ~~convincingly~~ shown that regular aerobic exercise significantly reduces the risk of premature death caused by the leading killers in our society, which include heart disease, cancer, and type 2 diabetes.[4, 7, 12, 15] For example, aerobic exercise has consistently been shown to lower cholesterol levels and blood pressure, to reduce body fat, and to improve insulin sensitivity. In addition, immune function is enhanced in individuals who perform aerobic exercise on a regular basis.[3,8,10] Based on these positive health outcomes, one might contend that aerobic exercise can slow the aging process and thereby extend the maximum lifespan potential. <u>However, this hypothesis is controversial, as witnessed by conflicting experimental results and theories on this topic.</u> Support for this hypothesis comes from a study conducted by Goldman,[5] who found that mice subjected to lifelong wheel-running

FIGURE 5.5 An effective revision of the problematic draft in Figure 5.4.

exercise lived beyond the maximum lifespan of normally active mice. <u>In contrast, Byrnes et al. found that wheel-running increased the average lifespan but not the maximum lifespan of mice.</u>[6] <u>Byrnes argued that Goldman's study was flawed by its failure to adequately control for the effects of diet.</u>

<u>In addition to conflicting findings from animal studies, there is debate about how aerobic exercise influences the biological processes of aging and longevity. Theoretically, exercise might slow aging through its effects on</u>~~A possible casual factor could be the positive effect that exercise has on~~ heat shock proteins, which are formed in response to physical exertion. These proteins play roles in maintaining cellular homeostasis and rebuilding cellular structures ~~Some have argued that Goldman's study was flawed. In addition, there is an accusation that~~ <u>On the other hand, some researchers believe that</u> exercise may accelerate the aging process by creating free radicals which damage cell membranes and mitochondrial DNA.[13] However, this <u>argument</u> has been countered by research ~~showing~~<u>suggesting</u> that exercise ~~actually~~<u>leads</u> to increased production of antioxidants, which neutralize the potentially harmful effects of free radicals.[6, 9] <u>Nevertheless, the debate persists because the overall effects of free radical accumulation versus antioxidant production in exercising humans have not been definitively determined.</u>

FIGURE 5.5 *continued.*

Let's focus on a few noteworthy improvements in Derek's revision. In the first sentence, Derek deleted the word "convincingly," which he originally used to describe the extent to which previous studies have supported arguments for positive effects of exercise on health and longevity. It's just one word, but the revision has a notable effect. By deleting the word "convincingly," Derek avoids giving readers the initial impression that his introduction is intended to argue for positive effects of exercise (when his actual goal is to present different sides of the argument). Next, in the middle of the first paragraph of Derek's revision, there is a pivotal new sentence: "However, this hypothesis is controversial, as witnessed by conflicting experimental results and theories on this topic." This sentence conveys a message that effectively grounds readers to Derek's rhetorical goal. Another highlight of the revision is that the content clearly presents a more balanced view on the issue. For example, Derek refers to a study by Byrnes et al. in which the results suggest that exercise does not increase the maximum lifespan. In the revision's second paragraph, Derek added content to better achieve his strategy for presenting the theoretical rationale that supports different sides of the issue. In addition, the last sentence of the revised version effectively captures the unresolved nature of the issue—it's another idea that successfully hits the bull's-eye on Derek's rhetorical goal.

We will close this section with a very important point: All flaws in draft content ultimately limit writers from accomplishing their rhetorical goals. This is true for the content-level problems that we have already discussed (missing content, ambiguous content, and inaccurate content). It's also true for all of the content-level problems that the rest of the chapter covers. I raise this point to emphasize the importance of taking a goal-direct approach to revising content. To be successful at this approach, you really have to know a lot about conventional rhetorical goals and strategies for scientific papers. We have touched briefly on these conventions so far in the book. I want to remind you, however, that Chapter 8 presents the conventions in great detail.

Content That Fails to Adequately Address Concerns of Audience

In a moment we are going to eavesdrop on a group of students in a scientific writing class. They are discussing a peer's draft, completed for an assignment to write a 10-page paper on any life science topic that is relevant to society. The professor gave explicit instructions to write the paper for a popular audience rather than for the scientific community. The students are talking about a draft authored by a classmate named Rosie, a senior majoring in biology, who wrote about the use of supplemental growth hormone as an antiaging therapy. Figure 5.6 is an excerpt from Rosie's introduction section.

In studies designed to elucidate the mechanisms of aging, life scientists have discovered very few therapies with the potential to increase the maximum life span, which is defined as the average age of the longest-lived decile of a given cohort. One of the most controversial anti-aging therapies is exogenous growth hormone. There is some evidence that growth hormone is capable of prolonging longevity. Age-related declines in structure and function have been linked to deficiencies in signaling mechanisms of the somatotropic axis and to imbalances in the GH/IGF-1 system.[1, 4, 7] Therefore, the exogenous administration of growth hormone has been hypothesized to confer anti-aging effects.[7] In contrast is the

FIGURE 5.6 Draft content that fails to account for concerns of the intended audience.

paradoxical finding that hypopituitary and GH-resistant mice often live longer than 4 years.[3,5] In addition, some researchers have suggested that high levels of growth hormone are associated with oxidative phosphorylation and, therefore, oxidative stress that would speed the aging process.[2,8] Clearly, the issue of whether growth hormone can extend longevity is debatable.

FIGURE 5.6 *continued.*

In response to Rosie's draft, a student named Reece says, "This is really good, Rosie. It shows that you know a lot about your topic, which is impressive. It just sounds very scientific—like a published article." A classmate named Brandon joins in with more praise. "What I like is how you're showing an interesting controversy about growth hormone," he adds. "That's a great way to hook your audience and make them want to keep reading the whole paper."

A student named Cecilia has a different view on Rosie's draft. "I agree that the draft shows that you know your topic very well, Rosie," Cecilia says. "And it's good that your introduction presents a controversy, which I know you're going to analyze and make arguments about in the body of your paper. But an important problem in this section of your introduction is that the content isn't geared for your audience. Remember, the assignment is to write for a popular audience. These are people who aren't specialists on our topics. This section of Rosie's draft goes too far over the heads of a popular audience. Her main readers would probably be middle-aged and older people who are interested in antiaging therapies but who do not know much about the scientific details and research on growth hormone. This audience won't understand terms like *deficiencies in signaling mechanisms of the somatotropic axis*, *imbalances in the GH/IGF-1 system*, and *oxidative phosphorylation*. If readers don't know what these terms mean, they won't understand the controversy that Rosie is trying to get across about her research issue. I'm not saying that you should dumb-down everything you've written. Instead, I suggest that you define the very technical terms as you present them. Or, maybe you could explain your research issue in more general terms in the introduction and save the more technical explanations for the body of the paper."

Cecilia's comments are especially insightful. You will appreciate why by re-reading Rosie's draft from the perspective of the intended audience. First, consider all of the undefined scientific terms and concepts that Rosie's nonscientific readers likely will not understand. By taking an audience-centered view, you will also diagnose the problem that key information and ideas, which Rosie's readers will need in order to understand her main message, are omitted. Take the following sentence: "In contrast is the paradoxical finding that hypopituitary and GH-resistant mice often live longer than 4 years." Rosie intends for this sentence to convey evidence that supports one side of her research issue. Specifically, it is the view that supplemental growth hormone does *not* extend longevity. The evidence comes from studies on mice that are genetically altered to produce low amounts of growth hormone. These are animals that Rosie refers to as "hypopituitary and GH-resistant mice." As she states, these mice can live for more than 4 years. To understand how this evidence supports one side of the research issue, readers must already know that normal laboratory mice live for only 2 or 3 years. This background knowledge is necessary to comprehend Rosie's intended point, which is that some mice survive beyond the normal life expectancy without having high levels of growth hormone. It's safe to assume, however, that Rosie's readers do not already possess the missing link of knowledge.

This aspect of revision—evaluating whether draft content successfully addresses concerns of intended readers—warrants special attention and concentrated passes through drafts in which writers apply audience analysis. The challenges are to put yourself in your readers' shoes, assume their characteristics, and reflect on how they will respond to what you have written. Ask yourself questions like the following.

- Have I provided definitions of terms and explanations of concepts that my readers likely do not know?
- Have I *avoided* defining terms and explaining concepts that my readers know very well?
- Does the content omit information and ideas that my readers will need for comprehension?
- Does the content account for the values, preconceptions, and biases that readers might hold about my research issue?
- In writing about published research papers that my audience likely has not read, have I sufficiently explained details about the motivating issues, methods, and results?
- In making arguments, have I acknowledged the positions that my readers are likely to take?
- For school projects, does my draft's content meet the standards for excellence that my professor has set?

If you have the opportunity to run your draft by members of your intended audience, take it. What better way is there to figure out whether you are effectively accounting for readers' needs, expectations, and values? If members of your specific audience are not accessible, seek reviewers who can take their perspectives and provide feedback accordingly.

Saying Too Little or Too Much

The two problems that we will consider here involve a quality of writing called *development*. In general terms, we can say that a paper's content is appropriately developed when its ideas are sufficiently explained, its concepts are adequately illustrated, and its arguments are supported with a critical mass of detailed evidence and reasoning. Revising for matters of development thus entails evaluating drafts for two potential problems: underdeveloped content (saying too little) and overdeveloped content (saying too much). The evaluations demand problem solving because there are no completely objective tools for measuring how much content is, as Goldilocks would say, *just right*. In the methods section of a lab report, for example, how would you know for certain whether you have presented an adequate amount of detail in describing your experimental procedures and analyses? In an explanation of a complex theory, how can you be sure that you have elaborated sufficiently on its conceptual foundations? In assessing the strength of your arguments, how can you tell whether claims are backed with sufficient evidence and reasoning? While the successful development of draft content cannot be measured on precise scales, the revision process for this quality can be effectively guided by task and audience analyses. It's worth double-checking parts of your assignment that indicate the relative importance of selected subtopics or rhetorical goals. In addition, you should reflect again on content that your intended audience might need or expect you to develop extensively. This revision process can be quite manageable for writers who, prior to drafting, make key decisions about content to emphasize versus de-emphasize. Recall that guidelines for making these decisions were presented in Chapter 4 (see pages 170 to 171) as part of the process of organizing content.

Logical Fallacies in Scientific Arguments

In Chapter 3, after hailing argument as the pinnacle of scientific thinking and communication, I presented a systematic approach to identifying strengths and weaknesses of arguments in published scientific literature. The approach entails answering the following diagnostic questions.

1. To what degree are the lines of support relevant to the argument's claim?
2. If the relevance of an argument's lines of support might not be obvious to readers, how successfully has the author developed necessary warrants?
3. How convincing are the data-driven lines of support for the claim?
4. How convincing are the concept-driven lines of support for the claim?
5. How successfully does the author acknowledge and refute viable counter-arguments?

Chapter 3 demonstrated how to apply these questions for evaluating scientific arguments to *generating* draft content, specifically to develop our own arguments by critiquing arguments that other authors have raised about our research issues. The very same diagnostic questions and procedures can be applied to *revising* our own arguments in drafts. I thus encourage you to add the diagnostic questions and procedures that we covered in Chapter 3 to your independent revision toolbox, so to speak.

Here we add a new set of tools for revising scientific arguments, specifically those weakened by logical fallacies. These are deceptive and faulty lines of evidence and reasoning. If you have studied formal logic, you know that logical fallacies go by catchy names such as *red herring, straw man, false cause,* and *appeal to authority*. In some fields of communication, writers and speakers deliberately exploit these misleading tactics to win arguments. An example is a negative political campaign in which candidates take their opponents' quotations completely out of context, create smoke screens with irrelevant statistics, and attack adversaries on petty personal matters rather than on their political views and records. In advertising, a common exploitation of a logical fallacy is to sell a product by appealing to consumers' emotions rather than by demonstrating its actual merits. Through education and experience in science and scientific communication, you will quickly learn to avoid constructing arguments with blatant logical fallacies such as personal attacks and appeals to emotion. But a number of other logical fallacies, presented as follows, have subtle qualities that may enable them to creep into first drafts unnoticed.

Circular Reasoning

This is a special case of underdeveloped content in an argument. It's called circular reasoning because the argument's support keeps revolving back to its claim. In other words, instead of presenting convincing data and reasoning, the author restates his or her claim in different forms. A simple example of circular reasoning is presented in the following passage.

> **(1)** Proteolytic enzyme therapy is an efficacious method for preventing pancreatic cancer as well as for slowing its progression in those who have developed the disease. **(2)** The treatment works by inhibiting tumorigenesis, the events that cause normal cells to transform into cancer cells. **(3)** Furthermore, proteolytic enzyme therapy attenuates the proliferation of existing malignant cells. **(4)** Through these mechanisms, cancer development can be thwarted. **(5)** For individuals who have already developed tumors, proteolytic enzyme therapy can play a valuable healing role.

The first sentence presents the author's claim for the positive treatment and preventive effects of proteolytic enzyme therapy on pancreatic cancer. The argument begins well enough. But notice the circular reasoning as each sentence following the claim simply restates it in slightly different words. Take sentence 2, which conveys the message that proteolytic enzyme therapy inhibits tumorigenesis, or the development of cancer cells. This is just another way of claiming that the treatment prevents pancreatic cancer. Then, sentence 3 contains the phrase "attenuates the proliferation of existing malignant cells." Stated simply, this refers to inhibiting the growth of cancer cells. Of course, this message circles back to the author's original claim that proteolytic enzyme therapy is an effective method for slowing the progression of pancreatic cancer. Sentences 4 and 5 also spin back to the claim. (By the way, this logical fallacy is also called *begging the question*, which refers to the evasive nature of arguments with this flaw.)

To independently detect circular reasoning, you have to evaluate the sufficiency of support for each claim in your draft's arguments. The task involves testing each supporting idea for whether it is based on convincing evidence and reasoning or whether it simply restates the claim. The problem's obvious solution is to eliminate instances of needless repetition and to add convincing support. Take the the sample argument for proteolytic enzyme therapy. To back up the claim that it *prevents* pancreatic cancer, what sort of supporting data would be convincing? If you were revising this argument, you would certainly need to present data from studies in which healthy subjects who received proteolytic enzyme therapy had, over time, a significantly lower risk of developing pancreatic cancer than control subjects who did not receive the treatment. To argue for the claim that proteolytic enzyme therapy effectively *treats* cancer, you would need to present data from studies on patients who had already developed the disease. The data must show that these subjects experienced statistically and practically significant reductions in tumor size and other markers of cancer progression.

Red Herring

As the story goes, this logical fallacy is named for a prank played on fox hunters centuries ago in the United Kingdom. To fool the hunters, saboteurs would sneak into the fields before organized hunting events, dragging smelly smoked herring in all directions. Distracted by the strong fishy odor, the hunters' dogs would veer off course haphazardly. The foxes, it turns out, were spared. Like the sidetracked hunting dogs, scientific audiences can lose their trail trying to follow arguments that are flawed by the logical fallacy called red herring. Here the term refers to lines of evidence and reasoning that are irrelevant to the argument's claims. As an example, consider an author's claim that therapeutic applications of embryonic stem cells are not potentially effective methods for extending human life. In one line of support for the claim, the author contends that embryonic stem cell therapy is unethical. Any doctor prescribing the therapy for the purpose of life extension, says the author, is immoral and should not be allowed to practice medicine. This reasoning, in the context of the specific claim, is red herring. The ethical implications might be relevant as support in a political argument about whether embryonic stem cell therapy should be used to prolong life beyond its normal limits. However, the reasoning based on ethics is not at all relevant to the author's specific claim that embryonic stem cell therapy is not effective for extending human life. Relevant support for this argument must come from scientific research and reasoning that discount the viability of embryonic stem cell therapy in combating disease and slowing the physiological aging process.

In Chapter 3, we worked through steps for detecting and diagnosing irrelevant evidence and reasoning in published arguments. The same procedures apply to revising for red herring in our own drafts. For each claim in every argument that you are making, a productive approach is to put your draft aside and raise the following question: What are the most logically relevant lines of support? Your answers will form a list of think-ahead criteria that you can check against the evidence and reasoning that your draft actually presents. If you catch a blatant red herring, the obvious solution is to delete it. More often, however, you will come across support that you believe is relevant but that you suspect readers might view as slightly fishy. In this case, the solution is to develop the argument's warrants—that is, to explain how the support is truly relevant to your claim.

Hasty Generalizations

In science, *to generalize* means to reach a conclusion that extends, at least a small degree, beyond the boundaries of its supporting evidence. The most common form of scientific generalization, as discussed in Chapter 3, is making inferences about an entire population based on data derived from a sample (a subset of the population). Generalizations come with the territory of the scientific method and its reliance on inductive reasoning to support arguments. However, scientific arguments are weakened when authors commit the logical fallacy of *hasty generalizations*. These are quickly reached conclusions that extend *too far* beyond the logical boundaries of an argument's supporting evidence. Take the claim that a newly discovered hormone therapy "causes significant increases in muscle mass and strength in older adults." Suppose that the author supports this claim with data from studies on older males only. This argument raises the red flag of a possible hasty generalization because the support comes from a restricted sample, one of males only. The claim infers that the new hormone treatment would benefit the entire population of "older adults," which obviously includes females. At this point in our analysis, we can say only that the argument reflects a *possible* hasty generalization. If, however, sound scientific knowledge exists to support the view that females would respond *differently* to the treatment, the argument is definitely flawed by the logical fallacy. Perhaps the treatment would interact with sex-specific hormones in women to actually inhibit increases in muscle mass and strength. This example reflects a case of hasty generalization across the sexes. Other common cases are presented in Figure 5.7.

Revisions of hasty generalizations begin with evaluation of claims, focusing on whether they overstep the boundaries of supporting evidence and the methods from which the evidence was derived. Some hasty generalizations are misinterpretations of study data; an example is concluding that data are practically significant when rational support is simply not evident. Most cases of this logical fallacy, however, stem from authors' oversights of methodological shortcomings in the research that they use to support their arguments. Common oversights include failing to acknowledge the following: study samples that do not adequately represent populations of interest, biases in assigning subjects to experimental treatments, and a lack of precision in instruments for collecting data. The revision process thus entails double-checking the procedures and analyses of studies from which you have obtained data to support your claims. Chapter 3 presented a set of diagnostic questions for evaluating the methods of published studies in order to generate draft content (Figure 3.11, page 146). The same questions apply to revising our own drafts to determine whether the supporting evidence for our arguments comes from studies with strong methods. The questions thus apply to assessing whether our generalizations are hasty

Case 1: Generalizing across subject characteristics such as sex, age, physical conditions, and states of health and disease

Example: Using data from research on relatively lean individuals to support the claim that a popular diet promotes health and longevity in obese individuals.

Explanation: Metabolic and hormonal processes differ in lean versus obese individuals. These differences might interact with dietary manipulations to cause different effects on health and longevity across the two populations.

Case 2: Generalizing across species

Example: Using data from research on simple organisms such as nematodes (worms) to support the claim that resveratrol, an antioxidant found in plants including red grapes, can increase longevity in humans.

Explanation: Genetic and physiological differences exist between humans and simple organisms that are bred specifically for research. Chemicals such as resveratrol might thus have different influences on aging and longevity across species.

Case 3: Generalizing from part to whole—in other words, reaching conclusions about how an entire system functions based on the functioning of its individual elements

Example: Using data collected from single myocardial (heart) cells to reach conclusions about how the whole heart works.

Explanation: Cells isolated from their natural, supporting environments may behave differently from *in situ* cells.

Case 4: Generalizing across time periods

Example: Using data collected from a 2-month study to support the claim that a form of gene therapy can ultimately cure liver cancer—the data indicated a statistically significant reduction in tumor size.

Explanation: Linear extrapolations of results are problematic for reaching conclusions about nonlinear systems, or systems that do not change consistently over time. In this case, tumors of the liver might grow again.

Case 5: Generalizing across treatment doses and experimental conditions

Example: Using data from subjects' responses to only one dose of a drug to support the claim that the drug is generally effective and safe.

Explanation: Responses to the drug might vary across different doses. Sound conclusions must therefore address "dose-response" effects.

FIGURE 5.7 Cases of hasty generalizations in scientific arguments.

or sound. The following strategies can be used to resolve the problem of hasty generalizations.

1. **Adapt your argument's claim to bring it within logical boundaries.** Imagine that your draft contains a hasty generalization across the sexes. Your claim covers the entire population of men and women, but your supporting data come from studies conducted only on men. A simple and effective solution is to rewrite the claim so that it applies to men only.

2. **Search again for studies with stronger methods.** Let's say that your draft contains a hasty generalization that stems from weak methods in studies that you have presented to support your argument. Maybe your original search for literature overlooked studies with stronger methods. So another thorough literature search, perhaps using a wider range of databases, may turn up more convincing evidence. You might even find that since you conducted your original search, a brand new supporting study with flawless methods was published.

3. **Develop necessary warrants to convince readers that your generalization is not so hasty after all.** This strategy applies when readers might cry hasty generalization in response to your argument, but you strongly believe that it is valid. As an example, let's say that you are writing a position paper to argue that calorie restriction can increase the maximum lifespan in humans. Due to ethical and practical concerns, no long-term experiments on this issue have been conducted on humans. So, to support your argument, you have used data from studies on animals. One part of your draft presents data from studies in which mice subjected to calorie restriction outlived control mice that ate freely. You diagnose a hasty generalization across species: You are using data from studies on mice to support a conclusion about positive effects of calorie restriction in humans. Suppose that, based on your advanced knowledge of this research field, you truly believe that the generalization is valid. As your draft argument stands, however, skeptical readers will indeed cry hasty generalization.

To solve this problem, you will need to develop warrants that convince readers that data from studies on mice are indeed relevant. An essential warrant must compare the physiological processes of aging in rodents to those in humans. You will have to present sound evidence and reasoning to argue that (a) the mechanisms of aging are similar in the two species and (b) calorie restriction would likely influence the mechanisms by slowing aging at a comparable relative rate in mice and humans. You may also need to respond convincingly to the objections of scientists who have argued that the rodent model of calorie restriction and longevity is not generalizable to humans. This strategy—developing necessary warrants to explain *how* study data logically support a claim or conclusion—is fundamental to many revisions of weak scientific arguments.

Subjectivism and Personal Experience

When a writer has strong personal beliefs about an issue that lends itself to argument, he may be inclined to support a claim with the reasoning "because I believe that it's so" or "because my experiences support it." Our personal beliefs and experiences can be truly invaluable in successfully guiding many important decisions in daily life. In science, however, we must distinguish between knowledge gained and decisions made through experience versus experiments. Some people strongly believe, for instance, that vitamin supplementation promotes optimal health and prolongs life. They may support their claims with personal observations of vitality and long life in elderly family members who take vitamins. It's quite possible, however, that vitamin supplementation actually has nothing to do with their relatives' successful aging. Maybe their relatives just have good genes. Without convincing data obtained from well-controlled experiments on vitamin supplementation and longevity, we cannot make compelling scientific arguments on this issue.

You may already know very well that subjectivism and personal experience are logical fallacies in scientific arguments. If so, you naturally would not rely on this sort of reasoning to support your claims. But these logical fallacies can still have subtle influences on how we construct arguments. For issues that are near and dear to our hearts, we may sometimes assume that our audiences will feel the same way and, therefore, will automatically accept our claims. The consequence is that the evidence and reasoning we present for our claims may be underdeveloped and one sided. To detect this problem in your drafts, look for claims that you would feel strongly about even if no scientific evidence were available to support them. Then, take the perspective of the most skeptical members of your audience by envisioning readers who do not naturally share your strong beliefs. Will they criticize your argument for a lack of scientific merit? Will they be convinced by your supporting evidence and reasoning? Have you acknowledged and successfully refuted their preconceptions as well as the scientific support for their beliefs? Thoughtful answers to these questions will guide you to eliminate arguments based on subjectivism and personal experience.

False Cause

As the name suggests, this logical fallacy underlies flawed claims about cause–effect relationships. This is a case in which an author falsely concludes that one variable directly *causes* another, when the variables are related only by association. Consider the variables of exercise participation and health in older adults. In typical studies on this issue, researchers ask subjects to report details about their exercise habits and health conditions. Correlational analyses in these studies usually indicate that subjects who report higher levels of physical activity are healthier and live longer than sedentary subjects. Suppose that a writer used these study results to support the claim that exercise in old age *causes* good health and longevity. Given the circumstances, this argument is compromised by the logical fallacy of false cause. The reason involves the observational design of the studies from which the writer obtained supporting results. As explained in Chapter 3, observational studies cannot reveal whether cause–effect relationships exist between variables; they reveal associations only. In observational studies on exercise participation and health in old age, there is no way of telling whether exercise causes good health or whether the opposite relationship is true. Maybe good health in old age enables people to perform regular exercise. Elderly people who exercise regularly might enjoy good health as a consequence of other lifestyle behaviors, such as eating nutritious foods and participating in social activities.

To revise your arguments for false cause, begin by noting any claims and conclusions that directly indicate or infer cause–effect relationships between variables. Then, double-check your backing evidence and reasoning. Is the support based on associations derived from studies with observational designs? If so, you might need to rewrite your claim as demonstrated in the following revision.

> Problem: The results prove that regular exercise participation increases longevity.
>
> Solution: The results suggest that an association exists between regular exercise participation and longevity.

The conditional language ("The results *suggest* an association") is essential to avoid misleading readers about "proof" for a cause–effect relationship that cannot be confirmed through observational study data. Another strategy for solving the problem of false cause is to seek stronger support for claims of cause–effect relationships. This entails searching for published studies that were specifically intended to identify causal factors. As we discussed in Chapter 3, these are studies with well-controlled experimental designs.

False Dichotomy

On the whole, the scientific method teaches us to be vigilant in avoiding flawed reasoning in arguments. There is, however, a unique logical fallacy that can surface in arguments as a consequence of a traditional approach to science. The approach, called *reductionism*, entails divvying up a large, complex system into relatively small parts. Each part is then studied intensely. Reductionism enables scientists to understand how the workings of individual elements contribute to the integrated function of complex systems. Given the great complexity of living systems, reductionism can indeed be a fruitful approach in the life sciences. But it also has its shortcomings. Misleading information and ideas can be advanced, for example, when different teams of scientists focus too narrowly on their separate, favorite parts of a common complex system. Each team can get so entrenched in its reductionist camp that the scientists lose sight of the big picture. They thus fail to account for how their part serves the system's overall function. Scientists who operate strictly by reductionism may also develop completely exclusive theories about how a whole system works, disregarding alternatives developed in other reductionist camps.

A common consequence of reductionism in scientific argument is the logical fallacy called *false dichotomy*, which also goes by the names of *black-or-white reasoning* and *either-or reasoning*. As a hypothetical example, consider another argument about the causes of aging. Suppose that a research team has devoted many years to studying a free radical called the "black" molecule. Another team of scientists in a lab halfway around the world has focused narrowly on an antioxidant, which is a molecule that is capable of neutralizing the harmful effects of free radicals. They call the antioxidant the "white" molecule. The first team argues that aging is caused solely by an accumulation of black molecules. In sharp contrast, the second team argues that aging is caused solely by a reduction in the number of white molecules. Each team has strong evidence to support its claim, so each vehemently accuses the other of being wrong. But there is a logical fallacy at hand: The scientists have created a black-or-white argument when both sides might actually be right. In this example, aging may in fact be caused by mutual influences—specifically, the accumulation of black molecules *along with* reduced numbers of white molecules.

To detect instances of false dichotomy in your own arguments, look for places in drafts where you are reasoning that your view and an opposing view *both* cannot be correct. Then, reflect on whether the research issue is truly defined by black-or-white logic. Consider the current state of knowledge about the research issue. Is it possible that your view as well as the opposing view have elements of truth? Might valid support exist for both sides? Through rethinking the candidate arguments, you might very well change your mind, eventually accepting and arguing for a more inclusive view. It might help to know that revisions of arguments flawed by false dichotomy reflect the wisdom of veteran scientists. They know that the truth in science, which aims to mirror the truth in nature, is often painted in shades of gray rather than black and white. The best answers to complex scientific questions are often multisided and dependent on the circumstances. So you might find yourself revising an argument by changing your claim from, "It has to be black because it cannot possibly be both black and white" to "It depends on the circumstances: Sometimes it's black, sometimes it's white, and sometimes it can be black and white simultaneously."

Appeal to Ignorance

Before the development of technologies that enabled scientists to identify individual genes and their functions, one might have argued against the existence of genes that regulate aging. "There cannot possibly be any aging genes," the claim would go, "because no material evidence of them has ever been found." This is an example of a logical fallacy called *appeal to ignorance*. The arguer reasons that a claim is not viable because it lacks direct supporting evidence, such as hard, cold data derived from research. But arguments based only on appeals to ignorance are problematic because they ignore indirect, conceptual evidence that may reasonably support claims. In an argument for genes that regulate aging, for instance, several indirect lines of support should not be ignored. One is that people in long-lived societies share similar genetic characteristics. Moreover, longevity tends to run in families. Appeals to ignorance may crop up in scientific arguments when authors overlook inherent limitations in developing research fields, such immature technologies for collecting and analyzing data. The lack of precise instrumentation for discovering and describing phenomena in nature does not, in itself, rule out the existence of the phenomena.

To check your drafts for appeals to ignorance, locate content inspired by the rhetorical goal to acknowledge and refute arguments that oppose yours. Detecting the flaw is a simple matter of noting claims that a counterargument is not valid simply because it lacks direct evidence. If you find such claims, think about whether there might actually be viable indirect, conceptual support for the counterargument. This reflection might entail reevaluating the evidence and reasoning that has been advanced by proponents of arguments that oppose yours. Suppose that, after opening your mind in this way, you determine that the counterargument is indeed valid. Then you will have some tough choices to make in rewriting your draft. The revision might require a major overhaul of your argument, perhaps even a 180 degree turnaround of your claim and support. (If you happen to completely reverse your position in an argument, consider yourself a card-carrying member of the scientific community. Veteran scientists occasionally change their views after reassessing the support for opposing arguments.) In a different scenario, however, you might acknowledge the indirect support for a counterargument but contend that it is problematic. In this case your revision will involve refuting the opposing conceptual evidence and reasoning.

Appeal to Authority

Committing this logical fallacy, the arguer reasons that a claim is valid because a so-called expert backs it. Take a hypothetical argument for the life-extending effects of a popular diet. An appeal to authority is demonstrated as follows.

> The Never-Die Diet, which was discovered by Dr. D.O. Dazzleus, has a significant influence in slowing the aging process and enhancing quality of life in old age. Dr. Dazzleus, a world's leading expert on nutrition and aging, advises all his patients to go on the revolutionary Never-Die Diet. Through extensive clinical observations, Dr. Dazzleus has witnessed profound anti-aging effects in his patients who have maintained the diet.

In nonscientific circles, this sort of reasoning would immediately lead skeptics to justly demand proof that the "expert" is indeed a credible authority. Good critical thinkers in a general audience might be inclined to accept the argument if Dr. Dazzleus is endorsed by a reputable medical organization, if he has written acclaimed articles and books on nutrition and aging, and if he has a sound reputation in the public eye. In the scientific community, however, the standards are considerably stricter. Well-trained scientists generally do not accept claims supported by appeals to authority, even if the authority figure has a stellar reputation in the research field. The reason is that appeals to authority are irrelevant and insufficient lines of support for scientific arguments. For example, the hypothetical argument for the antiaging effects of the Never-Die Diet is seriously flawed because it completely omits supporting evidence from peer-reviewed studies and a sound rationale for how the diet could possibly slow the aging process.

Bandwagon

In this logical fallacy, arguments are supported by the claim that something should be done or accepted because *everybody* does it and accepts it. Here's an example:

> The herb *Gotu Kola* promotes well-being in old age and can extend the lifespan. This is known because, for more than 5,000 years, many millions of people in China have taken this herb with successful results.

Perhaps *Gotu Kola* actually does promote good health and longevity. However, the claim for these positive outcomes cannot logically be supported by how many people have taken the herb, even if its popularity spans many millennia. As is true for appeals to authority, bandwagon should not be used to support claims in scientific arguments.

Straw Man

It's easy to build a man made of straw and then knock him over. This is the analogy behind the logical fallacy called *straw man*. The author raises only the flimsiest aspects of an opposing argument and then proceeds with the relatively simple task of dismissing them. Take an argument to support the claim that hormone replacement therapy (HRT) is an effective and safe treatment for reversing bone loss in postmenopausal women. In reality, one of the strongest lines of support for the opposing view is that HRT increases risks for cancer and heart disease. But rather than acknowledging and responding to this formidable premise, the author builds a straw man, as follows.

> In arguments against the efficacy and safety of HRT, my opponents are quick to contend that physicians administering the treatment have great difficulty determining the appropriate dose for each patient. Thus, my opponents argue, women are highly inconvenienced by having to visit their doctors frequently to adjust the dose of HRT. This argument is not convincing, however, because proper adjustments are typically made within two or three office visits, which should not pose great inconveniences. Moreover, any inconvenience associated with adjusting the dose of HRT pales in comparison to the debilitating symptoms of menopause and the risk of losing bone mass through the horrifying disease, osteoporosis.

The argument is problematic because it misrepresents the major premises of the opposing argument—that is, the argument against using HRT. In fact, those who oppose the author's argument will not emphasize the inconvenience associated with adjusting the dose of HRT. Instead, they will argue that HRT is not safe because it increases risks for cancer and heart disease. Of course, the author conveniently overlooks this argument.

To check for the straw man fallacy in your drafts, focus on content in which your rhetorical goal is to acknowledge and refute arguments that run counter to yours. Then, contemplate the strongest lines of support for the counterarguments.

What evidence and reasoning will your opponents raise to back their claims? These are the lines of support that your paper should acknowledge and refute.

REVISING GRAPHICS

In Chapter 4, I explained that graphics (tables as well as figures of various types) serve the same overall functions that text serves in scientific papers. Like text, graphics are vehicles for accomplishing rhetorical goals. It follows that the same procedures we have covered for revising text-based content generally apply to revising graphics. For example, graphics in drafts should be evaluated for whether they omit important content, include accurate information, address concerns of readers, and accomplish their targeted rhetorical goals. The process of revising graphics should be guided by the key principles for graphical design that we covered in Chapter 4 (see pages 195 to 199). As an example, one of the principles is to exclude embellishments that do not directly convey essential information. Examples of embellishments include shaded backgrounds, heavy gridlines, garish colors, and unnecessary 3-D effects. In the revision process, key design principles for graphics can be converted into very useful diagnostic questions, like the following: Do my graphs contain unnecessary embellishments that distract readers from grasping the most important information? Guided by a list of diagnostic questions that you can easily derive from the design principles covered in Chapter 4, you may find areas for improvement in your draft's tables and figures.

EXCELLING AT COLLEGIAL PEER REVIEW

As emphasized throughout this chapter, many writing problems are ideally solved through a combination of independent revision and feedback from reviewers. As an author, perhaps you have received peers' insightful critiques and helpful suggestions for improving your early drafts. If so, you know first-hand the very positive influence that effective peer review has on the quality of writing projects. As is true for gifts on birthdays and holidays, however, the act of *giving* feedback can be even more rewarding than receiving it. Through developing advanced skills for peer review, you will be highly valued and sought after by coworkers for your abilities. In addition, you will learn many important lessons that will ultimately enhance the quality of your own writing. This last section of the chapter presents guidelines for excelling at peer review. We will focus on *collegial* review, which differs from the competitive review process that scientific papers undergo in grant funding and publication endeavors. Collegial peer review is feedback from friendly coworkers or classmates who are not charged with making final decisions about grades, publication, and funding. Instead, collegial peer reviewers offer constructive critiques and suggestions on drafts to help authors produce final papers that shine in the competitive review process.

Apply Key Methods of Independent Revision to Guide Your Peer Review

The same principles and practices for revising independently apply in peer review. For example, peer review should engage problem solving, critical thinking, and the organized method of "revision by division," which I discussed earlier in the chapter. Reviewing a peer's draft for the quality of its content, you should focus on the same features that we have covered for revising content independently. Applying the diagnostic procedures presented earlier, you will evaluate the draft for accuracy, clarity, development, strength of argumentation, success in accomplishing rhetorical goals, and so on. In addition, the same routines for revising paragraphs and sentences independently—routines that you will learn in the following two chapters—also guide the process of reviewing peers' drafts for matters at these two levels of discourse.

Take a Goal-Directed Approach to Generating Feedback

Both the scientific writing and peer review processes share the fundamental challenge of generating content, or figuring out what to say. Of course, peer review demands figuring out what to say about someone else's draft. This is the core task no matter whether you are giving the author feedback in a casual conversation over coffee, in scribbles in the draft's margin, or in a well-developed document or e-mail message. How should you go about generating the content of your peer reviews? Why not take a goal-directed approach? My point is that the feedback you generate about a peer's draft should accomplish key rhetorical goals. It makes good sense because, like a scientific paper, a peer review is a form of communication intended to influence the knowledge, beliefs, and actions of audiences. As a collegial peer reviewer, your overall communication goal is to help your audience, the draft's author, write a successful final paper. As follows, I present three general rhetorical goals to guide peer reviewers in generating, organizing, and delivering their feedback.

Rhetorical Goal 1: Highlight the strengths of the author's draft. It's especially important to comment on effective features of a peer's draft for two main reasons. First, positive comments will help the writer reinforce strengths through independent revision. Second, positive comments soften the blow of feedback that focuses on weaknesses. As a case in point, let's recall some of the feedback that a student named Cecilia gave to her classmate, Rosie, about a paper on supplemental growth hormone as an antiaging therapy. (Cecilia was commenting on draft material presented in Figure 5.6 on pages 234–235.) Here's how Cecilia began her comments:

"I agree that the draft shows that you know your topic very well, Rosie. And it's good that your introduction presents a controversy, which I know you're going to analyze and make arguments about in the body of your paper. But an important problem in this section of your introduction is that the content isn't geared for your audience. Remember, the assignment is to write for a popular audience...."

Notice that the first two sentences of Cecilia's review focus on positive aspects of Rosie's draft. These comments set up the crux of the review, in which Cecilia explains how the content does not adequately address concerns of Rosie's intended audience.

Rhetorical Goal 2: Directly point out specific weaknesses in the author's draft and, if necessary, explain them in sufficient depth. This goal generates the constructive criticism that helps writers revise successfully. The first part of the goal calls for conveying each problem in a very direct statement. An effective example is the third sentence of the preceding excerpt of Cecilia's review. A concise and specific statement about a problem may be all the writer needs to work out successful solutions. Sometimes, however, elaborate explanations are necessary. As a writer, how would you respond to the following comments about one of your drafts?

- I just don't get your point here.
- This argument isn't convincing at all.
- These ideas go way off-tangent.
- This paragraph just doesn't flow.

Of course, you would want to know more.

- What *exactly* do you think is unclear about my point?
- Why *precisely* are you saying that this argument isn't convincing?
- Which ideas are going off-tangent, and why do you think so?
- What specific ideas in the paragraph aren't flowing well, and why?

To explain problems that are not obvious, peer reviewers must carry out diagnostic routines. If, for example, you detect a weak argument in a peer's draft, the challenge is to explain the specific underlying causes. Is the argument weak because the claim lacks a critical mass of supporting data? Has the author failed to develop necessary warrants? Or, do particular logical fallacies compromise the author's reasoning? As another example, suppose that part of a peer's draft elicits the vague sense of an underdeveloped idea. Your first impression is that the content is just plain skimpy. You will help the writer most by digging deep in your evaluation. *Why* does the content seem underdeveloped? What essential information, illustrations, and elaborations are missing? Well-developed explanations of problems will make all the difference in guiding the writer toward an effective revision.

Rhetorical Goal 3: Offer specific and detailed suggestions for solving problems. As is true for independent revision, some problems identified through peer review lead to straightforward solutions almost spontaneously. For more challenging problems, however, solutions will not be so obvious. Or, in some cases, a number of candidate solutions may be viable. In offering suggestions for solving difficult problems, peer reviewers should address all viable options and justify them with specific details. Go beyond giving vague and underdeveloped feedback, as in the following problematic comments.

- I think that you need to strengthen your argument.
- You should develop your ideas more.

- Give stronger examples to clarify concepts.
- Make your paragraphs flow better.

To elaborate on vague solutions, you may need to provide specific information and ideas for strengthening arguments, to suggest detailed examples for clarifying concepts, or to talk authors though effective patterns of idea development.

Make It *Constructive* Criticism

For career scientists with aims on publication and grant awards, the peer review process is extremely competitive and often quite harsh. Publication and funding rates are typically below 10% for high-quality journals and grant agencies. Peer reviewers, in their role as gatekeepers charged with ensuring quality control, are by no means obliged to offer collegial and constructive criticism. When they find major faults in the scientific merit and writing quality of submitted documents, peer reviewers often do not hold back their negative criticism. Sometimes their feedback is downright brutal.

In contrast, the aim in collegial peer review is to offer constructive criticism. It can be a formidable challenge when you are reviewing a draft with many major flaws. In this situation, one can easily slip into the role of a harsh judge, pointing out only what is wrong without helping the author understand and resolve the weaknesses. This is not to suggest that you should withhold comments about major problems in peers' drafts. In addition, you certainly do not need to sugarcoat your feedback. To offer honest constructive criticism, however, you will need to stay focused on the ultimate goal of helping authors improve the quality of their papers so that they succeed in upcoming competitive reviews. With this goal in mind, you will naturally generate helpful feedback concerning problems. Moreover, you will deliver the feedback in a positive manner. As I recommended in the previous guideline, it's a good idea to preface comments about problems by highlighting strong points. If you are giving feedback orally, be aware of the tone of your voice. Even a little nervous laughter, when you do not have any criticism in mind, can come across to an author as negative and patronizing. Another key to offering constructive criticism is to remind the author about the helpful intentions that underlie your feedback.

Take on the Role of Writing Teacher

Sometimes the best approach to supporting authors in peer review is to help them detect, diagnose, and solve problems independently. In this approach, the peer reviewer plays the role of a writing teacher. Suppose, for example, that in a classmate's draft you have identified the logical fallacy of red herring, which is irrelevant support for a claim. You could, of course, tell the author straightaway that the argument is weak because its evidence and reasoning are beside the point. You could also simply give the author your ideas for what constitutes relevant support. Consider, however, the option of guiding the author through an independent revision of the logical fallacy. Instead of conveying the problem directly, ask the author to set his draft aside and brainstorm the most relevant lines of support for the claim. Then, instruct the author to compare his think-ahead criteria to the actual supporting evidence and reasoning in the draft. Help the author see for himself where the argument goes off track. By guiding the author through the process of resolving the logical fallacy independently, you will help him improve the current draft as well as avoid the problem in future papers.

Encourage Dialogue

At its best, peer review should unfold as a rich dialogue between reviewer and author. After providing feedback, the reviewer can encourage dialogue by prompting the author to respond to questions like these: Are my comments and suggestions clear? What *don't* you understand about any ambiguous feedback? How might you actually apply my feedback to your revision? The resulting dialogue is essential for ensuring that your conscientious efforts as a peer reviewer will actually pay off for the author. In sustaining the dialogue, the author also has certain responsibilities. When the feedback is crystal clear and immediately useful, it is more than a nice gesture for the author to let the peer reviewer know. By learning what helps the author most, the reviewer can adapt future critiques and suggestions for revision accordingly. Authors should make a special effort to raise points about feedback that is not clear and that they do not agree with. In this scenario, authors may be inclined to keep quiet in order to be polite and avoid conflict. But these responses limit the potential for peer review to help. Conscientious peer reviewers, those who genuinely want their feedback to be understood and useful, do not want a passive response from authors.

Avoid Giving Feedback Based on Personal Preferences and Pet Peeves

Skillful peer reviewers justify their feedback with sound principles of communication rather than personal preferences and pet peeves. Imagine that you are reviewing a draft in which the author did not define key terms in his introduction section. You tell the author, "I suggest that you define and distinguish *longevity, survival,* and *maximum lifespan potential.* These terms have specialized meanings in research on aging and longevity, which your readers—who are clinicians rather than scientists—might not understand." This is sound advice because it is based on the important principle to consider the needs and expectations of readers. Suppose, however, that you responded this way: "I suggest that you define and distinguish *longevity, survival,* and *maximum lifespan potential* because introductions to scientific papers always include a bunch of definitions. I always start my introductions by defining terms, and it works great for me." This, of course, is a case of problematic feedback based on personal preferences.

Another problem is feedback based on pet peeves: "I scratched out the word *because* whenever you used it at the beginning of a sentence," a peer reviewer might say. "I just don't like it when sentences begin with *because*. Because no sound principle or rule forbids beginning complete sentences with *because,* the reviewer's feedback is unfounded. The take-home message is to think carefully about whether your criticisms and suggestions concerning peers' drafts are indeed informed by sound principles and established knowledge about written communication.

SUMMING UP AND STEPPING AHEAD

Revision means looking again at draft material for the ultimate purpose of reaching our greatest potential as writers. In covering the process at the levels of document design, global structure, and content, this chapter highlighted the expert practice of approaching revision as problem solving. We have covered procedures for detecting, diagnosing, and resolving numerous problems that can occur in drafts of scientific papers. It's worth considering, however, that your own drafts will not contain *all* of the flaws described in this chapter. So as you reflect on the chapter's lessons, you will benefit from identifying specific areas for improvement in your writing, to form a personalized plan for revising efficiently and successfully. You will also benefit from noting your drafts' strengths in the features that we have covered. Don't forget that revision is also a process of reinforcing and capitalizing on what we do well in drafts.

In our divide-and-conquer approach to the overall revision process, we focus first on large units of discourse, such as global structure and content. Then, step-by-step, we work our way through smaller and more refined units of discourse. The next logical step, which we will take in the upcoming chapter, is revising for matters at the paragraph level.

CHAPTER 6

Revising Paragraphs

Mind–Body Interactions in Health and Disease

Let's begin this chapter with a simple experiment—you'll be both scientist and subject. Start by bringing to mind the memory of an exceptionally pleasant day. Where were you and who was in your company? What were you doing? And what were you thinking and feeling? Take a minute or two to immerse yourself in this fantasy before reading on.

In the next stage of our experiment, we'll gather some physiological data. If you recalled a peaceful and relaxing day, you might notice your muscles loosening and your breathing and heart rates slowing. If we analyzed your blood, we would likely find high concentrations of feel-good hormones such as serotonin and beta-endorphin. How interesting that the purely mental activity of imagining a pleasant experience can induce a harmonious set of calming physiological responses. If, however, we altered the experiment by having you recall a most unpleasant day, your body would react quite differently. If the day were extremely stressful, for example, we would detect measurable increases in your respiration and heart rates, discernible tension in your muscles, and elevated blood concentrations of stress hormones such as adrenaline and cortisol.

Our simple experiment reveals intriguing mind–body interactions, the sort that scientists study in disciplines such as psychobiology, physiological psychology, and psychoneuroimmunology. It might seem a bit of a stretch but this chapter's focus on revising paragraphs has something in common with studying mind–body interactions. Both pursuits engage their practitioners in deeply analyzing interrelationships between the parts of truly complex systems in order to enhance their form and function.

INTRODUCTION

To better appreciate the analogy likening minds and bodies to paragraphs, consider your experiences and observations of these entities when they are in particularly good health and order. They elicit first impressions of integrity, unity, coherence, and efficiency. Looking below the surface, however, we find considerable complexity in the parts composing the whole. For example, the mind–body system includes literally trillions of connections between nerve cells in the brain. The projections of these neurons form nerve fibers that connect with hundreds of peripheral organs and tissues to produce thousands of molecules that determine our state of mind–body health. While nowhere near as complex, paragraphs are still inherently intricate. To describe the underlying qualities of a well-written paragraph, we would have to account for numerous relationships between its words, phrases, sentences, and ideas. It follows that the successful revision of disordered and dysfunctional paragraphs demands deep analyses of their many interacting parts.

Chapter 4 introduced the major qualities that define effective paragraphs: strong unity, well-crafted topic sentences, and impeccable coherence, or flow. The present chapter adds a few more defining features, including smooth cohesion, sentence variety, and sound design. For each of these paragraph-level qualities, you will learn routines for detecting, diagnosing, and solving common problems. The routines can be complex because problems at the paragraph level may have any number of underlying causes. Suppose, for example, that one of your draft's paragraphs is not clearly conveying your intended message. You have the vague sense that readers just won't get the main point. You begin revising by diagnosing the paragraph's topic sentence. But it's not a simple diagnosis, because effective topic sentences are defined by a number of interacting features:

- They are positioned strategically in their paragraphs.
- They contain specific words and phrases that reveal their paragraphs' central themes, main points, and rhetorical goals.
- They relate logically to their supporting sentences.
- They forecast the structure of ideas in their paragraphs.
- They reflect their paragraphs' relationships to surrounding paragraphs.

Testing your topic sentence for these defining features, let's say that you find no problem. If the topic sentence is not responsible for the paragraph's weakness, what is? To answer this question, you will carry out even more complex diagnoses for qualities of unity and coherence. At its best, your revision process will be comprehensive and systematic. These are the hallmarks of the revision process that we will cover in this chapter.

ABOUT THE PROCESS

In our overall approach to the scientific writing process, as diagrammed in Figure 6.1, the stage of revising paragraphs immediately follows that of revising

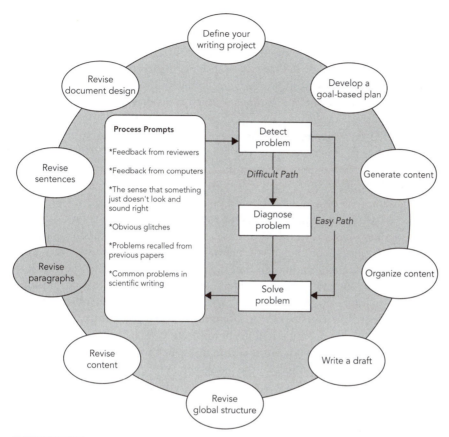

FIGURE 6.1 Process map for *revising paragraphs*.

content. At the paragraph level, we are no longer concerned about whether ideas are accurate and whether arguments are convincing and devoid of logical fallacies. These content-level matters should be resolved through the revision strategies that were presented in the previous chapter. There are, however, some overlaps between revising content and paragraphs. At both levels of discourse, we evaluate how effectively ideas relate to one another and accomplish rhetorical goals. At the paragraph level, the focus on these features is more narrow and refined, emphasizing sentence-to-sentence relationships in meaning, logic, structure, and function.

In our divide-and-conquer approach to the writing process, revising paragraphs is distinct from writing a draft. Separating these two stages makes good sense because, as just discussed, paragraph-level revision is a truly complex process. Even for the most experienced writers, it's too mentally taxing to account for *every* defining quality of effective paragraphs while drafting. Revising at the paragraph level, expert writers thus concentrate on individual features or small sets of related features. Another reason for separating the stages of revising paragraphs and drafting is to allow sufficient time to pass and to thereby gain a more objective view of our own writing.

Objectivity is especially important for successfully revising paragraphs for their coherence and unity. Perhaps, with insights gained through allowing time to pass between drafting from revising, you've reversed your assessment of these two qualities in your drafts' paragraphs. Immediately after drafting a paragraph, you congratulate yourself because it seems to flow beautifully and to convey a totally unified message. You can't imagine the paragraph giving your audience any trouble at all. When you read the paragraph again a few days later, however, your evaluation changes drastically. It becomes obvious that readers will justly complain that the paragraph's ideas are actually disjointed and that there is no unified message or goal. What could possibly have happened over the passing days? It's what happens to the best of writers. While drafting, you may have tangled up ideas in your mind with those in your paper. For missing gaps and disorganized concepts on paper, you naturally substituted ideas that were fresh in mind. As the mental reminiscence of the ideas wanes with time, the gaps and disconnects in your draft become evident. By separating the stages of drafting and revising paragraphs, you will find it easier to gain this essential objective view and to take your readers' perspective. Doing so will help you more easily identify problems to solve, along with strengths to reinforce.

Figure 6.1 includes a set of process prompts, which were defined in Chapter 5 as the cues that initiate revision. At the paragraph level, a common prompt is the vague sense that something just isn't quite right. It might be the feeling that your ideas are not flowing smoothly or that they are leading readers in too many different directions. You might sense that transitions across sentences seem jumpy or that a clear focus is lacking. If you can gain a sufficiently objective view of your draft's paragraphs, you may detect and solve these sorts of problems easily on your own. Consider, however, that the ideal sense of objectivity will come naturally to your peers, professors, and editors. The take-home message is that feedback from external reviewers is especially important for initiating paragraph-level revision.

In keeping with our overall approach to revising scientific papers, we will take on paragraph revision as a problemsolving process. As introduced in Chapter 5 and illustrated again in Figure 6.1, the major steps involve detecting, diagnosing, and solving problems. Recall that we are defining detection as initially sensing that a problem exists, without yet understanding its underlying causes. Many paragraph-level problems are not so simple to solve. To figure out the best solutions, we often must diagnose the underlying causes of problems. In other words, we have to take the "difficult path" to revising paragraphs. This chapter presents many examples of paragraph-level flaws that demand advanced diagnostic routines and problem-solving procedures. The chapter's sections are organized by challenging paragraph-level problems that commonly occur in first drafts of scientific papers. Outlined in Figure 6.2, the problems are grouped by six defining features of paragraphs: unity, topic sentences, coherence, cohesion, sentence variety, and design.

Revising for Unity

Fractured unity (264)
Faded unity (267)
Frazzled unity (271)

Revising Topic Sentences

Missing topic sentences (when they're needed) (278)
Misplaced topic sentences (281)
Topic sentences as broken promises (282)
Vague topic sentences (283)
Topic sentences that are *too* specific (286)

Revising for Coherence

Disordered ideas (287)
Missing knowledge links (when they're needed) (291)
Oversights of readers' expectations (292)
Lack of parallel structure (when it's needed) (293)

Revising for Cohesion

Missing cohesion cues (298)
Misplaced cohesion cues (301)
Unnecessary cohesion cues (302)

Revising for Sentence Variety

Lack of variety in sentence length (304)
Lack of variety in sentence beginnings (306)
Lack of variety in grammatical structure (309)
Lack of variety in tone (310)

Revising for Paragraph Design

Paragraphs that are too long (312)
Paragraphs that are too short (312)
Paragraphs that lack variety in length (312)

FIGURE 6.2 Common paragraph-level problems in scientific papers. The pages on which the problems are introduced are shown in parentheses.

REVISING FOR UNITY

In Chapter 4, we defined unity as the extent to which a paragraph's ideas do the following:

(1) Adhere to and develop one central topic.

(2) Convey and support one main message.

(3) Contribute consistently to achieving one focused rhetorical goal or one related set of strategies.

Unity is thus the quality of *oneness* in a paragraph's topic, message, and goal. Well-written paragraphs elicit the sense of unity for each of these three qualities almost instantaneously. So a practical approach to evaluating a paragraph's overall unity is to time how long it takes to identify the qualities. It's a diagnostic routine that we will call *the stopwatch test*. Here's how it works: After reading a paragraph, start a stopwatch. If more than 5 seconds pass and you cannot clearly identify the central topic, main message, and focused rhetorical goal, the paragraph lacks strong unity. Let's try the stopwatch test on the paragraph in Figure 6.3, which comes from a published review paper on the effects of humor and laughter on

---

(1) In recent decades, a major impetus for the increased popularity of humor and health was the publication of Norman Cousins's (1976) article "Anatomy of an Illness" in the New England Journal of Medicine, which was expanded into a best-selling book (Cousins, 1979). (2) The story of how Cousins recovered from ankylosing spondylitis, a progressive and painful rheumatoid disease involving inflammation of the spine, through laughter (and massive doses of Vitamin C), has become part of contemporary folklore. (3) Cousins claimed that 10 min of hearty laughter had a reliable analgesic effect, providing 2 hr of pain-free sleep. (4) In addition, he reported that episodes of laughter reliably resulted in reductions in the sedimentation rate, the rate at which red blood cells descend in a test tube, which is a measure of inflammation. (5) These observations have given rise in particular to the idea that laughter reduces pain, perhaps by stimulating the production of endogenous opioids such as beta-endorphin, and also enhances immune system functioning. (6) Although the case of Norman Cousins is widely cited as evidence for the health benefits of laughter, it is of course only anecdotal and suggestive at best. (7) It is unknown whether Cousins' recovery can be attributed to the laughter, or to the Vitamin C, or to particular personality traits such as optimism or a will to live, or to some totally unrelated factor, or indeed whether the disease may have been misdiagnosed in the first place.

FIGURE 6.3 A paragraph with excellent unity from a published review paper authored by Martin (2001). Reproduced with permission from the American Psychological Association.

physical health. (You might be surprised to learn that quite a lot of research has been published on this issue in top-quality life science journals.)The paragraph in Figure 6.3 was written by Rod Martin, a professor of clinical psychology at The University of Western Ontario in Canada.

After just one reading, Martin's paragraph reveals its central topic without delay. The paragraph is clearly about the observations and claims of physician and author Norman Cousins. Specifically, the paragraph summarizes Cousins's claims about how laughter contributed to his recovery from an immune-related disease called ankylosing spondylitis. Martin's paragraph also quickly conveys its main message, which is that Cousins's "evidence" for the curative effect of laughter is subjective and perhaps insufficient as scientific support. And even though Figure 6.3 takes Martin's paragraph out of context—it's actually the fourth paragraph of the paper's introduction section—we can easily derive its rhetorical goal, which is to introduce the debatable research issue that motivates Martin's review of the literature on humor, laughter, and health. Martin's specific strategy is to present a case that exemplifies the issue: the case of the physician Norman Cousins. Because we can effortlessly identify the central topic, main message, and rhetorical goal of Martin's paragraph within only a few seconds, it passes our stopwatch test with flying colors. In other words, we can say that the paragraph has strong unity.

A deeper diagnosis of unity requires parsing and analyzing the topics, messages, and goals of a paragraph's consecutive sentences. The following questions guide this diagnosis.

- **Do the topics of consecutive sentences relate logically to one another? In addition, do the sentences' topics adhere to and develop the paragraph's overall topic?**
  A sentence's topic is usually its grammatical subject—in other words, what the sentence is about, who or what its principal actor is, or whose story the sentence tells.
- **Do the messages of consecutive sentences relate logically to one another? In addition, do the sentences' messages support the paragraph's overall message?**
  A sentence's message may be a point about its topic, the action carried out by its principal actor, or what happened in the story being told. The message typically comes in the sentence's predicate, which is the main verb and the word groups that complement it.
- **Do the goals of consecutive sentences relate logically to one another? In addition, do the sentences' goals serve the paragraph's overall rhetorical goal?**
  A sentence's goal is its reason, or purpose, for being in the paragraph. When analyzing paragraphs that we have not authored, we usually have to infer the purposes of sentences.

A useful framework for answering these unity-diagnosing questions is presented in Table 6.1 (page 262). The table's format makes it easy to examine the

TABLE 6.1 A Framework for Diagnosing Paragraph Unity (Applied to the Paragraph in Figure 6.3)

	Topic	Message	Goal
Overall	Norman Cousins's personal observations and claims about the effects of laughter on health	Cousins's evidence for the curative effect of laughter is somewhat subjective and perhaps insufficient as scientific support	To introduce the debatable research issue—involving the effects of humor and laughter on health—that motivates Rod Martin's review paper. The specific strategy is to present a case that exemplifies the issue: the case of the physician Norman Cousins.

Sentences

	Topic	Message	Goal
1	A major impetus for the popularity of humor and health	The major impetus was Norman Cousins's article	To introduce Norman Cousins's article
2	The story of how Dr. Norman Cousins used laughter to recover from a serious disease, ankylosing spondylitis	Cousins's story is part of contemporary folklore about the effects of laughter on health	To set up the author's explanation of Cousins's anecdotal evidence for the curative effects of laughter
3	Cousins's observations and claims about hearty laughter	Hearty laughter lessened Cousins's pain and enabled pain-free sleep	To present an example of Cousins's anecdotal evidence for the curative effects of laughter
4	Cousins's observations and claims about laughter	Laughter reduced something called the *sedimentation rate*, a measure of inflammation	To present another example of Cousins's anecdotal evidence for the curative effects of laughter
5	Cousins's observations and claims about laughter	Cousins's observations and claims have led to ideas about specific mechanisms by which laughter reduces pain and favorably influences immune function	To explain how Cousins's observations and claims have influenced ideas about the mechanisms by which laughter reduces pain and favorably influences immune function
6	The case of Norman Cousins	The case is only anecdotal and suggestive	To raise a concern about whether Cousins's observations and claims are scientifically valid
7	Cousins's recovery from ankylosing spondylitis	We don't know whether Cousins's recovery was really due to laughter, to personality traits, or to misdiagnosis	To raise further, specific concerns about whether Cousins's observations and claims are scientifically valid

relationships among the topics, messages, and goals of a paragraph's sentences. This example is fleshed out with details from Rod Martin's paragraph in Figure 6.3. To diagnose the paragraph's topical unity, let's compare the contents in the table's *Topic* column. Notice that the main subjects of sentences 2–7 all refer, in one way or another, to Norman Cousins's story or his observations and claims about laughter. The paragraph's excellent unity is partly explained by the topical consistency and the relevance of each sentence's topic to the overall topic. Notice again that the format of Table 6.1 facilitates these comparisons.

The successive messages and goals of the sentences in Martin's paragraph also relate logically to each other and support the paragraph's overall message and goal. Consider, for example, the messages of sentences 3, 4, and 5. Sentences 3 and 4 convey points about the specific health benefits that Cousins claimed were caused by laughter. Then, sentence 5 conveys the message that, inspired by Cousins's observations, specific physiological mechanisms have been proposed to explain the analgesic and health-restoring effects of laughter. Martin's goals for sentences 3, 4, and 5 are also in accord: The author clearly intends for the sentences to present Cousins's observations as support for the "contemporary folklore" of how laughter benefits health.

In highlighting the consistent themes in Martin's paragraph, I am not suggesting that, to have strong unity, all of a paragraph's sentences must focus on the same topic, convey a similar message, and achieve an identical goal. Imagine the redundancy and monotony that would result if they did. These elements can certainly vary from sentence to sentence without undermining the overall unity. However, drastic and unexpected shifts in topics, messages, and goals always weaken paragraph unity. To avoid the confusion that can result from shifts across sentences, skilled writers purposefully alert readers to them and then reinforce the overall themes of the paragraph after a shift occurs. These techniques are nicely demonstrated in the transition across sentences 5 and 6 in Rod Martin's paragraph.

> **(5)** These observations have given rise in particular to the idea that laughter reduces pain, perhaps by stimulating the production of endogenous opioids such as beta-endorphin, and also enhances immune system functioning. **(6)** Although the case of Norman Cousins is widely cited as evidence for the health benefits of laughter, it is of course only anecdotal and suggestive at best.

Like sentences 3 and 4, sentence 5 conveys an idea about positive and plausible effects of laughter on health. Specifically, it's the idea that Norman Cousins's observations have led to suggestions of physiological mechanisms that might underlie the association. Then, sentence 6 raises a contrasting idea, which is that Norman Cousins's case is "anecdotal and suggestive at best." Of course, in response to this shift, we do not feel disoriented. Moreover, we do not accuse Martin of changing the subject and contradicting himself. The paragraph's unity is maintained because the author alerts us to the contrasting idea and shows us how it relates to the paragraph's overall topic, message, and goal. The first signpost comes way back in sentence 3, which conveys the idea that Cousins's story is "part of contemporary folklore" regarding views on the curative effects of laughter. The word *folklore* grounds us to the idea that Cousins's claims are not scientific facts. Another signpost comes in the subordinate clause that begins sentence 6: "Although the case of Norman Cousins is widely cited as evidence for the health benefits of laughter." The subordinating word *although* alerts us to a contrasting idea, thereby smoothing the sentence-to-sentence transition and maintaining the paragraph's global unity. Martin begins sentence 7 with the idea that "it is unknown whether Cousins's recovery can be attributed to...laughter," which reinforces the main theme of the paragraph after the shift in ideas across sentences 5 and 6. The words and phrases that Martin uses to maintain paragraph unity are examples of what we will call *unity-grounding cues*. More examples of them are presented just around the corner.

You might not need to create a formal unity-diagnosing table, like the example in Table 6.1, to diagnose the unity of every paragraph in your drafts. For diagnosing straightforward paragraphs, the table's framework might just serve as a mental guide. However, for diagnosing complicated paragraphs that elicit a strong sense of disunity, creating the formal table on paper might be just the trick to revealing underlying problems and their best solutions. In upcoming pages, I will present more examples that apply the unity-diagnosing framework to resolving paragraph disunity as well as flaws in coherence.

Our diagnosis of Rod Martin's paragraph in Figure 6.3 reveals its excellent features of unity. When your own revision process uncovers strengths in unity, use them as guides to revise other paragraphs in your draft. It's a matter of capitalizing on what you are doing well, which might be maintaining a consistent focus, preparing readers for necessary shifts in topics and messages, or using unity-grounding words and phrases. It's likely, however, that you will come across at least a paragraph or two that requires revision for improving unity. So we turn next to problems of disunity that are fairly common in drafts of scientific papers. We will call these flaws *fractured unity*, *faded unity*, and *frazzled unity*.

Fractured Unity

I could use a little help revising the paragraph in Figure 6.4, which we'll say comes from a draft of a review paper that I am writing on laughter and pain perception. Take a few minutes to get a general sense of the paragraph's unity.

(1) Research on the effects of laughter on pain perception is characterized by extremely conflicting findings. (2) This situation has significantly limited progress in the research field; thus, it is important to analyze the existing research and explain the inconsistencies. (3) The contradictory outcomes may

FIGURE 6.4 A case of fractured unity.

be due to differences in experimental design, varying subject characteristics, and discrepancies in the methods used to measure pain threshold. **(4)** One hypothesis to explain why laughter has been shown to increase pain tolerance is that laughter has a distracting effect, which diverts attention from a painful stimulus. **(5)** This hypothesis contrasts arguments that laughter reduces pain through direct physiological effects, which have been proposed to include the increased secretion of endorphins and other analgesic chemicals from parts of the brain that regulate pain perception.

FIGURE 6.4 *continued.*

Reading the paragraph closely from start to finish, you will likely reach the end asking, "How did I get here from there?" This is a sign of fractured unity, a case in which the topics, messages, and/or purposes of sentences abruptly change without reason and warning.

Let's begin to diagnose the problem by applying the stopwatch test: In 5 seconds or less, can you identify my draft paragraph's central topic, message, and goal? At first glance, the overall topic appears to address the effects of laughter on perceptions of pain. But across the sentences you may sense a problematic shift in topics, which takes more than 5 seconds to sort out. Parsing the sentences' topics, we have to ask whether the paragraph is about *research on the effects of laughter on pain perception* (sentence 1) or *hypotheses to explain why laughter increases pain tolerance* (sentence 4). The main message of the paragraph also does not reveal itself straight away. Is the main message that inconsistencies in research findings must be explained (sentence 2)? Or is it that inconsistencies in hypotheses are attributable to contrasting psychological and physiological explanations (sentences 4 and 5)?

When I first planned the paragraph in Figure 6.4, my overall goal was to explain why existing studies on laughter and pain threshold have yielded contradictory findings. My intended overall message was that the inconsistencies may be attributable to methodological factors, specifically the three factors listed at the end of sentence 3. Sentence 1, which introduces the idea of conflicting findings from previous research, lines up perfectly with my intended overall goal and message. The idea in sentence 2, which addresses the importance of resolving the contradictory outcomes, follows aptly. Sentence 3 also maintains sound unity by speculating that the disagreements in study outcomes may be attributable to differences in specific aspects of their methods. But notice how sentence 4 breaks the paragraph's unity. It does so by raising a completely new idea without any alerting signpost. The divergent idea concerns a hypothesis for a mechanism that might explain positive associations between laughter and pain tolerance. To achieve my goal of explaining why research outcomes on this issue are contradictory, I do not need to present hypotheses about any underlying mechanisms. The same goes for conveying my intended message about how methodological differences may contribute to contradictory study outcomes. The paragraph goes fully off course in sentence 5, which presents a contrasting hypothesis for why laughter increases pain tolerance. This diagnosis reveals a complete fracture in paragraph unity across sentences 3 and 4.

Abrupt breaks in unity commonly occur when writers focus too narrowly on local topical cues. These are words and phrases that reflect what a sentence (or a group of related sentences) is about. The problem emerges during the drafting process when the writer relies too heavily on an individual sentence's topical cues to generate ideas for the next sentence(s). The narrow focus obscures the writer's view on the paragraph's bigger picture. So the writer fails to check for whether the ideas in successive sentences adhere to the paragraph's overall topic, support its main assertion, and contribute to achieving its focused rhetorical goal. It's a classic example of getting lost in the trees and losing sight of the forest. It's also a case of forgetting to apply the helicopter thinking method that we discussed in earlier chapters. This is what caused the unity flaw in my draft paragraph—I just forgot to do the necessary helicopter thinking while drafting. In sentences 1 through 3, I focused my attention too narrowly on the topical cue "laughter and pain threshold" and on all the phrases reflecting the contradictory research in the field. I figured that it made sense, while I was talking about contradictory research, to raise the contradictory hypotheses regarding the mechanisms by which laughter might increase pain tolerance. The problem is that I lost sight of my original goal for the paragraph, which was to explain how specific methodological limitations have caused contradictory outcomes in previous studies on the issue. An insightful diagnosis of fractured unity requires a sentence-by-sentence analysis, with regular reflections on a paragraph's overall themes, as demonstrated in the preceding analysis of my draft paragraph in Figure 6.4.

Figure 6.5 presents general strategies for solving unity problems. To choose optimal solutions, you must determine whether the sentences that are causing disunity truly belong in their paragraphs or not. If they do belong, the solutions involve repairing fractures by (a) adding ideas that clearly establish logical relationships, (b) integrating reinforcing cues, and (c) smoothing out troublesome transitions. As outlined in Figure 6.5, strategies 5, 6, and 7 serve these purposes. As we have diagnosed, the unity flaw in my preceding draft paragraph is caused by sentences that do not belong; these are sentences 4 and 5 in Figure 6.4 (the sentences that bring up the contrasting hypotheses about how laughter affects pain tolerance). So my best solutions are to delete or move these sentences (strategies 2 and 3 in Figure 6.5). Once the errant sentences are eliminated from the paragraph, I will have a new problem to solve: The paragraph will have only three sentences and its content will be clearly underdeveloped. So, in keeping with the true iterative nature of the revision process, I will need to revisit the planning process of generating content. That is, I will have to develop new ideas to elaborate on how the methodological limitations that I listed in sentence 3 have actually led to contradictory findings in studies on laughter and pain perception. It's a worthwhile revision that will markedly improve the quality of my paper.

1. **Strengthen your topic sentences.** The overall topics, messages, and purposes of unified paragraphs are conveyed by their topic sentences.

2. **If the sentences that weaken a paragraph's unity are not essential, delete them.** This strategy entails assessing whether each sentence in a paragraph is absolutely necessary for developing its central topic, supporting its main message, and accomplishing its overall rhetorical goal.

3. **If the sentences that weaken a paragraph's unity would fit better elsewhere, move them accordingly.** This might involve moving the sentences to an existing paragraph or using them to construct a new paragraph.

4. **Begin successive sentences with the same topic.** This solution corrects the problem of drastic shifts in topics; however, as we will discuss later in the chapter, the solution can also create a problematic lack of variety in sentence beginnings.

5. **Use transitional words, phrases, and sentences to signal and smooth out potentially troublesome shifts in topics, messages, and purposes.** Sometimes, just a simple transition across two disjointed sentences can resolve a major case of disunity.

6. **Start new sentences with words and phrases that link back to the topics, messages, and purposes of previous sentences.** This solution applies the advice from Gopen and Swan's classic article on scientific writing, which I introduced in Chapter 4.

7. **Use *unity-grounding cues*.** These are words, phrases, and sentences that refer to the overall topic, message, and rhetorical goal of a paragraph.

8. **For paragraphs that completely lack unity, take a do-over.** This strategy entails reoutlining and redrafting hopelessly disunified paragraphs from scratch.

FIGURE 6.5 Strategies for solving problems that weaken paragraph unity.

Faded Unity

We have defined unity as the quality of *oneness* in the topic, message, and purpose of a paragraph. When paragraphs have the qualities *two-ness*, *three-ness*, *four-ness*, or more, they can overwhelm readers by pushing and pulling them in too many different directions. This is certainly what happens when writers do not draw the logical connections between more than one disparate topic, message, or goal. However, disunity also results when paragraphs have the quality of *less than oneness*, so to speak. A common case is a paragraph that begins with a unified focus, but the focus blurs progressively across successive sentences. This flaw is more subtle and difficult to detect than a single clean fracture. An example of the problem, which we will call *faded unity*, is presented in Figure 6.6 (page 268).

(1) Future studies should examine health-related physiological measures in subjects who are exposed to humor over longer durations than researchers have previously investigated. **(2)** In typical laboratory experiments in this field, subjects watch timed segments of comedic television shows or humorous movies. **(3)** Before and after the intervention, the researchers measure outcomes such as blood pressure, stress hormone concentrations, and immune system activity. **(4)** A common finding in these studies is that the experimental treatment has no influence on health-related physiological measures. **(5)** For example, Browne et al.[7] found that a 15-minute exposure to humorous movies had no significant effect on markers of immune function, including immunoglobulin A activity and T-cell helper-suppressor ratio. **(6)** However, subjects might not be receptive to humorous stimuli because they are uncomfortable or anxious in the unfamiliar laboratory setting. **(7)** Future studies are needed to address this limiting factor, which may have led to skewed results in previous studies.

FIGURE 6.6 A case of faded unity.

It's a draft paragraph from a review paper that a student named Justin wrote for a physiological psychology class. Justin's topic concerns the effects of exposure to television and movie humor on health-related physiological functions. The paragraph comes from the end of Justin's paper, where his rhetorical goal was to suggest ideas for future research. Justin's strategy was to suggest new studies in which subjects are exposed to humorous stimuli over long durations of time—specifically, longer durations than have been previously studied.

Justin's paragraph begins in sharp focus. Sentence 1 directly reflects his goal to suggest future studies in which researchers expose subjects to humor for relatively long durations. In sentence 2, notice Justin's reference to "timed segments" of exposure to comedic stimuli. This two-word phrase grounds readers to the topical theme of *duration of humor exposure*, thereby maintaining the paragraph's unity. Even the phrase that begins sentence 3, "Before and after the intervention," helps to keep the unity in focus. From this point on, however, the unity fades progressively.

In sentences 3 through 5, Justin's goal was to explain how the short exposure to humor in previous studies might have led to flawed outcomes; this is the basis for Justin's argument for new studies in which subjects are exposed to humor over longer durations. To achieve the goal for sentences 3 through 5, Justin appropriately explains the methods and results of typical previous studies. But his explanation lacks sufficient references to the previous studies' duration of exposure to humor. In other words, the sentences do not ground readers to the paragraph's unity of purpose. Sentence 5 refers to "a [short] 15-minute exposure to humorous movies" in Browne et al.'s study, which revealed no effects of humor on immune function. While it refers to the core topic of study duration, sentence 5 does not reflect ideas that explain *how* the short-duration exposure might have led to flawed results and conclusions. In addition, the sentence does not directly support Justin's argument for studies with longer exposures. Justin's original plan was to explain that participants in the short-duration studies might not have had sufficient time to adjust to unfamiliar laboratory settings. The subjects, he reasoned, might have been too anxious to experience physiological responses to humorous stimuli in such short exposures. Justin is trying to get these ideas across in sentence 6. But that sentence is still out of focus because it does not refer directly to the methodological shortcoming—the limited exposure to humorous stimuli. The same problem characterizes sentence 7, with its vague reference to "this limiting factor."

Faded unity is a fairly common occurrence in drafts of scientific papers. It can be a consequence of our losing sight of our main themes and objectives for paragraphs. In some cases, faded unity is the result of our mistaken assumptions that readers will grasp paragraph topics, messages, and goals without explicit references to them. Once the problem is detected and diagnosed, the solution is to regularly refocus readers' attention by integrating unity-grounding cues at strategic points in paragraphs. As introduced earlier, these are words, phrases, and sometimes even entire sentences that keep readers focused on a paragraph's overall topic, message, and goal. A productive revision strategy is to put your draft aside in order to brainstorm unity-grounding cues. Justin's refocusing efforts and unity-grounding improvements are evident in his revision in Figure 6.7.

(1) Future studies should examine health-related physiological measures in subjects who are exposed to humor over longer durations than researchers have previously investigated. **(2)** In typical laboratory experiments in this field, subjects watch segments of comedic television shows or humorous movies over periods lasting only 15 to 30 minutes. **(3)** Before and after the intervention, the researchers measure outcomes such as blood pressure, stress hormone concentrations, and immune system activity. **(4)** A common finding in these studies is that the experimental treatment has no

FIGURE 6.7 A revision of the draft paragraph in Figure 6.6.

influence on health-related physiological measures. **(5)** For example, Browne et al.[7] found that exposure to humorous movies had no significant effect on markers of immune function, including immunoglobulin A activity and T-cell helper-suppressor ratio. **(6)** However, the participants in this study viewed the movies for only 15 minutes. **(7)** This exposure may not have been long enough for enabling health-related physiological responses to occur. **(8)** In addition, in typical short-duration experiments subjects may not have time to adjust to feelings of discomfort or anxiousness in the unfamiliar laboratory settings. **(9)** Thus, one suggestion for future studies is to extend the laboratory exposure to humor to periods of an hour or more.

FIGURE 6.7 *continued.*

Let's take a close look at Justin's use of unity-grounding cues to sharpen the focus of his original paragraph. As follows, we will compare key sentences across his draft and revision. The added unity-grounding cues are in bold font.

Draft: In typical laboratory experiments in this field, subjects watch timed segments of comedic television shows or humorous movies.

Revision: In typical laboratory experiments in this field, subjects watch segments of comedic television shows or humorous movies **over periods lasting only 15 to 30 minutes**.

In the revision, Justin replaces a vague reference to "timed segments" of comedy exposure with a phrase that more specifically reflects his message that the duration of exposure to humor in previous studies was relatively short.

Draft: For example, Browne et al.[7] found that a 15-minute exposure to humorous movies had no significant effect on markers of immune function, including immunoglobulin A activity and T-cell helper-suppressor ratio. However, subjects might not be receptive to humorous stimuli because they are uncomfortable or anxious in the unfamiliar laboratory setting.

Revision: For example, Browne et al.[7] found that exposure to humorous movies had no significant effect on markers of immune function, including immunoglobulin A activity and T-cell helper-suppressor ratio. However, the participants in this study viewed the movies **for only 15 minutes. This exposure may not have been long enough for enabling health-related physiological responses to occur**.

In his draft, Justin mentioned the 15-minute duration of exposure in Browne et al.'s study, but the idea is hidden in the middle of the sentence. Moreover, the idea is not developed at all in the draft's following sentence. The revision resolves both of these flaws. The last sentence in the revised text strongly grounds readers to Justin's goal to explain how previous studies may have had questionable outcomes.

> Draft: Future studies are needed to address this limiting factor, which may have led to skewed results in previous studies.
>
> Revision: **Thus, one suggestion for future studies is to extend the laboratory exposure to humor to periods of an hour or more**.

In the revision, notice the direct reference to the need for future studies that extend the duration of exposure to humor. This is another cue that grounds readers to Justin's rhetorical goal for the paragraph and thereby maintains its focus and strengthens its unity.

As you might have recognized, there is a potential drawback to continually interjecting unity-grounding cues into paragraphs with faded unity. If the writer goes overboard with the technique, the reader will respond, "Okay, I get your point already. You don't have to hit me over the head with it!" In your own iterative revision process, this is something worth double-checking. It would be good, for example, to ask peer reviewers for specific feedback on whether your unity-grounding efforts go too far. But when you are in doubt, it's always better to err on the side of nudging readers to see your ideas in clear focus.

Frazzled Unity

So far we have covered cases of paragraph disunity that are not tremendously complicated and difficult to fix—that is, if you have a sound approach to detecting and diagnosing the problems. For example, revising paragraphs with fractured unity is often a relatively simple matter of identifying clearly defined breaks in ideas across successive sentences and then reconnecting the pieces. While the challenge is greater for revising paragraphs with faded unity, at least some of their ideas begin in focus, which aids the revision process. Now we take on the supreme challenge of revising paragraphs that can be described as totally *frazzled*. These are the sorts of paragraphs in which every sentence seems like a tattered, disconnected thread. Maybe you won't ever have to deal with this major paragraph-level flaw in your own drafts. But if you do, or if you come across it in reviewing peers' drafts, you will need a skillful and effective approach to resolving it.

A case of frazzled unity is demonstrated in Figure 6.8 (page 272), which is a draft paragraph from the introduction section to a conceptual review paper authored by a student named Fran. The assignment, for a psychology class, was to argue for an original hypothesis in the research field of mind–body health. Fran's hypothesis addresses the role that laughter might play in preventing diseases involving the immune system.

(1) The common cold, asthma, and rheumatoid arthritis are all examples of immune-related diseases that could theoretically be prevented by laughter. **(2)** Negative emotional states such as anxiety, hostility, anger, and feelings of personality disintegration have been shown to have an adverse effect on health. **(3)** Studies reveal that frequent laughter is more often associated with positive psychological states such as joy, optimism, and compassion. **(4)** Diminished proliferation and activation of immune cells, such as neutrophils and helper T-lymphocytes, have been observed in subjects who scored high on tests of negative emotions. **(5)** High concentrations of hormones such as cortisol and adrenaline may play a role. **(6)** Thus, it is conceivable that laughter can prevent immune-related diseases.

FIGURE 6.8 A case of frazzled unity.

Running the stopwatch test on Fran's paragraph, you will discover that it lacks a unified topic. From the ideas in the first and last sentences, you might guess that the paragraph is mainly about laughter and immune-related diseases. But the ideas in the intervening sentences do not clearly relate to this overall topic. The ambiguity is partly due to the abrupt changes in topics from sentence to sentence. The drastic shifts are highlighted in Table 6.2, which presents another example of the unity-diagnosing framework that I introduced earlier. Take a minute to examine the subjects of the paragraph's sentences as listed in the table's *Topic* column. While sentences 3 and 6 are both about laughter, the topics of the other sentences are mismatched. As mentioned earlier, topical shifts across sentences are not automatically problematic; instead, problems occur when the shifts are not closely preceded or followed by unity-grounding cues. If, for example, the message of a given sentence sufficiently prepares readers for a topical shift in the following sentence, paragraph unity will be maintained. (One other requirement is that the following sentence's topic is still relevant to the paragraph's overall topic.)

The bare-bones analysis of Fran's paragraph, as presented in Table 6.2, reveals that the sentences' messages do not sufficiently prepare readers for upcoming topical shifts. Consider the transition across sentences 1 and 2. Sentence 1 conveys the message that the three conditions defining its topic (*the common cold, asthma, and rheumatoid arthritis*) are immune-related diseases that laughter might prevent. Then, sentence 2 introduces the topic of *negative emotional states*. Do sentences 1 and 2 convey a unified idea? In other words, is it clear how *negative emotional states* relate to *immune-related diseases that might be prevented by*

TABLE 6.2 A Framework for Diagnosing Paragraph Unity (Applied to the Paragraph in Figure 6.8)

	Topic	Message	Goal
Overall	?	?	?
	Sentences		
1	The common cold, asthma, and rheumatoid arthritis	…are immune-related diseases that might be prevented by laughter	To present a hypothetical connection between laughter and disease prevention
2	Negative emotional states	…have an adverse effect on health	To present information about the effects of negative emotions on health
3	Frequent laughter	…is associated with positive psychological states	To present information about the association between laughter and positive psychological states
4	Diminished proliferation and activation of immune cells	…have been observed in subjects who scored high on tests of negative emotions	To summarize a research finding concerning immune function in subjects who experience negative emotions
5	High concentrations of hormones	…may play a role (in what specific process, the writer doesn't clarify)	To present information about the role that certain hormones play (in some unclear process)
6	Laughter	…conceivably can prevent immune-related diseases	to present a hypothesis concerning the relationship between laughter and diseases that involve the immune system

laughter? Nothing in the first sentence's message logically connects with the topic of sentence 2. In addition, the message of sentence 2 does not reinforce the unity. Sentence 2 makes a point about some vague adverse health effects rather than referring specifically to immune-related diseases.

Fran's paragraph gets us back on track at the start of sentence 3. Its main topic, *frequent laughter,* is familiar because laughter was part of the message in sentence 1. But sentence 3 conveys a somewhat incongruous message. Notice that sentences 1 and 2 address *negative* emotions and *negative* health effects. Without any warning, sentence 3 expresses a contrasting message about *positive* effects, specifically positive effects of laughter on psychological states. At this point in the paragraph, we have to stop and ask, "Is this about the effects of laughter on immune-related diseases? Or, is it about the effects of laughter on psychological and emotional states? If it's about the latter, does the topic concern negative or positive psychological and emotional states?" As you will see shortly, Fran actually intended for the paragraph to present all of these topics in a unified way. But the unity is obviously frazzled in her draft.

At the start of sentence 4 of Fran's draft paragraph, we look for unity-grounding cues to resolve the confusion caused by shifts in the topics and messages of the first three sentences. But the topic of sentence 4, *diminished proliferation and activation of immune cells*, diverges yet again. Sentence 4 expresses the message that these immune responses occur in subjects who demonstrate negative emotions. Given the focus on *positive* emotions in sentence 3, we experience another 180 degree turn. Table 6.2 highlights yet another problematic topical shift across sentences 4 and 5, which further frazzles the unity of Fran's draft paragraph.

It's worth emphasizing that Fran's ideas for the paragraph are actually very strong. So we are not dealing strictly with underlying content-level problems. Instead, it's a case in which strong ideas simply are not expressed and structured in unity. It happens all the time in first drafts. For this flaw, however, there is no simple fix. It cannot be resolved by deleting a sentence or adding a transitional phrase or two. The solution usually requires a major overhaul of the paragraph—a complete *do-over*. A productive approach is to rebuild the paragraph from scratch by reoutlining the goals and rough ideas of its successive sentences. Fran's reoutlined paragraph is shown in Table 6.3.

Frazzled unity often results when we lose sight of our goals for paragraphs during the drafting process. So a useful first step to reoutlining a paragraph is to refocus on its rhetorical goal. It's simply a matter of asking yourself, "What was I trying to do and say in this paragraph in the first place?" You may not actually realize the truest answer to this question until you reach the revision process, when you have gained wisdom through the hindsight of drafting and when you can better concentrate your thinking. For Fran, refocusing on her goal naturally revealed her paragraph's unity of purpose. As Fran noted at the top of her outline (Table 6.3), the rhetorical goal for her paragraph was to introduce the hypothesis on which her conceptual review paper is based. In addition to introducing the hypothesis, Fran intended for the paragraph to provide an overview of its supporting evidence, which she elaborated on in the paper's body. To better understand the details in Fran's outline, recall that she is writing a paper to argue for an original hypothesis, which is that laughter can play a role in preventing diseases that involve the immune system.

The common, novice approach to drafting paragraphs is to wrestle immediately with what to say in their sentences. Often, a more productive method begins with figuring out what to *do* in successive sentences. In other words, the

TABLE 6.3 A Demonstration of the Paragraph Reoutlining Technique

Rhetorical Goal: To introduce my hypothesis and present a brief overview of its supporting evidence, to prepare readers for the detailed arguments in the body of my paper.

Sentence (or sentence sets)	Goal	Rough idea
1	To introduce my hypothesis and the overall purpose for my paper	Purpose is to advance hypothesis about laughter and immune-related diseases
2	To introduce the premise for my hypothesis	Premise is that laughter might deter negative emotions and their harmful effects on immune system
3	To introduce two research areas that offer support for my hypothesis	Hypothesis hasn't been tested directly, but it has indirect support through two research areas
4	To summarize first line of support	Studies showing that people who laugh a lot don't have negative emotions
5	To summarize second line of support	Studies showing that negative emotions weaken immunity
6	To explain how negative emotions can weaken immunity	Negative emotions increase release of hormones like adrenaline and cortisol, which prevent immune cells from working right
7	To present an overview of my plan for arguing for my hypothesis, showing readers how I'll use the two lines of support	Going to argue that laughter can prevent immune-related diseases by steering people away from negative emotions and their harmful effects on immune system

expert approach is to focus first on the goals that you want your sentences to achieve. This is how Fran developed her paragraph outline in Table 6.3. In the table's second column, Fran lists goal statements for her revised paragraph's sentences. Once you have set goals for your sentences, you will find that it is relatively easy to figure out what to say in them. To avoid getting bogged down by trying to write complete perfect sentences in your paragraph outline, you might try noting your ideas in phrases or short and simple sentences. That's what Fran does in the last column of her outline. Taking this same approach, you will find that your ideas flow smoothly and logically into your paragraph outlines, which will result in revised paragraphs with strong unity and coherence.

As she began reoutlining her paragraph, Fran realized that its first sentences needed to introduce her hypothesis and its premise directly, which is something that her draft paragraph did not do. Fran recorded these goals, along with corresponding rough ideas, in the rows for sentences 1 and 2 in her paragraph outline. As noted in the last column of row 2, the premise for Fran's hypothesis is that laughter may "deter negative emotions and their harmful effects on the immune system." Reflecting on her paragraph's overall rhetorical goal and the first two rows of her outline, Fran realized that she next needed to summarize the supporting evidence for her hypothesis. This, you can see, was Fran's plan for sentences 3, 4, and 5 of her revised paragraph. As indicated by the goals and rough ideas for these sentences, Fran's supporting evidence comes from two areas of research on negative emotions: (a) studies which have revealed that laughter deters negative emotions and (b) studies showing that negative emotions impair immune function.

Paragraph reoutlining is a powerful revision tool. By rethinking what we actually intended to do and say in a draft paragraph sentence by sentence, we may identify glitches that compromise unity and flow. This method enabled Fran to clearly see disjointed ideas and gaps in her draft. One insight came when Fran remapped her plan to summarize research on how negative emotions are related to laughter and to immune function. Fran realized that her first draft's reference to the association between laughter and *positive* psychological states (see sentence 3 in Figure 6.8) contributed to the original paragraph's frazzled unity. The process of reoutlining revealed another major gap in Fran's draft: She never directly introduced her plan to synthesize the two research areas on negative emotions in order to support her original hypothesis. Fran's revised plan is outlined in the row for sentence 7 in Table 6.3. Using her outline to guide the way, Fran wrote a second draft of her paragraph, which follows (Figure 6.9).

(1) This paper is intended to advance the hypothesis that regular bouts of laughter can play a role in preventing immune-related diseases such as the common cold, asthma, and rheumatoid arthritis. **(2)** The premise for this argument is that laughter may deter negative emotional states, which have been shown to weaken the immune system. **(3)** Although the hypothesis has not been tested experimentally, it has indirect backing from two lines of psychophysiological research. **(4)** First, support comes from observations that subjects who reported frequent laughter tended to avoid experiencing negative emotions such as anxiety, hostility, anger, and feelings of personality disintegration. **(5)** Second, studies have revealed that negative emotions are associated with diminished immune responses, including decreases in the proliferation and activation of neutrophils and helper T-lymphocytes. **(6)** Negative emotions adversely affect immunity by stimulating the release of hormones that directly inhibit immune cell function. **(7)** These

FIGURE 6.9 A revision of the draft paragraph in Figure 6.8.

hormones include adrenaline and cortisol. **(8)** Through synthesizing the two supporting lines of research, I will argue for the hypothesis that laughter can prevent immune-related diseases by steering people away from negative emotional states and their adverse influences on immune function.

FIGURE 6.9 *continued.*

Several features of Fran's revised paragraph are noteworthy. First, it obviously corresponds to her outline in Table 6.3. So Fran's redrafting process essentially involved translating the outline's goals and rough ideas into fully formed, coherent sentences. The unity of the revised paragraph is strikingly improved over the draft. If you take a few minutes to compare the two versions, you will agree. Fran's revision gets an "A" on the stopwatch test. The topics, messages, and goals of the revised paragraph's individual sentences relate clearly to their overall counterparts. A final point is that Fran's revision reflects key strategies for improving unity that were presented earlier in Figure 6.5 (page 267). One of the strategies is to use unity-grounding cues that reinforce the defining themes of a paragraph. The unity of Fran's paragraph would be effectively reinforced by regular references to her hypothesis, which is the paragraph's central topic. Notice the unity-grounding cues—the specific references to the hypothesis—in sentences 1, 2, 3, and 8 of Fran's revision.

Another unity-improving strategy is to use transitions to smooth out potentially troublesome shifts in topics, messages, and goals. In sentences 4 and 5 of her revision, Fran uses transitions (the words *first* and *second*) to maintain a unified sense of her purpose, which is to summarize two areas of research that support her hypothesis. Notice also that several sentences in Fran's revision have similar topics, which further reinforces the unity. Moreover, Fran's revision nicely demonstrates the strategy of starting new sentences with words and phrases that link back to the topics, messages, and goals of previous sentences. As one of many examples, consider the beginning of sentence 7. The noun phrase "these hormones" links back to the previous sentence's message about how negative emotions affect immunity. These improvements in unity were facilitated by Fran's use of the paragraph reoutlining method.

REVISING TOPIC SENTENCES

Back in Chapter 4, in its section on hints for drafting paragraphs, I praised the virtues of well-written topic sentences. Considering their many useful functions, I even suggested renaming them as topic-message-goal-structure sentences (if the term were not so clumsy). It's always worthwhile to evaluate draft paragraphs for qualities of their topic sentences. The main problems to look for, which we cover in this section of the chapter, include missing topic sentences, misplaced topic sentences, topic sentences that are in discord with their supporting sentences, vague topic sentences, and topic sentences that are too specific.

Missing Topic Sentences (When They're Needed)

Topic sentences are not mandatory for all paragraphs in scientific papers. This point is demonstrated in Figure 6.10, which comes from the body of an informative review paper on meditation and vascular function.

No single sentence captures the paragraph's central topic, main message, rhetorical goal, and overall structure. But the lack of a topic sentence is not problematic because these elements are conveyed implicitly. In sentence 1, we learn that the paragraph's overall topic concerns research on meditation and flow-mediated brachial artery vasodilation (FMBAV), which assesses the functional health of blood vessels. Sentence 1 also indicates the specific focus on a study conducted by researchers named Whitney and Stein. Because the paragraph comes in the body of an informative review paper, readers will intuitively grasp its rhetorical goal, which is to summarize studies on the topic. Sentences 2 and 3, which review the methods and results of Whitney and Stein's study, maintain our sense of the paragraph's topic and goal.

The main message and structure of the preceding sample paragraph are conveyed in the relationships between its ideas. Consider, for example, the transition of ideas across sentences 3 and 4. The subject that begins sentence 4, *Similar outcomes to those observed by Whitney and Stein*, further reinforces our sense of what the paragraph is about and what rhetorical goal it serves. From the information leading up to sentence 5, we know that Whitney and Stein found a significant increase in FMBAV among subjects who performed meditation. We also know that other studies have revealed similar outcomes. Beginning sentence 5 the transitional phrase, *in contrast*, alerts us to contradictory findings. This transition

> **(1)** Whitney and Stein[12] conducted a 16-week study to examine the effects of meditation on flow-mediated brachial artery vasodilation (FMBAV), a marker of vascular health. **(2)** The participants, 32 females and 30 males between 22 to 36 years of age, were randomly assigned to a meditation group and a control group that viewed informational videos about meditation. **(3)** The results showed a significant 18% increase in FMBAV ($P < .001$) among participants in the meditation group; no significant change in FMBAV occurred in the control group. **(4)** Similar outcomes to those observed by Whitney and Stein have been reported. [3,7,8]
>
> **(5)** In contrast, Levinson[6] and other researchers[2,9] have reported that meditation had no effect on FMBAV.

FIGURE 6.10 A paragraph that does not need a topic sentence.

implicitly conveys the paragraph's take-home message. If the paragraph had an explicit topic sentence, it might read something like this:

> Studies on the effects of meditation on flow-mediated brachial artery vasodilation (FMBAV) have produced conflicting results.

But it's not essential to add this topic sentence. The reason is that the paragraph's topic, message, goal, and structure are easily surmised in the context of the entire paper and through the relationships between ideas at the sentence level.

In contrast to the preceding example, some paragraphs that lack topic sentences would be much improved by their addition. For a case in point, we will examine a paragraph from an Honors thesis authored by a student named Yvonne. Her faculty advisor was the principal investigator for a study on meditation and health. The study's specific aim was to determine whether a technique called positive imagery meditation, or PIM, has any clinical value. Specifically, Yvonne's advisor wanted to know whether PIM could meaningfully improve health status to a degree that doctors would recommend the technique to their patients. Yvonne was assigned a small part of the project that focused on subjects at risk for cardiovascular disease. She collected and analyzed data on measures of blood pressure; HDL and LDL cholesterol, which are the healthy and unhealthy forms, respectively; and serum concentrations of C-reactive protein and homocysteine, both of which contribute to the disease condition of atherosclerosis. Yvonne's written thesis took the form of an IMRAD-structured research paper. For the discussion section, Yvonne intended the first paragraph to achieve the rhetorical goal of presenting and supporting her conclusions. Figure 6.11 shows Yvonne's first draft of the paragraph.

> The results of the present study showed that subjects who performed positive imagery meditation (PIM) experienced statistically significant changes in cholesterol levels. From baseline to the end of the 3-month study, the mean HDL-cholesterol concentration increased from 39 to 44 mg/dl ($P < .001$). According to the American Heart Association (AHA), HDL-cholesterol concentrations must be above 60 mg/dL to provide significant protection against heart disease (http://www.americanheart.org). Over the study period, LDL-cholesterol decreased from 164 to 160 mg/dL ($P < .001$). AHA guidelines indicate that LDL-cholesterol concentrations below 100 mg/dL are optimal for preventing heart disease. Statistically significant changes in a favorable direction were also observed for blood pressure. Mean blood pressure values fell from

FIGURE 6.11 A paragraph that needs a topic sentence.

144/93 mm Hg to 141/87 mm Hg. AHA guidelines indicate that significant reductions in risks for cardiovascular disease occur when blood pressure falls below 120/80 mm Hg. No changes were observed over the study period for concentrations of C-reactive protein and homocysteine.

FIGURE 6.11 *continued.*

As was true for the sample paragraph in Figure 6.10, no single sentence in Yvonne's paragraph qualifies as a bona fide topic sentence. All of the sentences either list individual results or comment on their clinical implications. The lack of a topic sentence in this case, however, is definitely problematic. Given its context in the discussion section of a research paper, as well as its rhetorical goal, the paragraph will naturally lead readers to ask, "So what's the point of all these results and implications?" In other words, readers will justly expect a take-home message. They will be looking specifically for Yvonne to interpret her results and to present a conclusion that answers her study's motivating question. Recall that Yvonne sought to determine whether PIM would lead to clinically significant health improvements, such that doctors would recommend the technique to their patients. Through revision, Yvonne diagnosed the problem of a missing topic sentence when one is actually needed. Her solution was to write a topic sentence that conveyed her conclusion and summarized the support for it. Presented as follows, Yvonne's topic sentence turned out to be two related sentences that function together.

Despite its statistically significant effects on cholesterol concentrations and blood pressure, positive imagery meditation (PIM) does not induce improvements that are great enough to support its recommendation as a clinically relevant treatment for cardiovascular disease. This conclusion is further supported by the lack of a significant effect of PIM on C-reactive protein and homocysteine concentrations.

To best appreciate Yvonne's effective revision, re-read her draft paragraph with the preceding tandem topic sentence placed at its beginning. You will see that the addition gives readers the take-home message they are looking for. In addition, the tandem topic sentence sheds a whole new light on the paragraph's other sentences. They are no longer simply a list of different results and implications. Instead, the sentences now clearly serve the higher purpose of supporting Yvonne's conclusion, as presented in her newly added tandem topic sentence.

If you come across draft paragraphs that lack topic sentences, it's certainly worth asking whether they are needed. Think about it from the perspective of your readers. Would added topic sentences enhance their understanding and give them a stronger sense of paragraph unity? Good candidate paragraphs for this revision are ones described as follows.

(1) **Long paragraphs that are packed with extensive details and complex ideas**. These are the sorts of paragraphs that lose readers and raise that all-important question, "So what?"

(2) **Paragraphs that are intended to make arguments.** Any claim in a scientific argument is ideally conveyed by a topic sentence.

(3) **Paragraphs with subtle rhetorical goals.** Given their context and content, some paragraphs naturally reflect their authors' rhetorical goals, even without topic sentences. Other paragraphs, however, may not express their goals so explicitly. If you are concerned that readers might not get your specific rhetorical aims for a paragraph, add a topic sentence with goal-grounding cues. This strategy will be demonstrated later.

Misplaced Topic Sentences

Misplaced topic sentences can weaken paragraph structure, impair readability, and even limit comprehension. The most common flaw is placing the topic sentence too late in a paragraph. Imagine a massive 15-sentence paragraph in which the writer's goal is to convince readers to accept his novel hypothesis. The first 14 sentences present and explain the supporting data, which are extremely complex. Sentence 15, which is the topic sentence, contains the actual hypothesis, presented as the argument's claim. In this position at the paragraph's end, the topic sentence is like the punch line of a joke. The writer reasoned that readers would happily work their way through the first 14 sentences in suspenseful expectation of the punch line. From the writer's perspective, the bottom-up paragraph structure is not problematic at all because he knows what's coming at the end and how it relates to the previous sentences. But consider that the audience is clueless (through no fault of their own). In the middle of the paragraph, as they are struggling to understand the complex data, readers will raise pressing questions for the writer: *Why are you telling me this? Why do I need to know all this stuff? What's your point, anyway?* In frustration, some readers might stop reading the paragraph in midstream. Those who hang in until the end might not extend much effort to understand and remember the information in sentences 1–14 because its relevance is unclear. So when they finally get to the topic sentence, readers might not understand its rationale. A deep understanding might require re-reading the whole paragraph again with the ending topic sentence in mind. Who wants to do that?

There is, of course, a simple solution to the problem of topic sentences that come too late in lengthy, complex paragraphs: Begin with the punch line, so to speak. In other words, the solution is to create a *top-down* structure by placing the topic sentence early in the paragraph—if not the first sentence, at least one of the first few sentences. If you are concerned that readers might lose sight of an early topic sentence's main message as a paragraph develops, integrate unity-grounding cues at strategic points along the way. Another useful technique is to reintroduce the topic sentence at the end of the paragraph, varying the phrasing from its initial presentation. This technique gives long, complex paragraphs effective closure.

Some topic sentences are placed too early in their paragraphs. This is a problem when readers need a little contextual information before a paragraph's main point. Maybe a few terms need to be defined or a concept needs to be explained. In this situation, it's best to hold off on the topic sentence until the foundation is set for it. In another troublesome case, the early-arriving topic sentence is a claim in an argument that readers are likely to react against strongly. If the claim comes immediately in the paragraph intended to support it, readers might be immediately turned off to the writer's argument. If, however, the writer eases into the claim by first presenting a manageable set of supporting details, there is a better opportunity to convince inherently biased readers.

Topic Sentences as Broken Promises

Placed strategically at the start of paragraphs, topic sentences make promises about the content, structure, and functions of their paragraphs' other sentences, giving audiences useful advance information and frames of reference. As promise statements, well-written topic sentences help us read, understand, and recall the content of paragraphs. However, these desirable outcomes depend on whether paragraphs actually keep the promises that their topic sentences make. A case of a broken promise is demonstrated in the following sample paragraph (Figure 6.12), which comes from a research proposal to study the effects of meditation on brain wave activity.

Sentence 1 is clearly the topic sentence, given its placement in the paragraph and its bold claim. The sentence promises that the rest of the paragraph will

> **(1)** The results of this proposed study will provide exciting new information that can be directly applied in clinical settings. **(2)** We propose to investigate the influence of meditative states on electroencephalography (EEG) activity in the prefrontal cortex. **(3)** Specifically, we will examine the EEG activity of alpha and theta waves while subjects perform 30-minute sessions of guided meditation. **(4)** We will also investigate patterns of neural synchrony in the prefrontal cortex during the meditation sessions. **(5)** Increases in the amplitude of slow alpha and theta waves have been associated with acute changes in parasympathetic neural output, which may influence cardiovascular and metabolic function.[3,4,6,8] **(6)** Given society's widespread interest in mind-body health interventions, knowledge about the effects of meditation on brain activity is especially clinically relevant.

FIGURE 6.12 A paragraph that breaks the promise of its topic sentence.

address the clinical applications of the proposed study's potential outcomes. We would thus expect the author to explain how clinicians might apply plausible findings from the study to advise and treat patients. The paragraph's content, however, does not meet our expectations. Sentences 2, 3, and 4 present the proposed aims and methods for the study, specifically describing how brain wave activity will be measured and analyzed. Given the topic sentence's promise, we want to know how clinicians will actually benefit from learning about the electroencephalography (EEG) activity of alpha and theta waves in the prefrontal cortex of meditating subjects. But the paragraph does not follow through on the promise. While sentence 5 contains an idea about the relationship between changes in brain wave activity and health-related physiological functions, it still does not address potential *clinical applications* of the proposed study. Neither does sentence 6.

The first step to revising the problem at hand is to isolate your topic sentences and view them as promise statements. Then think ahead about the expectations that your readers will form based on the promises that your topic sentences make. Use these expectations as a checklist to assess whether your paragraphs are following through. It's a revision process similar to evaluating whether the supporting ideas for an argument's claim are relevant. If you find broken promises in your drafts, you must decide what to rewrite: the topic sentence or the rest of the paragraph. Did you overstate your topic sentence, making a promise that could not be kept? If so, rewrite the topic sentence accordingly. Or did you make an appropriate promise but overlook the necessary content to keep it? If so, rewrite the body of the paragraph in accord with the topic sentence. For the author of the preceding example, the latter solution applies. A successful revision challenges the author to develop specific and convincing ideas about how the proposed study's potential outcomes will actually be useful in clinical settings.

Vague Topic Sentences

Let's say that I'm writing a research proposal to study how meditation influences the body's response to stress. My hypothesis is that the regular practice of meditation over many years enhances the body's resistance to stress-related diseases. In my proposal's introduction section, I have devoted a paragraph to briefly describing the novel and unique features of my study. My goal is to convince readers that the study will surpass previous ones in revealing useful insights about how meditation affects resistance to stress-related diseases. Suppose that all of the previous studies on this issue have focused only on the immediate effects of meditation. My plan is to investigate the long-term effects. In addition, the previous studies did not adequately account for potentially confounding influences of lifestyle behaviors such as diet and exercise. My study will be the first to directly control for these extraneous factors. In a draft paragraph that introduces these novel and unique features of my proposed study, I began with the following topic sentence.

> The proposed study will take a novel and unique approach to examining meditation and responses to stress.

It's an example of a vague topic sentence and what writing instructors call *deadwood*. It takes up space without conveying an essential idea. Even if the rest of the paragraph does an excellent job of describing the study's novel and unique features, it all begins on a very dull note. Unlike my draft sentence, the best topic sentences engage readers. The main quality that elicits such a favorable response is *specificity*. Great topic sentences contain specific words and phrases that convey their paragraphs' subjects, points, and purposes.

To diagnose my example topic sentence, look for whether it contains specific words and phrases that reflect the overall topic of its paragraph, which concerns the extraordinary features of my proposed study. While the topic sentence says that the study is *novel and unique*, it does not refer to any specific features. You might justly assert that the specifics should come in the body of the paragraph. But at least an overview of them will make all the difference in perking up the draft topic sentence. What exactly makes my study novel and unique? Recall that it will be the first study to focus on long-term, chronic effects of meditation on stress-related diseases. Moreover, it is designed to control for extraneous variables that could have limited previous studies. Did I even refer to these exciting features in my original topic sentence? Take another look for yourself:

> The proposed study will take a novel and unique approach to examining meditation and responses to stress

While you are evaluating the sentence's specificity, check for whether it conveys my intended message for the paragraph, which is that the proposed study will go beyond previous studies in adding significant knowledge about how meditation affects resistance to stress-related diseases. The sentence falls short of conveying this message.

Solving the problem of vague topic sentences first involves eliminating the deadwood—that is, getting rid of language that does not reflect specific aspects of the paragraph's overall topic, message, and goal. Then, it's a matter of brainstorming and adding the specifics. Taking this approach, I realized that my draft topic sentence should better focus readers' attention on my rhetorical goal to introduce the novel and unique features of my proposed study. The sentence should also give advance information about those features. My revision, another case of a two-sentence tandem topic sentence, follows.

> The proposed study affords the first opportunity to determine the effects of meditation on resistance to immune-related disease in individuals who have meditated over long periods of their lives. In addition, the study is designed to control for extraneous factors that may have led to flawed results and conclusions in previous research on this issue.

After these two sentences, which would begin the paragraph, I would then describe the novel and unique features of my proposed study in detail.

More examples of vague topic sentences, along with specific revisions, follow. The examples come from a coauthored research paper on the effects of hypnosis on sleep quality in people who suffer from insomnia.

Example 1: For a paragraph in the paper's introduction section, the authors' goal is to convince readers that researchers have not sufficiently determined whether hypnosis improves sleep quality in people with insomnia. The authors' main message is that a limiting factor in previous studies has been a lack of control for subjects' use of sleeping pills.

Vague topic sentence: Previous studies on the effects of hypnosis on insomnia have been limited by confounding variables.

Revised, specific topic sentence: Whether hypnosis improves sleep quality in people with insomnia is uncertain, partly because previous studies have not adequately controlled for the confounding influence of sleeping pills.

Example 2: For a paragraph in the methods section, the authors' goal is to justify their use of EEG, which measures the electrical activity of neurons in the brain. The authors' main message is that the EEG method was necessary to determine the underlying mechanisms by which hypnosis affects sleep quality.

Vague topic sentence: EEG data were collected while subjects underwent hypnosis, allowing vital information to be obtained regarding the treatment.

Revised, specific topic sentence: We collected EEG data in order to determine whether hypnosis influences sleep quality by inducing changes in the brain's electrical activity.

Example 3: For a paragraph in the discussion section, the authors' goal is to explain the study's results in terms of underlying physiological mechanisms. The results indicated favorable effects of hypnosis on sleep quality. In the paragraph, the authors speculate that the outcome may have been attributable to the effects of hypnosis on the parasympathetic nervous system.

Vague topic sentence: Physiological effects may have been responsible for the observed results.

Revised, specific topic sentence: In subjects who underwent hypnosis, the improvement in sleep quality may have been due to increased activation of the parasympathetic nervous system.

Example 4: For a paragraph in the discussion section, the authors' goal is to suggest future research on hypnosis and sleep quality, specifically in individuals over 70 years of age. The authors' main message is that the new research is necessary because this population is especially prone to insomnia.

> Vague topic sentence: New studies on the elderly are needed to elucidate and expand on the findings of the present study.

> Revised, specific topic sentence: Given the high incidence of sleep disorders such as insomnia among individuals over 70 years of age, future studies should investigate the effects of hypnosis on sleep quality in this population.

Topic Sentences That Are *Too* Specific

Because it's such an important point, I have to raise it again: Readers benefit greatly from topic sentences that contain specific cues about the topics, messages, and goals of paragraphs. But sometimes writers can go too far with specificity. Revising the vague topic sentence from my hypothetical research proposal on meditation and stress-related diseases, I happened to get a little carried away. Here's how the sentence turned out:

> The proposed study on the long-term effects of concentration-based and mindfulness meditation on physiological responses to psychological stress is unique compared to previous studies, which have focused exclusively on short-term acute effects, because it affords the first opportunity to determine how individuals who have meditated regularly for long-term periods of at least 6 months to more than 5 years respond physiologically, as will be determined by changes in serum glucocorticoid and epinephrine concentrations, to psychologically stressful conditions such as timed mathematical tests, exposure to video of gruesome traffic accidents, and exposure to noxious noises.

Now that's just way *too* specific! It's an overload of advance information that will unquestionably overwhelm readers. You likely will not have to deal with such an extreme case in your own drafts, but you should still check your topic sentences for excessive specificity. First, look for details that are not essential to present in the topic sentence because they will play a supporting role in the paragraph's body. For instance, the preceding example does not need the details about how long individuals have meditated ("periods of at least 6 months to more than 5 years") and about the psychological stressors that I am proposing to study ("timed mathematical tests, exposure to video of gruesome traffic accidents, and exposure to noxious noises"). I will include that information after the topic sentence. Second, look for terms that readers might not have the background knowledge to understand. Suppose that I am planning to submit my research proposal to a funding organization that does not specialize in physiological research. How might reviewers respond to the undefined phrase "serum glucocorticoid and epinephrine concentrations"? For such an audience, I had better use a more general term, like "stress hormones" in the topic sentence. Then, in the body of the paragraph, I will introduce and define the more specific terms.

REVISING FOR COHERENCE

As I was watching the World Series on television one late-October evening several years ago, it occurred to me that pitching baseballs is nothing at all like writing paragraphs in scientific papers. The baseball pitcher's goal is to trick the batter by varying the ball's speed and trajectory from throw to throw. A fastball might be followed by a curve, a slider, a change-up, or a knuckleball. Only the pitcher and catcher know what's coming up next. If the batter cannot anticipate how fast and where the next ball will arrive at the plate, a swing-and-miss is in the offing, which is naturally what the pitcher desires. The situation is completely different for skilled scientific authors. In crafting their paragraphs, they want their pitches—that is, the ideas in consecutive sentences—to be highly predictable based on the meaning, logic, and structure of previous sentences. When ideas come at readers erratically from sentence to sentence, the result is a breakdown in the sense of paragraph flow. As defined in Chapter 4, this is the quality that writing teachers call *coherence*. In response to paragraphs that lack coherence, readers must take repeated swings in order to resolve unexpected change-ups and to figure out how the ideas in successive sentences relate to one another. The effort wastes mental energy and obscures a sense of overall paragraph unity.

So a commendable aim for writing paragraphs is to throw home-run pitches with every sentence. The reality, however, is that even the best writers inadvertently throw a few strikes every now and again in first drafts. The results are gaps and breaks in coherence. This section of the chapter focuses on revising common coherence problems. They include disordered ideas, missing knowledge links, oversights of readers' expectations, and ineffectively structured sentences. We have already addressed the most common coherence flaw, which is a sudden shift in the topic, message, or purpose across two sentences. As demonstrated earlier in Figures 6.4 (pages 264–265) and 6.8 (page 272), such shifts are responsible for cases of fractured and frazzled unity. As you will see in the upcoming sections, many of the same solutions for resolving paragraph disunity also apply to improving paragraph coherence.

Disordered Ideas

Some paragraphs have all the right ideas but lack coherence because their ideas are not effectively ordered. To demonstrate how to revise paragraphs with disordered ideas, let's use the example in Figure 6.13 (page 288). It's a draft paragraph from a student's critical review paper on behavioral therapies for attention-deficit hyperactivity disorder (ADHD) in children and adolescents. The author, named Roberta, was writing to an audience of family physicians who treat ADHD primarily with conventional drugs. Roberta assumed that her readers are not particularly familiar with the research literature on alternative behavioral therapies. In her draft paragraph, presented as follows, Roberta reviews studies on music therapy.

(1) Controls for confounding effects of medications are essential in studies on behavioral treatments for attention-deficit hyperactivity disorder (ADHD) in youth. (2) Lynch's apparently positive results should be viewed with caution because the study lacked a control group. (3) The study was the first to investigate effects of calming music as a treatment for ADHD in a younger pediatric population. (4) Several studies have been conducted to determine the effects of calming background music on ADHD symptoms in teenagers. (5) At baseline and after the therapy, tests were administered to measure the subjects' attention span and activity level. (6) The results indicated favorable effects of calming music in this teenage population. (7) Thirty children (mean age = 8.7 years) with physician-confirmed diagnoses of ADHD participated in eight weekly sessions of music therapy, with each session lasting 35 minutes. (8) Compared to baseline measures, the post-treatment results indicated that subjects demonstrated significant increases in attention span and decreases in disruptive classroom behaviors, as well as improved scores on math tests. (9) These results led Lynch to conclude that the behavioral intervention was highly efficacious. (10) However, because Lynch's subjects were on Ritalin, the most commonly prescribed medication for ADHD, it is not possible to conclude that the positive outcomes were attributable to the music therapy.

FIGURE 6.13 A case of disordered ideas.

As in many cases of paragraph incoherence, this one elicits the initial response, "It just doesn't flow very well." The main problem is that key ideas are not ordered effectively from sentence to sentence. Roberta diagnosed the problem by taking the perspective of her audience and analyzing the flow of ideas. She narrowed in on several specific places where the paragraph's coherence is disordered.

- A few words into sentence 2, Roberta's readers will stop and ask, "*Which apparently positive results from Lynch's study are you referring to? And what was Lynch's study about, anyway?*" (Given that Roberta's readers do not specialize in alternative therapies for ADHD, she realized that Lynch's study would likely be unfamiliar to them.) The answers to these questions do not come until much later in the paragraph, beginning in sentence 7.

- Readers will have difficulty getting through sentence 3 because they will be carrying a lingering question about sentence 2: "Why does the lack of a control group in Lynch's study warrant caution in viewing the results?" Roberta does not answer this vital question until sentence 10, where she alludes to the problem that Lynch's subjects were taking the ADHD drug Ritalin.
- In the middle of sentence 5, readers will likely be confused about whether the tests for attention span and activity level were administered in Lynch's study or in the previous studies on the effects of calming background music on ADHD symptoms in teenagers. The confusion does not get resolved until the end of sentence 6, where Roberta casually refers to "this teenage population."
- Readers might not fully grasp Roberta's conclusion in sentence 10—the conclusion that it is not possible to attribute Lynch's observed improvements in ADHD symptoms directly to the music therapy. Even though Roberta suggests that the results might have been due to Ritalin, a very important idea to complete her argument is presented way back in sentence 2—it's the idea that Lynch's study did not include a control group.

After identifying the disordered ideas in her draft, Roberta proceeded to rearrange them. She relied on several of the principles for organizing content that were presented in Chapter 4 (Figure 4.5, page 161) and demonstrated, for revising the global structure of scientific papers, in Chapter 5. Recall that one of the principles is to adopt conventional approaches to organizing content. In her revision process, Roberta applied this principle by examining published critical review papers in which authors sought to accomplish the same rhetorical goal that she aimed for in her draft paragraph. The goal is to summarize and critique study methods and results. Roberta looked for effective patterns in how experienced authors ordered their ideas. For example, it's customary (and plain sensible) to begin reviewing a study by describing key details about its methods. Next, a summary of the study's most relevant results is in order. This essential background information naturally sets up the author's ensuing critique of the study. In several places, Roberta's draft paragraph did not follow this conventional and common sense order. So Roberta used the conventional pattern to guide her revision. She also applied the organizing principle to rely on audience analysis to make sound decisions about ordering ideas. Roberta reorganized her draft paragraph by reconsidering the order of ideas that would best serve her readers' needs and expectations. Roberta's revised paragraph is presented as follows (Figure 6.14).

(1) Controls for confounding effects of medications are essential in studies on behavioral treatments, including music therapy, for attention-deficit hyperactivity disorder (ADHD) in youth. **(2)** Several highly controlled studies have been conducted to determine the effects of calming background music on ADHD symptoms in

FIGURE 6.14 A revision of the draft paragraph in Figure 6.13.

teenagers. **(3)** At baseline and after the therapy, tests were administered to measure the subjects' attention span and activity level. **(4)** The results indicated favorable effects of calming music in this teenage population. **(5)** A study conducted by Lynch was the first to investigate calming music as a treatment for ADHD in a younger pediatric population. **(6)** Thirty children (mean age = 8.7 years) with physician-confirmed diagnoses of ADHD participated in eight weekly sessions of the music therapy, with each session lasting 35 minutes. **(7)** Compared to baseline measures, the post-treatment results indicated that the children demonstrated significant increases in attention span and decreases in disruptive classroom behaviors, as well as improved scores on math tests. **(8)** These results led Lynch to conclude that the behavioral intervention was highly efficacious. **(9)** Lynch's apparently positive results should be viewed with caution because most of the subjects were taking Ritalin, the most widely prescribed drug for ADHD; in addition, the study lacked a control group. **(10)** Thus, it is not possible from Lynch's study to conclude with certainty that the positive outcomes were attributable strictly to the behavioral therapy.

FIGURE 6.14 *continued.*

The improved paragraph begins with a topic sentence that grounds readers to the essential point of Roberta's critical review of research on music therapy for ADHD. In sentences 2 through 4, Roberta briefly reviews the studies on music therapy and ADHD symptoms in teenagers. Recall that in her draft, the reference to those studies got tangled up with Roberta's summary of Lynch's study on younger children. In sentence 5, Roberta begins presenting the key background information about Lynch's study. To address the needs of audience members who have never read the study, Roberta describes its key methods and results *before* bringing up the possible flaw involving the study's lack of control for ADHD medications. After presenting Lynch's conclusion in sentence 8, in sentence 9 Roberta raises her criticism that Lynch's results should be viewed with caution because the study lacked controls for subjects who were taking Ritalin. Finally, Roberta reaches her own conclusion about Lynch's study in sentence 10. To best appreciate the improvement in coherence, compare the revised paragraph to the draft in Figure 6.13.

Missing Knowledge Links (When They're Needed)

To understand the collective meaning of consecutive sentences in paragraphs, readers must connect their ideas with knowledge that may range from simple facts to well-developed ideas and concepts. Gaps in paragraph coherence occur when writers omit *necessary* knowledge links—that is, the background information that readers will need in order to understand relationships in meaning between successive sentences. Here's a case in point:

> Studies show that individuals who are prone to anger and hostility produce abnormally high concentrations of homocysteine. This may explain why individuals with these personality traits have a high risk of suffering heart attacks and strokes.

At least on the surface, the basic message of the first sentence is clear: Plenty of something called homocysteine is present in the bodies of people who tend to be angry and hostile. The surface meaning of the second sentence is also fairly straightforward: A high risk of heart attacks and strokes might be explained by the high concentrations of homocysteine in people who have the traits of anger and hostility. Notice, however, that the deeper meaning of the two sentences combined is not so immediately apparent. To understand the collective meaning, readers must link the two sentences with knowledge that the writer has omitted. Specifically, readers must know about homocysteine and its role in cardiovascular disease. A vital fact is that homocysteine is an amino acid that, when produced by the body in high concentrations, can damage the linings of coronary blood vessels and thereby contribute to cardiovascular disease. Readers who do not possess this knowledge might complain that the two sentences in the example above are incoherent. Their complaint would indeed be justified if the readers are part of the writer's intended audience. This is the key point about missing knowledge links: They are problematic only if the writer's intended readers actually need them.

Catching omissions of essential knowledge links in drafts requires audience analysis. As presented in Chapter 1, a guiding question for this process asks, "What do your readers know and need to know about your topic?" Applied to paragraph revision, the answers to this question will help you identify terms and concepts that you should define and explain to maintain sentence-to-sentence coherence. This revision strategy is demonstrated as follows.

> Studies show that individuals who are prone to anger and hostility produce abnormally high concentrations of homocysteine. This amino acid contributes to cardiovascular disease by damaging the inner linings of the coronary arteries and promoting blood clots. Thus, high concentrations of homocysteine may explain why individuals with the personality traits of anger and hostility have a high risk of suffering heart attacks and strokes.

Oversights of Readers' Expectations

As demonstrated by the case of missing knowledge links, paragraph coherence is weakened when writers overlook the needs of their intended readers. A closely related flaw results when a sentence leads readers to form an expectation that is not fulfilled by the next sentence(s). See if your expectations are satisfactorily met in the following paragraph (Figure 6.15).

(1) There have been many anecdotal reports of health benefits associated with emotional states such as joy, optimism, and compassion. **(2)** Proponents of this supposed connection have hypothesized that positive emotions stimulate the brain to transmit neural signals that activate the peripheral immune system. **(3)** However, based on experimental findings, the effects of positive emotions on the immune system and clinical health outcomes are extremely controversial. **(4)** Many studies have reported that high levels of job satisfaction and strong social networks are associated with enhanced immune function. **(5)** To determine whether there is a true cause-effect relationship, researchers must assess specific immune responses, such as *in vivo* T-cell proliferation and natural killer cell activity.

FIGURE 6.15 A case of overlooking readers' expectations.

Did you notice the gap across sentences 3 and 4? The topic of sentence 3 concerns the effects of positive emotions on the immune system and clinical health outcomes; the sentence's message is that these effects are *extremely controversial*. Reflecting on this idea, what do you expect sentence 4 to be about? You might logically anticipate that sentence 4 will elaborate on the extreme controversy regarding the research on positive emotions, the immune system, and clinical health outcomes. Or you might expect sentence 4 to present some of the controversial experimental findings that sentence 3 refers to. Sentence 4, however, does not meet these natural expectations. Its topic concerns job satisfaction and social networks, which are not directly related to the topic of positive emotions. So, across sentences 3 and 4, there is a drastic shift in topics. Moreover, sentence 4 does not meet our logical expectations for elaboration on the extreme controversy regarding previous studies on positive emotions, immune system function, and clinical health outcomes. Neither does sentence 5.

To detect and diagnose oversights of readers' expectations independently, you have to put yourself in your readers' shoes and exercise your imagination. For any given sentence, what ideas might readers expect based on the ideas in the previous sentence (or the previous set of related sentences)? A useful tactic is to focus attention on word groups that convey the topic and message of the previous sentence(s). These word groups are often positioned at the beginnings and endings of sentences, respectively. It makes sense that in any given sentence, readers will expect the writer to address the topic and to elaborate on the message of the previous sentence(s).

The best solutions to the problem at hand depend on our original intentions for sentences that fall short of meeting readers' expectations and thereby create gaps in coherence. In one scenario, we deliberately intend for the (errant) sentences to meet readers' expectations, but we miss the mark in drafting them. As a case in point, suppose that the author of the preceding sample paragraph (Figure 6.15) had originally planned for sentence 4 to elaborate on the extreme controversy involving research findings on positive emotions, the immune system, and clinical heath outcomes. The best solution would be to smooth out the coherence gap by adding essential ideas that better address readers' expectations. In another scenario, we send readers off track by inadvertently setting up misleading signposts. For example, maybe the author of our sample paragraph did not intend for sentence 4 to elaborate on the *extreme controversy* involving positive emotions, the immune system, and clinical health outcomes. Perhaps the author brought up the controversy as an inconsequential bit of background information. The solution then is to eliminate the misdirecting cues and to redraft the misleading sentence to redirect readers' expectations appropriately.

Lack of Parallel Structure (When It's Needed)

Included in the upcoming chapter is a demonstration of how to revise a sentence-level problem called *lack of parallel structure*. Here's a sentence that contains the flaw:

> Antidepressant drugs may reduce symptoms of depression by inhibiting the neuronal reuptake of serotonin, inducing gene expression of neurotrophic factors, and the production of corticotropin-releasing hormone can be decreased.

Something is awkward and out of place, isn't it? That something is the independent clause at the sentence's end: *the production of corticotropin-releasing hormone can be decreased.* (An independent clause has a subject and verb, and it expresses a complete idea.) The sentence lacks parallel structure because the independent clause is not consistent in grammatical form with the two word groups that precede it: *inhibiting the neuronal reuptake of serotonin* and *inducing gene expression of neurotrophic factors.* These word groups are participial phrases. (A phrase does not express a complete idea.) Notice that the three word groups have the same function in the sentence, which is to describe how antidepressant drugs may reduce symptoms of depression. When word groups within a sentence are coordinated in function, they should usually be expressed in parallel grammatical form. The following revision applies this useful guideline and thereby resolves the awkwardness of the original sentence.

> Antidepressant drugs may reduce symptoms of depression by inhibiting the neuronal reuptake of serotonin, inducing gene expression of neurotrophic factors, and decreasing the production of corticotropin-releasing hormone.

The guideline for parallel structure within sentences has a counterpart at the paragraph level: When a paragraph's ideas are coordinated in function, the word groups expressing those ideas should be structured consistently across sentences. Here's a draft paragraph that lacks parallel structure (Figure 6.16):

> **(1)** In many studies on both animals and humans, researchers have determined the effects of various physical and psychological stressors on the immune system and immune-related diseases. **(2)** For example, reductions in the rate of lymphocyte proliferation in the thymus gland of dogs have been observed in response to exposure to noxious sound. **(3)** Electric tail-shock suppresses normal natural-killer (NK) cell activity in mice. **(4)** Tumor development is an outcome of these immune-weakening effects. **(5)** Additional evidence on spousal caregivers of Alzheimer's disease patients reveals correlations between periods of heightened psychological stress and an increased risk of infection in response to influenza vaccination caused by reduced antibody responses. **(6)** Finally, serum concentrations of glucocorticoids increase with the stress of social isolation in primates. **(7)** This effect reduces the proliferation and migration of macrophages, which increases susceptibility to various pathogenic diseases.

FIGURE 6.16 A paragraph that lacks parallel structure.

On first read, the paragraph elicits a general sense of incoherence. A deeper diagnosis requires analyzing the meaning, structure, and function of the paragraph's sentences. Sentence 1, the topic sentence, conveys the main message that researchers have determined how, in animals and humans, the immune system and immune-related diseases are influenced by various forms of stress. As we might expect, the following sentences summarize the research that has been conducted on this topic. Notice that sentences 2 through 7 contain information about four aspects of the research:

(1) Who or what the subjects were—dogs, mice, humans, and primates.

(2) The type of stressor studied, which included exposure to noxious sound, electric tail-shock, the psychological stress of caring for Alzheimer's patients, and social isolation.

(3) How some part of the immune system is affected by the stressor—for example, reductions in the rate of lymphocyte proliferation, suppressed natural-killer cell activity, and reduced antibody responses.

(4) The implications for disease outcomes, which include tumor development and an increased risk of infection.

In summarizing studies on stress and immune system responses, sentences 2 through 7 serve a unified purpose. Their ideas, specifically the four aspects of the studies summarized, are inherently coordinated. If the ideas were expressed in parallel structure, they would be positioned consistently across consecutive sentences. But they are not. For example, in sentence 2 the word group conveying information about the type of stressor studied (*exposure to noxious sound*) comes at the very end. In contrast, in sentence 3 the stressor (*electric tail-shock*) is placed at the very beginning. Then, in sentence 5 the stressor (*heightened psychological stress*) is placed in the middle. The positioning of word groups conveying information about the studies' subjects, as well as information about the immune responses to stress, also varies unpredictably from sentence to sentence. You can check it out for yourself. The lack of parallel structure of the coordinated ideas underlies the sample paragraph's lack of coherence. To better appreciate this point, compare the draft in Figure 6.16 to the following revision (Figure 6.17).

(1) In many studies on both animals and humans, researchers have determined the effects of various physical and psychological stressors on the immune system and immune-related diseases. **(2)** For example, studies on dogs have shown that exposure to noxious sound reduces the rate of lymphocyte proliferation in the thymus gland. **(3)** In mice, electric tail-shock suppresses normal natural-killer (NK) cell activity. **(4)** These immune-weakening effects have been shown to promote tumor development. **(5)** Additional evidence reveals that in spousal caregivers of Alzheimer's disease patients, periods of heightened psychological stress are correlated with reduced antibody responses to influenza vaccination, leading to an increased risk of infection. **(6)** Finally, in

FIGURE 6.17 A revision of the draft paragraph in Figure 6.16.

> primates the stress of social isolation is associated with increases in serum concentrations of glucocorticoids. **(7)** This effect reduces the proliferation and migration of macrophages, which increases susceptibility to various pathogenic diseases.

FIGURE 6.17 *continued.*

Compared to the draft, the revision is considerably more coherent; consequently, it is much easier to understand and remember. Sentence for sentence, the ideas are the same in the draft and the improved paragraph. The only difference is the positioning of word groups that express those ideas. The revised paragraph in Figure 6.17 is, for the most part, structured in parallel form. For example, sentences 2, 3, 5, and 6 have a similar grammatical pattern for expressing their naturally coordinated ideas. At or close to the start of each sentence, the animal or human studied is named in a prepositional phrase:

studies on dogs (sentence 2), *in mice* (sentence 3), *in spousal caregivers of Alzheimer's patients* (sentence 5), and *in primates* (sentence 6).

These phrases are closely followed by the main grammatical subject, which names the stressor studied:

exposure to noxious sound (sentence 2), *electric tail-shock* (sentence 3), *periods of heightened psychological stress* (sentence 5), and *the stress of social isolation* (sentence 6).

After a verb or verb phrase that describes an action or condition about the stressor, each sentence ends with a noun phrase that expresses the aspect of immune-related function that is affected:

the rate of lymphocyte proliferation in the thymus gland (sentence 2), *normal natural-killer (NK) cell activity* (sentence 3), *reduced antibody responses to influenza vaccination, leading to an increased risk of infection* (sentence 5), and *increases in serum concentrations of glucocorticoids* (sentence 6).

As you were reading the revised paragraph in Figure 6.17, perhaps you recalled former English teachers urging students to spice up their writing by varying sentence structure. As addressed shortly in this chapter, sometimes added variety is indeed essential for improving weak paragraphs. But when the ideas in successive sentences are naturally coordinated, the repetition of structure (a deliberate lack of variety) actually enhances readability and comprehension.

REVISING FOR COHESION

Technically, the term *paragraph coherence* refers to how sentences relate to each other at the relatively deep levels of meaning, logic, and structure. However, coherence can be influenced by simpler features that are closer to the surface of sentences—that is, by the effects of single words and short phrases that serve as transitions and links (such as the word "however" at the start of this sentence and the phrase "that is" after the dash). Writing instructors use the term *cohesion* to describe the sentence-to-sentence flow that is aided by surface transitions and links, or what we will call *cohesion cues*. Figure 6.18 presents common cohesion cues, which are grouped by the functions they serve.

Qualifying or clarifying cues signal that a sentence explains, clarifies, or exemplifies an idea in the previous sentence.

> Examples: for example, for instance, to illustrate, specifically, that is, indeed, of course

Cause-effect cues signal that a sentence conveys a result, or an outcome, associated with an idea in the previous sentence.

> Examples: therefore, thus, consequently, because, as a result, hence, accordingly, thereby

Contrast cues signal that a sentence conveys an idea that opposes, or conflicts with, an idea in the previous sentence.

> Examples: however, but, whereas, although, nevertheless, yet, still, conversely, in contrast, on the other hand, on the contrary, notwithstanding, even so

Addition cues signal that a sentence conveys an idea that has a cumulative or collective relationship with an idea in the previous sentence.

> Examples: in addition, also, and, moreover, furthermore, further, next, besides, again

Comparison cues signal that a sentence conveys an idea that agrees with an idea in the previous sentence.

> Examples: similarly, likewise, also, in agreement, in the same way

Time cues signal the chronological relationship between ideas in successive sentences.

> Examples: before, earlier, until now, formerly, recently, immediately, presently, now, meanwhile, after, later, next, following, thereafter

Summary cues signal that a sentence conveys a synopsis or conclusion based on ideas in the previous sentence or set of related sentences.

> Examples: In conclusion, in summary, to summarize, therefore, taken together, in other words, on the whole

Meaning links are words and phrases in a sentence that repeat key ideas from the previous sentence or set of related sentences. Meaning links can be verbatim repetitions of key words and phrases, or they can be synonyms, pronouns, or similar ideas that are expressed in different words. In the following paragraph, meaning links are highlighted in bold font; the ideas they repeat, from the previous sentence, are underlined.

FIGURE 6.18 Common cohesion cues.

Researchers have shown that depression is associated with reduced levels of serotonin, dopamine, and norepinephrine in the brain. **The deficiency of these chemical signaling molecules** may cause depressive symptoms by impairing normal synaptic transmission in the brain. **This abnormal condition** can be reversed by various pharmacological treatments. **The most common treatments** are categorized as monoamine oxidase (MAO) inhibitors, tricyclic drugs, or selective serotonin reuptake inhibitors (SSRIs). Ultimately, **these drugs** reverse depressive symptoms by increasing the availability of serotonin, dopamine, and norepinephrine to brain cells.

FIGURE 6.18 *continued.*

Cohesion cues function somewhat like bridges and signposts. In this analogy, a cohesion cue that serves as a bridge is word or phrase that connects the topic, message, or goal of a sentence to a counterpart in the previous sentence or set of related sentences. More formally, these bridges in paragraphs are called *meaning links*. As defined in Figure 6.18, they are repetitions of key words and phrases across sentences. Meaning links are synonymous with the unity-grounding cues that were described earlier in the chapter. In contrast to bridges, which link backwards, signposts are words and phrases that give readers a heads-up about the direction of upcoming ideas. In Figure 6.18, the examples of contrast cues, addition cues, and comparison cues are analogous to signposts. When used properly, cohesion cues can greatly enhance paragraph flow. But as we consider next, cohesion cues are not always used properly.

Missing Cohesion Cues

If cohesion cues are like helpful bridges and signposts, their absence at critical junctions in paragraphs can cause readers to lose their way by missing important points, making wrong turns in interpreting meaning, and stumbling across sentences. To appreciate the advantages of appropriately used cohesion cues, you might try removing them from successful paragraphs and then assessing the effect on readability and comprehension. This is just what I have done with a well-written paragraph from a published research paper about molecular markers of major depressive disorder (MDD). The paper reports a study conducted by Salvatore Alesci and coworkers, who measured plasma (blood) concentrations of a molecule called interleukin-6 (IL-6) in patients with MDD over a 24-hour period (Alesci et al., 2005). To understand Alesci et al.'s paragraph, which comes at the end of the research paper's introduction section, you will need to know that concentrations of IL-6 have a circadian rhythm, rising and falling predictably over the course of each day. In mentally healthy individuals, IL-6 concentrations are low during the morning hours. You will also need to know that the researchers measured concentrations of a hormone called cortisol. When we experience stressful situations, cortisol is released via a physiological pathway called the hypothalamic-pituitary-adrenal axis, or what Alesci et al. refer to as the *HPA axis*. With my apologies to the authors, Figure 6.19 presents a version of Alesci et al.'s paragraph from which I have removed all of the word groups that originally

---

(1) Elevated levels of plasma IL-6, assessed by single morning measurements, have been previously reported in subjects with MDD (34-36). **(2)** The contribution of potential confounders, such as age, gender, and body composition, was overlooked. **(3)** A single-time-point measurement of IL-6, undermining important temporal considerations, was used. **(4)** We compared plasma IL-6 levels, measured hourly around the clock, in patients with MDD and healthy controls, individually matched by gender, age, and BMI. **(5)** Considering the relevance of HPA axis dysregulation in patients with MDD, and the activating role of IL-6 on the HPA axis, we analyzed the relationship between IL-6 and cortisol secretion in our study groups.

FIGURE 6.19 A paragraph with its cohesion cues removed. As explained in the text, the paragraph comes from a well-written published research paper authored by Alesci et al. (2005). Copyright The Endocrine Society.

served as cohesion cues. (After removing the cues, I had to make a few minor structural changes to ensure that the sentences were grammatically complete.)

It's a bumpy read, requiring many stops and starts to figure out how the sentences' ideas relate to one another. Midway into sentence 2, for example, you might stop to ask, "What idea does *the contribution of potential confounders* relate to in sentence 1?" Then, at the end of sentence 2, you might wonder *who* overlooked the contribution of potential confounders. The topics of sentences change sharply without any warning, which breaks our reading momentum. An example is the topic of sentence 3: *A single-time-point measurement of IL-6*. To find a connected thread, we have to look all the way back to the start of sentence 1, which refers to IL-6. Another rough transition occurs across sentences 4 and 5, the latter of which abruptly raises a new idea about *the relevance of HPA axis dysregulation in patients with MDD*. From sentence to sentence, the ideas are obviously not well connected. The following paragraph (Figure 6.20) (page 300), taken verbatim from Alesci et al.'s paper, adds back the authors' original cohesion cues, which I have underlined for emphasis.

(1) Elevated levels of plasma IL-6, assessed by single morning measurements, have been previously reported in subjects with MDD (34-36). **(2)** However, these reports overlooked the contribution of potential confounders, such as age, gender, and body composition, to this biological abnormality. **(3)** Furthermore, they relied on a single-time-point measurement of IL-6, undermining important temporal considerations. **(4)** To overcome these limitations, we compared plasma IL-6 levels, measured hourly around the clock, in patients with MDD and healthy controls, individually matched by gender, age, and BMI. **(5)** Additionally, considering the relevance of HPA axis dysregulation in patients with MDD, and the activating role of IL-6 on the HPA axis, we also analyzed the relationship between IL-6 and cortisol secretion in our study groups.

FIGURE 6.20 Cohesion cues added back into Alesci et al.'s (2005) original paragraph. Copyright The Endocrine Society.

Just a few added words and phrases appreciably improve the paragraph's coherence, in addition to its unity. Take the transitional word *however* at the start of sentence 2. It's a signpost that alerts readers to a caveat concerning the main message of sentence 1. With just one word, the authors tell us, "We want you to notice something important about the previous reports of elevated levels of plasma IL-6 in subjects with major depressive disorder—look out for a contrasting message coming up in the present sentence (sentence 2)." The authors' advance warning gives us confidence in the direction we are heading and deepens our understanding of the relationship in meaning between the first two sentences.

In addition to its opening signpost, sentence 2 has two bridges. The first is the meaning link, *these reports*. This phrase clearly refers to the previous sentence's summary of the research finding about elevated levels of plasma IL-6. To appreciate how this meaning cue enhances the paragraph's flow, go back to Figure 6.19 and read the paragraph with the cue removed. The second bridge in sentence 2 comes in its ending phrase, *this biological abnormality*. It's a helpful link back to a key idea in sentence 1, which is that in depressed patients, levels of plasma IL-6 are elevated during the morning hours—this is the biological abnormality (because, as mentioned earlier, in mentally healthy individuals, plasma IL-6 levels are low in the morning). The bridge at the end of sentence 2 is critical for conveying the main message of sentences 1 and 2 combined: Previous reports of the abnormal condition in MDD patients should be viewed with caution because the studies overlooked potential confounders. This message does not come through clearly in the version of Alesci et al.'s paragraph from which the cohesion cues are removed.

In Alesci et al.'s original paragraph (Figure 6.20), notice the signpost at the start of sentence 3—it's the word *furthermore*. This cue alerts readers to an upcoming idea that logically adds on to previous ideas. The effect, however, runs deeper than the surface link. The word *furthermore* reinforces the paragraph's unity of purpose, or its rhetorical goal. After reading sentences 1 and 2, we have a fairly strong sense that the authors' goal is to discuss methodological limitations (the *potential confounders*) to previous studies on plasma IL-6 levels in subjects with major depressive disorder. From the combined message of the first two sentences, we know that one methodological limitation is that the previous studies have overlooked potential extraneous variables, including age, gender, and body composition. Reading just one word at the start of sentence 3, the word *furthermore*, we instantly know that another methodological limitation will be presented. This advance notice very effectively reinforces our sense of the paragraph's goal. As specified in the rest of sentence 3, the additional methodological limitation is that previous researchers have measured IL-6 only at one point in time during the day.

Take a look at the phrase that begins sentence 4 in Alesci et al.'s original paragraph: *To overcome these limitations*. This phrase serves the combined role of signpost and bridge. The bridge is the phrase *these limitations*, which directly links back to the methodological shortcomings that Alesci et al. wrote about in sentences 2 and 3. The signpost is the phrase *to overcome*, which hints at the authors' intention to describe their study's improved methods. The rest of sentence 4 provides the details, informing readers that Alesci et al. measured IL-6 levels every hour and that they controlled for extraneous variables by matching subjects for gender, age, and body mass index (BMI). Then, the last sentence of the paragraph begins with the cohesion cue *additionally*, giving advance notice of another method that the authors employed to overcome limitations of previous studies. Again, to best appreciate the effects of the cohesion cues, compare Alesci et al.'s original paragraph (Figure 6.20) with its cue-less version (Figure 6.19).

The problem of missing cohesion cues—that is, when their inclusion would be helpful for readers—can be difficult to detect in independent revision. How do you revise something that is *not* in your draft? The trick is to pay attention to potentially shaky transitions in ideas across your draft's sentences. It might help to read out loud, listening for slowing and hesitation in your voice. Another useful approach is to parse the topics, messages, and purposes of successive sentences in especially complex paragraphs. If you detect sharp deviations across consecutive sentences, try adding appropriate cohesion cues to improve the flow.

Misplaced Cohesion Cues

In popular and nontechnical literature, cohesion cues are commonly placed in the middle of sentences or toward their endings. In these positions, the cues may serve stylistic and rhetorical purposes such as balancing out a sentence's word groups and giving readers a chance to pause and reflect. You will see many places in this book, for example, where cohesion cues are embedded in the middle of sentences, as in this sentence and the next one. In scientific writing, however, cohesion cues more commonly show up at the very beginning of sentences, which is fitting. There they best serve their most important functions, which are to indicate relationships in meaning between successive sentences and to orient readers to upcoming changes in direction. When a paragraph's ideas are especially complex, the positioning of a cohesion cue anywhere *but* the start of a new sentence is like placing a road sign for a sudden turn long after drivers have passed the turn.

In the following excerpt from Alesci et al.'s paragraph, I have moved two cohesion cues away from the start of their sentences, where they originally appeared.

> Elevated levels of plasma IL-6, assessed by single morning measurements, have been previously reported in subjects with MDD. These reports overlooked the contribution of potential confounders, however, such as age, gender, and body composition, to this biological abnormality. They relied on a single-time-point measurement of IL-6, furthermore, undermining important temporal considerations.

To appreciate the problem of misplaced cohesion cues in the preceding excerpt, compare it to the original version, as follows.

> Elevated levels of plasma IL-6, assessed by single morning measurements, have been previously reported in subjects with MDD. However, these reports overlooked the contribution of potential confounders, such as age, gender, and body composition, to this biological abnormality. Furthermore, they relied on a single-time-point measurement of IL-6, undermining important temporal considerations.

It's a minor revision, moving the signposts *however* and *furthermore* from the middle to the start of their sentences; however, the effect on readability is not trivial by any means.

Unnecessary Cohesion Cues

From grade school to graduate school, some teachers forbid the use of *any* transitional words and phrases in student papers. In red ink, these teachers strike through all instances of *however, for example, in contrast, therefore,* and so on. They do so probably to prevent students from writing paragraphs like the following one (Figure 6.21).

> **(1)** Whereas numerous studies have shown that anxiety weakens the immune system, others, in contrast, have revealed that anxiety can actually enhance immunity. **(2)** For instance, anxiety may have this positive effect to aid us in maintaining health during life situations that are especially stressful and threatening. **(3)** In addition, people also commonly become sick with immune-related

FIGURE 6.21 A case of overused cohesion cues.

illnesses, such as the common cold, when anxiety-provoking situa-tions in daily life are diminished. **(4)** Therefore, anxiety may have a protective effect against disease. **(5)** On the other hand, there is a substantial amount of contrasting research evidence to indicate otherwise. **(6)** However, future studies are necessary to determine the degree to which individual differences might account for the effects of anxiety on the risk for developing immune-related dis-eases. **(7)** Furthermore, these future studies will offer insights into whether individual differences are due to genetic factors.

FIGURE 6.21 *continued.*

Notice that every sentence has at least one cohesion cue, which should indeed raise a red flag during the revision process. Even if all of the transitional words and phrases reflect logical relationships between the ideas in successive sentences, their excessive use can distract readers and undermine the sense of paragraph flow. But no good reason exists for eliminating cohesion cues altogether. When used appropriately they are very helpful devices, especially for smoothing transi-tions across sentences that have complex relationships in meaning and logic. So the best advice for revision is to think carefully about whether your readers will truly need and benefit from the cohesion cues in your draft. If any cues are not essential, delete them.

REVISING FOR SENTENCE VARIETY

If your former writing teachers harped on adding variety to spice up academic prose, you might find something curious about the paragraph-level problems that we have covered so far in this chapter: Many of the problems are attribut-able to *too much* variety. For example, paragraph disunity is often caused by variations in the topics, messages, and goals of successive sentences. In addi-tion, when a paragraph's ideas are coordinated in function, coherence is com-promised if they are structured inconsistently. The main strategy for solving these sorts of paragraph-level problems is to use repetition, or to eliminate vari-ety. For example, we have seen cases in which unity is strengthened by making the subjects of consecutive sentences the same. Moreover, paragraph flow can be enhanced by organizing successive sentences by the same (parallel) gram-matical structure. The problem of excessive sameness—in other words, a lack of sentence variety—is evident in the following draft paragraph (Figure 6.22) (page 304) from the introduction section to a review paper on the association between religious belief and physical health.

> Religion is very important to many people in our society. Participation in religious groups and activities has increased recently. Marsh (2006) investigated interests in religion and spirituality. The study found 84% of Americans reporting a strong interest. This significantly surpasses the 56% reported only five years earlier (Marsh, 2001). Religious practice is recognized as important for social and mental well-being. The physical health benefits of religion and spirituality are controversial. Some studies have shown less disease in religious versus non-religious individuals. Their results are typically flawed by extraneous variables. A critical review of the literature is necessary to resolve this issue.

FIGURE 6.22 A paragraph that lacks sentence variety.

Reading the paragraph out loud, you will clearly hear the problem: The short, jerky sentences create a tedious rhythm. The effect reminds me of the schoolroom scenes in television episodes of the cartoon *Peanuts*. The students drift off into space as Charlie Brown's teacher drones on endlessly in sentences that sound to the class like this: *Wah wha-wha wah. Wah wah wha-wha. Wha-wha wah wah.* The children never grasp the content of the lessons because all they hear is the monotonous rhythm of the teacher's voice. While a problematic lack of sentence variety is usually fairly easy to detect, the best solutions are not always obvious. Presented as follows, the diagnostic routines demand parsing the elements of discourse—words, phrases, ideas, and grammatical structures—that are repeated excessively in paragraphs and that elicit the Charlie-Brown-teacher effect.

Lack of Variety in Sentence Length

In the preceding example of a paragraph that lacks variety (Figure 6.22), all of the sentences contain between 8 and 12 words. This narrow range contributes the paragraph's dullness and choppiness, despite the fact that its ideas are really quite interesting. A lack of variety in sentence length underlies the reader's sense of tedium regardless of whether a paragraph's sentences are all short or all long. Imagine, for example, the dreary chore of reading a 10-sentence paragraph in which each sentence is between 45 and 50 words long. The most common flaw in scientific writing, however, is stringing together too many short sentences in a row. Novice writers may do this purposefully, misapplying the rule that scientific writing must be succinct. (Years ago, one of my students told me about a science professor who commanded his class to write simple sentences "like a fourth grader would." To this day, I still cringe at the thought!). To detect paragraphs that lack variety in sentence length, read your drafts out loud and listen to the rhythm of successive sentences, noting any choppiness or monotony.

In addition, you can use your word-processing program to count the words in consecutive sentences. As a general guideline, paragraphs with at least three consecutive sentences ranging between approximately 5 to 12 words are worth examining for insufficient variety in length. For solving the most common flaw, too many short sentences in a row, several useful techniques apply. They are demonstrated in the following paragraph (Figure 6.23), a revised version of the problematic draft in Figure 6.22.

> Religion is very important to many people in our society, and participation in religious groups and activities has increased recently. As revealed by Marsh (2006), 84% of Americans reported a strong interest in religion and spirituality, which significantly surpasses the value of 56% that the same author reported only five years earlier (Marsh, 2001). Religious practice is recognized as important for social and mental well-being. However, the physical health benefits of religion and spirituality are controversial. Whereas some studies have shown a lower incidence of cancer and cardiovascular disease in religious versus non-religious individuals, their results are typically flawed by extraneous variables. A critical review of the literature is necessary to resolve this issue.

FIGURE 6.23 A revision of the draft paragraph in Figure 6.22.

The revision eliminates the choppy rhythm that characterized the draft. Extending the draft's restricted range of 8 to 12 words per sentence, the revision adds variety by interspersing relatively short sentences with longer ones. The revised paragraph's sentences range from 11 to 34 words. The author's main revision strategy was to combine sentences using the techniques of *coordination* and *subordination*. Coordination entails combining sentences in which the ideas are equal in importance or similar in function. The main tools for combining these sorts of sentences are coordinating conjunctions: *and, but, or, nor, yet, for,* and *so.* In the revised paragraph in Figure 6.23, the author used coordination to combine the first two sentences of the draft paragraph:

> Religion is very important to many people in our society, and. Pparticipation in religious groups and activities has increased recently.

In this simple revision, the coordinating conjunction *and* joins two independent clauses containing ideas that the writer viewed as serving a similar function, which is to convey a message about the popularity of religion in our society.

In contrast to coordination, subordination entails combining sentences in which some ideas are relatively less important than others. In the revised paragraph in Figure 6.23, the writer used subordination as follows.

Draft: Some studies have shown less disease in religious versus non-religious individuals. Their results are typically flawed by extraneous variables.

Revision: Whereas some studies have shown a lower incidence of cancer and cardiovascular disease in religious versus non-religious individuals, their results are typically flawed by extraneous variables.

In the revision, the subordinating conjunction *whereas* sets up a dependent, or subordinate, clause (the word group leading up to the sentence's comma). The dependent clause conveys the idea that studies have revealed a positive association between religion and specific health outcomes. The writer makes this idea subordinate to the idea conveyed in the revised sentence's independent clause (everything after the comma). The independent clause conveys the idea that the results of previous studies are characteristically flawed. The first idea is subordinate, or relatively less important, because the writer's goal for the paragraph is to introduce the controversy and problems that are associated with research on religion and physical health. Instead of using the subordinating conjunction *whereas,* the writer could have substituted *although* or *while.*

The preceding revision demonstrates another strategy for adding variety to paragraphs composed of many short sentences in succession. In the original paragraph, the writer referred to studies showing "less disease in religious versus non-religious individuals." The revision refers to studies showing "a lower incidence of cancer and cardiovascular disease in religious versus nonreligious individuals." The revision adds 6 words to the original sentence by providing details about the diseases studied and by summarizing the results more precisely. The change opposes the common advice to say things as concisely as possible in scientific writing. In this case, however, the added detail effectively enhances the precision of ideas while adding variety to sentence length.

Depending on the logical relationship between sentences combined through subordination, the appropriate subordinating word might be one of the following: *as, because, if, since, unless, when, whether,* or *which.* In the following revision, the author used two subordinating conjunctions (*as* and *which*) to form one long sentence from three short ones.

Draft: Marsh (2006) investigated interests in religion and spirituality. The study found 84% of Americans reported a strong interest. This significantly surpasses the 56% reported only five years earlier (Marsh, 2001).

Revision: As revealed by Marsh (2006), 84% of Americans reported a strong interest in religion and spirituality, which significantly surpasses the value of 56% that the same author reported only five years earlier (Marsh, 2001).

Lack of Variety in Sentence Beginnings

As presented earlier, a strategy for improving paragraph unity and coherence is to begin successive sentences with the same subject. Of course, this strategy backfires when writers apply it too liberally, as in the following example (Figure 6.24).

> Corticotropin-releasing hormone (CRH) is a neuropeptide composed of 41 amino acids. CRH initiates the stress response through two pathways, the HPA axis and the sympathetic nervous system. CRH acts on the HPA axis when it is secreted from the hypothalamus. CRH is then transported to the anterior pituitary through a system of capillaries. The anterior pituitary responds by secreting adrenocorticotropin hormone (ACTH), which enters the systemic circulation and triggers the release of cortisol from the adrenal glands. CRH influences other brain regions in addition to the anterior pituitary. CRH can bind to the locus coeruleus and the amygdala, which are parts of the sympathetic nervous system. CRH thus stimulates activity of the sympathetic nerves through binding at these regions.

FIGURE 6.24 A paragraph that lacks variety in sentence beginnings.

One thing is for certain: We don't have to ask what the paragraph is about. It's obviously a well-unified definition of corticotropin-releasing hormone. However, the subject's repetition at the start of so many sentences, 7 of 8 sentences in all, is rather excessive. The lack of variety in sentence beginnings elicits the sense of monotony that can distract readers from understanding a paragraph's underlying meaning. This problem is not limited to consecutive sentences that begin with the exact same subject. It can also be caused by starting too many sentences with the same part of speech, the same word group, or the same grammatical structure. In the paragraph in Figure 6.24, compounding the glitch of a repeated verbatim subject, all of the sentences begin with a noun or a noun phrase that is followed immediately by a verb or a verb phrase. This overdone pattern contributes to the paragraph's monotonous tone. Helpful strategies for adding variety to sentence beginnings are presented as follows. In the sample revisions that follow, added text is underlined.

1. Use sentence-combining techniques (coordination and subordination) to eliminate instances of a verbatim simple subject that begins numerous consecutive sentences.

Revision through coordination: CRH acts on the HPA axis when it is secreted from the hypothalamus. ~~CRH~~ and is then transported to the anterior pituitary through a system of capillaries.

Revision through subordination: Corticotropin-releasing hormone (CRH), which is a neuropeptide composed of 41 amino acids, ~~CRH~~ initiates the stress response through two pathways, the HPA axis and the sympathetic nervous system.

2. Vary the literal subjects of successive sentences by using appropriate synonyms and pronouns.

Corticotropin-releasing hormone (CRH) is a neuropeptide composed of 41 amino acids. This hormone initiates the stress response through two pathways, the HPA axis and the sympathetic nervous system.

3. Rearrange word groups to form introductory phrases that delay a repeated subject.

Draft: Corticotropin-releasing hormone (CRH) is a neuropeptide composed of 41 amino acids. CRH initiates the stress response through two pathways, the HPA axis and the sympathetic nervous system.

Revision #1: Corticotropin-releasing hormone (CRH) is a neuropeptide composed of 41 amino acids. Through two pathways, the HPA axis and the sympathetic nervous system, CRH initiates the stress response.

Revision #2: Corticotropin-releasing hormone (CRH) is a neuropeptide composed of 41 amino acids. Acting on the HPA axis and the sympathetic nervous system, CRH initiates the stress response.

4. Form introductory transitional phrases that delay a repeated subject.

Draft: CRH influences other brain regions in addition to the anterior pituitary. CRH can bind to the locus coeruleus and the amygdala, which are parts of the sympathetic nervous system.

Revised: In addition to binding to the anterior pituitary, CRH binds to receptors in other brain regions. For example, it can bind to the locus coeruleus and the amygdala, which are parts of the sympathetic nervous system.

5. Create a bulleted list in which a sentence-beginning subject appears only once.

Corticotropin-releasing hormone (CRH)

- is a neuropeptide composed of 41 amino acids,
- initiates the stress response through the HPA axis and the sympathetic nervous system,
- acts on the HPA axis after secretion from the hypothalamus, and
- influences other brain regions in addition to the anterior pituitary.

Lack of Variety in Grammatical Structure

The draft paragraphs in Figures 6.22 and 6.24 demonstrate a lack of variety in sentence length and sentence beginnings, respectively. Both paragraphs also happen to lack variety in their grammatical structure. They are examples of parallel structure gone overboard. To describe the flaw, I will bring back the paragraph presented earlier in Figure 6.22 for a closer look. As follows, I have underlined the subjects of its sentences and used a bold font to highlight the verbs (or verb phrases).

> Religion **is** very important to many people in our society. Participation in religious groups and activities **is increasing.** Marsh (2006) **investigated** interests in religion and spirituality. The study **found** 84% of Americans reporting a strong interest. This significantly **surpasses** the 56% reported only five years earlier (Marsh, 2001). Religious practice **is recognized** as important for social and mental well-being. The physical health benefits of religion and spirituality **are** controversial. Some studies **have shown** less disease in religious versus non-religious individuals. Their results **are** typically flawed by extraneous variables. A critical review of the literature **is** necessary to resolve this issue.

All 10 sentences have a simple structure, which means that there is only one subject-verb unit per sentence. Paragraphs with one simple sentence after the next are usually monotonous to a fault. They are doubly dull when all of their verbs come immediately or soon after their subjects, and they are triply tedious when they also offer little or no variation in the actual verbs used. The preceding example paragraph demonstrates the triple curse. Notice how closely the verbs follow their subjects. Then check out how many sentences contain a similar verb—specifically, a form of the verb *to be* (*is* or *are*).

In scientific papers, long rows of simple sentences are the most common and troublesome cases of deficient variety in grammatical structure. But any pattern that is repeated exactly in more than two or three sentences can be problematic. Suppose that, in revising the preceding sample paragraph, the writer transformed the 10 simple sentences into five consecutive compound sentences. A compound sentence has two or more independent clauses that are joined by conjunctions. The result would still be monotonous for readers. There is, however, one exception to the rule for varying the grammatical structure of sentences in paragraphs. As explained earlier, the exception applies when the ideas in consecutive sentences are coordinated in function. The exception resolves the paradox about whether to emphasize variety versus sameness in paragraphs. For draft paragraphs with numerous consecutive sentences that lack variety in their grammatical structure, run a simple diagnostic test: Ask whether the structural sameness serves the useful purposes of highlighting coordinated ideas and enhancing coherence. If not, proceed with strategies for adding variety.

If the problem is a long row of simple sentences, the solutions are similar to ones that apply to resolving a lack variety in sentence length and beginnings. The sentence-combining techniques of coordination and subordination are effective for adding variety to simple sentences. The integration of an occasional introductory phrase or clause also solves the problem. Other strategies apply to revising numerous consecutive sentences with more complex structures. Take the case of a paragraph in which every sentence is composed of an introductory dependent clause followed by an independent clause. The necessary revision demands breaking down some of the complex sentences to form simpler ones to be interjected every now and again.

Lack of Variety in Tone

We commonly define the word *tone* by referring to qualities of sound. To many people, for example, the sound of a cat purring has a pleasant and calming tone. In contrast, the sound of a cat fight has a shrill and distressing tone. Writing can also be described by its tone, or generally by how it "sounds" to readers. More specifically, tone in writing refers to impressions that are elicited about the nature of author-reader relationships. The impressions fall somewhere along the following sorts of continua: from formal to familiar, from unapproachable to inviting, from objective to subjective, and from passive to active. Take an e-mail message that you might write to your best friend about plans for an upcoming weekend. On the unapproachable-to-inviting continuum, the tone of your e-mail would obviously be inviting and friendly. This quality of tone would emerge from your use of everyday language and maybe some playful slang. Your e-mail message would not contain only declarative sentences; instead, you would raise questions to engage your friend in a conversation. The friendly tone would also emerge naturally from the qualities of sentence variety that we have been discussing. For example, your sentences would vary in their length, beginnings, and grammatical structure. Some might not be sentences at all—instead, they might be phrases and fragments.

On the formal-to-familiar continuum, scientific writing obviously has a relatively formal tone, which emerges from the use of highly technical language and strict adherence to guidelines for academic grammar. In addition, scientific writing generally has an objective tone that tends to distance readers rather than to engage them directly. At the sentence level, a deliberate technique for creating an objective tone is the use of passive voice. Take the sentence, "Salivary cortisol was measured as an indicator of the stress response." The sentence is written in passive voice because the subject, *salivary cortisol*, does not name who or what carried out the action of the verb phrase, *was measured*. Passive voice naturally has an objective tone because it focuses on objects rather than who or what acts on them. An option is to write the sentence in active voice: "I measured salivary cortisol as an indicator of the stress response." It's a case of active voice because the subject, *I*, indicates who carried out the action of the verb, *measured*. In the upcoming chapter, you will learn how to resolve misuses of passive and active voice. Here, just take a moment to notice how active voice conveys a more subjective and engaging tone.

For good reasons, the tone of scientific writing should, for the most part, be formal and objective. A scientific paper is not an e-mail message to your best friend! Nonetheless, a common flaw in scientific communication is to restrict the tone excessively by exaggerating formality and objectivity. Reading top-quality articles authored by elite scientists in leading journals, you might be surprised by the relatively relaxed tone of writing. It's a refreshing change from conventions that are sometimes needlessly strict. As an example of scientific writing that relaxes qualities of tone very effectively, consider the paragraph in Figure 6.25. It comes from a review paper authored by Dr. Ma-Li Wong and Dr. Julio Licinio, both of the University of Miami School of Medicine (Wong & Licinio, 2001). The paper, which concerns research on treatments for major depressive disorder, was published in the first-rate journal, *Nature Reviews*. The authors' goal for the paragraph is to discuss the difficulty of diagnosing clinical depression. They attribute this difficulty to incomplete knowledge about the biological basis of the disease. As you are reading Wong and Licinio's paragraph, see if you can identify features that add sentence variety and thereby naturally relax the tone.

One feature that makes the paragraph's tone quite lively is variety in sentence length. After a 29-word opening sentence, the second sentence contains two short and snappy independent clauses (10 and 11 words, respectively) joined by a semi-colon. Sentence 3 is 25 words, setting up the paragraph's longest sentence, which is 45 words. The first four sentences explain the problem of diagnosing depression

(1) A key problem in diagnosis that has baffled nonpsychiatrists is the fact that the elaborate classification systems that exist today are solely based on the subjective descriptions of symptoms. **(2)** Such detailed phenomenology includes the description of multiple clinical subtypes; however, there is no biological feature that separates one subtype from another. **(3)** It is well known that different diseases can show similar clinical manifestations and also that the same disease can manifest itself differently in different patients. **(4)** For example, precocious puberty can be caused by various diverse and unrelated genetic defects and environmental factors, and hepatitis B can manifest itself in forms as diverse as acute hepatitis, chronic active hepatitis, cirrhosis of the liver or hepatocarcinoma, or it can be fully asymptomatic. **(5)** In clinical research on depression, some related fundamental questions remain unanswered. **(6)** Is each subtype of depression the result of different biological abnormalities? **(7)** Or are they merely different manifestations of the same underlying disease process? **(8)** Such basic aspects about the nature of depression have yet to be clarified because we still lack an unequivocal understanding of the biology of this disorder.

FIGURE 6.25 A paragraph with excellent variety of tone. From a published review paper authored by Wong and Licinio (2001). Reproduced with permission from Macmillan Publishers Ltd.

by using conventional classification systems, which are subjective in nature. The core of this problem, the authors explain in sentence 2, is that various clinical subtypes of the disease have not been linked to specific biological abnormalities. After presenting examples that illustrate the problem in diseases other than depression (in sentences 3 and 4), the authors begin to narrow in on the specific problem faced by researchers and clinicians in their field. Sentence 5 introduces the problem. The relative brevity of sentence 5, which contains only 11 words, accentuates the research problem. In addition, sentence 5 nicely sets up the very pointed questions presented in sentences 6 and 7, which are 11 and 12 words, respectively. The final sentence, of medium length, sums up the paragraph's overall point in 26 words. Wong and Licinio's paragraph demonstrates an interesting paradox and a lesson worth applying: Short sentences are dull and uninviting when they are all a paragraph has to offer; however, they can be emphatic and engaging when interspersed with longer sentences.

The relaxed tone of Wong and Licinio's paragraph is also a product of variety in sentence beginnings and structure. In addition, the direct questions that the authors raise in sentences 6 and 7 spice up the paragraph. So does the use of active voice and the first-person reference (the word *we*) in the last clause of sentence 8. These variations engage us and invite us to keep reading.

The take-home message is to evaluate your drafts' paragraphs for opportunities to broaden and enrich their qualities of tone. Enhancements may involve adding variety to sentence length, beginnings, and structure. In addition, you might try out techniques like raising direct questions and, when appropriate, using active voice.

REVISING FOR PARAGRAPH DESIGN

In a last pass through paragraph-level revision, it's a good practice to check for features of design. This pass through a draft is easy—in fact, you need not even read what you have written. Instead, just look at how your paragraphs appear on paper. Begin by identifying any excessively long paragraphs. Readers expect and need a break every now and again—in other words, they need indentations that signal new paragraphs at reasonable intervals. This is true regardless of whether your paragraphs are impeccably unified, coherent, and engaging. Readers use paragraph breaks to take a mental rest, to reflect on what they have just read and how it fits into the whole text, and to gear up for the next paragraph. If the text of your manuscript is double-spaced on 8.5 × 11 inch paper, a good general rule is to limit the length of individual paragraphs to approximately three-quarters of the page's length. Paragraphs that are very short can give readers the impression of skimpy, underdeveloped ideas and arguments. If your ideas are indeed well developed but they are spread out across several short paragraphs in a row, the solution is, of course, to combine the paragraphs.

Also check for variety in paragraph length. Many consecutive paragraphs that are equal in length can give manuscripts a boxy and bland look-and-feel. When the identically shaped paragraphs are short, readers may sense a choppiness and monotony, like they do with long stings of short sentences.

SUMMING UP AND STEPPING AHEAD

We began this chapter with an analogy likening paragraphs to minds and bodies in conditions of health and disease. The essential similarity is that these entities are complex systems that function through the interrelationships of many working parts. Like mind–body systems, when paragraphs are not functioning perfectly their rehabilitation requires applying comprehensive and systematic diagnostic procedures. We have seen many examples of these sorts of procedures applied to solving problems that compromise effective unity, topic sentences, coherence, cohesion, and sentence variety. Successful revision at the paragraph level is definitely painstaking and challenging. But the effort is always worthwhile because improvements in paragraph-level features directly enhance readability and comprehension.

Next up we take the last step in our approach to revision, further refining our level of analysis. In the present chapter, we revised primarily by looking *across* sentences. In the upcoming chapter, we will look *within* sentences to detect, diagnose, and solve problems involving matters of logic, clarity, grammar, punctuation, and word choice.

CHAPTER 7

Revising Sentences

The Science of Sleep

Reflecting on what we humans do in the course of a lifetime, you can't deny the curiosity sparked by realizing that we spend roughly one-third of our lives asleep. Why is sleep such an essential part of life? What life-sustaining functions does it serve? What's really going on in the body and brain while we are sleeping? How are daily cycles of sleep and wakefulness regulated? These are a just a few of the fascinating questions that inspire scientists who study sleep.

It's easy to think of sleep as a passive state. After all, when we are asleep our bodies are fairly motionless and we lack conscious awareness of internal and external experiences. The reality, however, is that sleep is an extremely active state. The brain is hard at work forming the neural synapses that underlie memory and learning. The body secretes hormones, maintains its immune system defenses, and restores its energy sources. Some physiological functions actually peak while we are asleep.

For many of us, the demands of daily life keep us awake when, for the sake of optimal health and performance, we should be asleep. Research shows, for instance, that students who pull all-nighters studying for tests are at a significant disadvantage compared to peers who distribute their studying over several days and get a full night's sleep before exam days. And health care professionals commit a greater number of mistakes when they are sleep deprived versus well rested. If you didn't get a full night's sleep, take a nap and then take on this chapter. The science of sleep provides the backdrop for the chapter's focus on revising sentences.

INTRODUCTION

As the old saying goes, "First impressions are the strongest." For audiences of scientific papers, first impressions are greatly influenced by qualities of writing at the sentence level—that is, by features of sentence punctuation, grammar, word choice, style, and clarity of expression. It's a bit paradoxical that the most effectively written scientific papers leave readers with little conscious impression of these sentence-level features. In other words, well-crafted sentences are transparent in the sense that they enable readers to see directly through to the deeper levels of content and meaning. It follows that readers may never have the opportunity to be impressed by a writer's insightful ideas and convincing arguments if a paper's sentences contain punctuation and grammar errors, illogical expressions, awkward phrases, and other sentence-level flaws.

Experienced scientific audiences—especially professors, journal editors, and grant officials—are strict critics of the quality of writing at the sentence level. With tall stacks of lengthy papers to evaluate, these readers often look for quick and easy reasons to deduct points, deny publication, and disapprove proposals. One glimpse of a few glaring sentence-level mistakes may immediately provoke strong negative impressions. The errors may dissuade these busy readers from investing the extra energy to work through the surface turbulence in order to reach the deeper levels. Moreover, many critical readers view sentence-level flaws as signs of the writer's lack of effort and motivation.

As the old joke goes, "It's never too late to make a good first impression." The take-home message nicely reflects the attitudes and approaches of accomplished writers, who work diligently and skillfully at revising their drafts' sentences. The outcomes are final papers that are easy to read and understand, even when the science is extremely complex, and that strongly impress audiences by their ideas and arguments. Toward helping you achieve these desirable outcomes for your scientific papers, this chapter presents a comprehensive approach to revising sentences.

ABOUT THE PROCESS

As illustrated in Figure 7.1, the stage of revising sentences follows stages of revising for features of global structure, content, and paragraphs. This order makes good sense because it's not worth wasting time correcting sentence-level mistakes if your draft still has room for improvement in its global unity, its development of ideas and arguments, its paragraph coherence, and so on. By taking on the bulk of sentence-level revision *after* your draft's global structure, content, and paragraphs are in great shape, you can devote the necessary brainpower to the process at hand.

Like the earlier revision stages, sentence-level revision is a problem-solving process. Some sentence-levels problems are easy to solve immediately, whereas others require more demanding diagnoses. Whether your revisions take the easy or difficult path in our process map depends partly on your knowledge

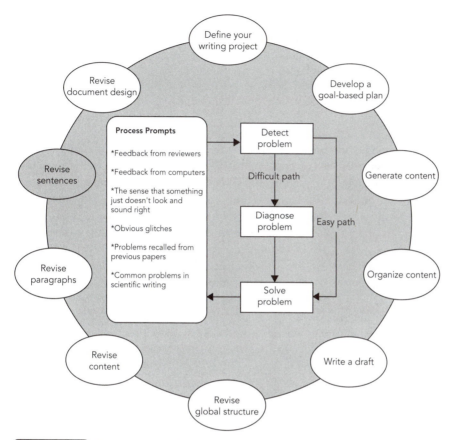

FIGURE 7.1 Process map for *revising sentences.*

of grammatical, structural, and stylistic conventions for crafting effective sentences. In addition, your path depends on your experience and skill in applying this knowledge to detect, diagnose, and solve problems. To better appreciate the previous points, see how easy it is for you to solve the problem in the following sentence.

Immediately after awakening from a full night of sleep, systolic blood pressure increases by 10 to 20 percent.

If the flaw isn't immediately obvious, focus on the sentence's introductory phrase: *Immediately after awakening from a full night of sleep.* This phrase serves to introduce, or modify, the sentence's subject. We would logically expect the subject to be a word or phrase conveying *who* or *what* might awaken from a full night's sleep. It makes sense, for example, to say that any group of people (women, men, children, older adults—or just people in general) can awaken after a full night's sleep. But in the preceding example, the subject is not a person or a group of people; instead, it's *systolic blood pressure*. Read literally, the sentence conveys the illogical idea that systolic blood pressure is able to sleep and to awaken. In the following sentence, the solution is in bold text.

> Immediately after awakening from a full night of sleep, **most people** experience a 10 to 20 percent increase in systolic blood pressure.

If you had good English teachers back in grade school, you quickly recognized the original sentence's flaw as a dangling modifier. If you have experience detecting and diagnosing dangling modifiers, you likely solved the problem automatically. If, however, you lack a strong background in grammar and the flaw is unfamiliar, you would need to diagnose the sentence somewhat systematically, applying conventions of grammar, to revise it correctly. My point is that knowledge, skill, and experience make all the difference in this stage of revision. This chapter is intended to help you gain these advantages so that, ultimately, your approach to sentence-level revision is manageable and successful.

Figure 7.1 shows six process prompts for initiating sentence-level revision. An especially helpful prompt is feedback from computers, specifically from computer-based grammar and spelling checkers. These programs are exceptionally good at catching misspelled words, sentence fragments, subject–verb disagreement, and incorrect punctuation. By relying on your computer to detect and resolve these sorts of relatively simple flaws, you can save mental energy for the more demanding challenges of revision. But even the best computerized spelling and grammar checkers do not always identify all problems correctly. In some cases, the programs give flawed advice for fixing problems. While I was writing this chapter, my word processor's grammar checker misinformed me a handful of times. For example, I wrote that a clause is "a construction that contains a subject and its predicate." My computer's grammar checker complained about the word *its*, telling me that I should have written *it's*. But this is not correct because the word *it's* is the contraction for *it is*. An apostrophe should not be used to indicate possessive case for the pronoun *its*. The take-home message is to double-check your computer's suggested revisions.

Another useful computer tool is the *Find* program that is included in word processing programs such as Microsoft Word. Let's say that, based on feedback I have received from peer reviewers, I am aware that I occasionally misuse the word *affect*, as in this example:

> In the brain, the main affect of caffeine is to promote wakefulness.

The sentence uses *affect* as a noun. Because the meaning of the noun *affect* is "a feeling or an emotional state," the sentence makes no sense. The correct noun for the sentence is *effect*, which means "an outcome or result." To check my drafts for this flaw, I could comb through every sentence looking for the word *affect*. A more efficient approach is to let my computer do the work. So I will type *affect* into my word processor's Find program. For each instance that it finds, I will carry out a diagnostic test to determine whether I have used the word correctly. Later in the chapter, when we focus on specific sentence-level problems (including the misuses of *affect* and *effect*), I will offer more tips on using computer-based Find programs to revise efficiently.

As is true for revising at all other levels of discourse, we cannot solve problems at the sentence level unless we first detect them. But some sentence-level problems are especially difficult to detect independently. They include flaws that compromise logic and clarity, that weaken the tone and directness of ideas, and that emerge when obscure rules of grammar are broken. This is why feedback from external reviewers is often essential for successful sentence-level revision. You may have opportunities to get feedback from your professors, classmates, and the staff of a writing center on campus. If so, ask your reviewers to help you identify specific strengths and weaknesses in your draft's sentences: Do my ideas come across clearly and logically? Are my sentences structured effectively? Do you see any problems with the grammar and punctuation? If professors or peers have alerted you to specific problems in the past, ask for targeted feedback accordingly. For example, if you are aware that your drafts' sentences tend to be "wordy," ask your reviewers for help revising this specific problem. In responding to feedback from your reviewers, ask for explanations of problems they have marked and the suggestions they have offered for revision. Don't automatically accept the feedback as valid.

My students often tell me that their friends, roommates, and family members helped them revise their drafts for matters of punctuation, grammar, and style. This is encouraging because these students are practicing a form of collaborative writing that is essential for success in school and the working world. But friends, roommates, and family members don't always give the most useful feedback. They may overlook major problems, turn small problems into big ones, and even create new problems. I certainly don't mean to dissuade you from seeking feedback from people who are close to you and have your best interests at heart. But I do offer the advice to solicit feedback from supportive reviewers who are especially knowledgeable and even critical about sentence-level matters.

Sentence-level revision can be an exacting and time-consuming process. Even a single flawed sentence can sometimes take what seems like forever to improve. Perhaps you have had the experience of detecting an awkwardly structured sentence in your draft—it's just not getting your point across directly. Setting out to revise, you might think, "This'll be a breeze. How long could it possibly take to fix one simple sentence? I'll rephrase the sentence and be done with it." But every time you reposition a phrase, a new structural glitch surfaces. Just as one part of the sentence becomes crystal clear, another part gets blurry. All of a sudden, 15 minutes have passed and the sentence still isn't quite right. Welcome to the world of expertise in revision! The good news is that regular and skillful practice at sentence-level revision greatly enhances our efficiency in the process. And, of course, with greater knowledge, skill, and efficiency in revising sentences, we can actually prevent future problems in the first place.

As a case in point, suppose that a science professor has marked your previous papers for "ambiguous passive voice." You suspect that your current draft contains instances of the problem. In sentences written in passive voice, the subject–verb unit does not explicitly name who or what carried out the action of the verb. Here's an example:

I Sleep deprivation is hypothesized to weaken immunity.

The subject–verb unit, *sleep deprivation is hypothesized*, does not name *who* hypothesized that sleep deprivation weakens immunity, so the sentence is constructed in passive voice. Whether scientific authors should use passive voice depends on the situation. When its use leaves readers confused about who carried out some action, passive voice is definitely a flaw. Depending on the content of sentences that surround the preceding example, the use of passive voice would indeed be problematic if it creates ambiguity about who raised the hypothesis concerning sleep deprivation and immunity. If you are not accustomed to evaluating sentences for problems involving passive voice, the diagnosis can indeed be exacting and time consuming. It would actually require parsing and testing every subject–verb unit in your draft. But after practicing the diagnostic routine a sufficient number of times, you will find that the process becomes automatic. Within a matter of seconds, you will detect passive constructions and decide whether they are appropriate. Eventually, you will carry out this process while you are drafting, so that the problem of ambiguous passive voice never even arises during formal revision.

The body of this chapter is organized by common sentence-level problems that warrant revision in drafts of scientific papers. Grouped in six categories, the problems are outlined in Figure 7.2. To understand the explanations of certain problems and the methods for solving them, you must know some basic terminology of English grammar. For example, to understand how to detect, diagnose, and solve several problems that compromise logic and clarity, you will need to know the definition of a *clause*: It's a construction that contains a subject and its predicate. Then, of course, you must know what subjects and predicates are. In addition, you will need to know the definitions of the two primary types of clauses: independent clauses and dependent clauses. When you come across unfamiliar grammatical terms in the pages that follow, check out their definitions in the glossary presented in Appendix B.

The chapter covers a fairly large number of sentence-level problems. To manage your revision process efficiently, take stock of ones that are most relevant to your concerns. Suppose that you are especially concerned about the structure and style of your draft's sentences. You might take separate passes through your draft to revise structural and stylistic problems such as wordiness, unnecessary jargon, and long noun trains. On the other hand, you might be very confident in your use of punctuation. If so, plan to take just one quick pass through your draft to double-check uses of commas, semicolons, apostrophes, and so on. Even if you have

Revising for Logic and Clarity

Illogical expressions and comparisons (323)
Anthropomorphism (325)
Dangling modifiers (327)
Vagueness (330)
Unclear pronoun reference (330)
Illogical tense shifts (332)
Problematic shifts in point of view (333)
Misplaced and awkward phrasing (333)
Inappropriate emphasis (335)

Revising for Style and Structure

Weak subjects and verbs (335)
Misuses of active voice and passive voice (339)
Wordiness (343)
Unnecessary jargon (345)
Excessive separation of subjects and verbs (348)
Long noun trains (350)
Lack of parallel structure (352)

Revising Basic Grammar Errors

Sentence fragments (354)
Subject–verb disagreement (355)
Noun–pronoun disagreement (357)

Revising for Word Choice

Affect, effect (358)
As, since, because (359)
Amount, number (359)
Compose, comprise (360)
Gender, sex (360)
Less, few, fewer (361)
Study, experiment (361)
That, which (361)
Than, then (363)
Who, whom (363)

Revising Punctuation and Mechanics

Commas (364)
Semicolons (368)

FIGURE 7.2 Common sentence-level problems in scientific papers. The pages on which the problems are introduced are shown in parentheses.

Colons (369)
Apostrophes (371)
Hyphens (373)
Quotation marks (374)
Capitalization (374)

Revising for Biased and Inadvertently Offensive Language

Sexist language (376)
Age-biased language (377)
Biased language involving ethnic and racial groups (377)

FIGURE 7.2 *continued.*

already developed advanced skills for crafting impeccable sentences, it doesn't hurt to learn the routines for detecting, diagnosing, and solving common flaws in scientific writing. When providing feedback on peers' drafts, you may have had the experience of knowing that a problem exists and knowing how to fix it, but *not* knowing how to explain the problem and its solution. By learning the routines for revising various sentence-level flaws, you will develop more advanced peer review skills, which your colleagues and coworkers will greatly value.

One last but important point before we move on: Even though it covers considerable ground, the chapter is not intended to teach all of the rules for English grammar and all of the conventions for writing effective sentences in scientific papers. Instead, the main goal is to help you learn and master a systematic problem-solving approach to revising sentences. So we will concentrate deeply on selected flaws—ones that are especially common in drafts of scientific papers. And you will learn detailed strategies for detecting these problems and for diagnosing them to derive the best solutions. You can apply the chapter's overall problem-solving approach to revising new sentence-level flaws that you learn about in educational resources that focus exclusively on sentence punctuation, grammar, structure, logic, and style.

REVISING FOR LOGIC AND CLARITY

The sentence-level problems in this category tend to be the most troubling of all in scientific writing because they ultimately cloud and confuse meaning. They are the sorts of problems that leave readers saying, "This sentence just doesn't make sense" or "I have no idea what the writer is trying to say here." Most of the flaws that compromise logic and clarity are not grammatical errors. Instead, they reflect a lack of sensibility and precision in expressing ideas.

Illogical Expressions and Comparisons

Creative writing and other nonscientific genres encourage figurative and embellished language that takes liberties with logic. In contrast, scientific writing is by and large characterized by its literal and logical precision—and for good reasons. A single word that is imprecise in meaning may express an illogical idea, as in the following sentence.

> Sleep apnea is a disease where affected individuals stop breathing for periods of 10 seconds or longer during sleep.

The problem involves the word *where*, which the dictionary defines as "at or in what place." Read literally, the sentence conveys the irrational idea that sleep apnea is actually a place, like Paris or Poughkeepsie. You might say that it's not such a big deal. After all, readers will still get the sentence's main message. But to experienced readers such as professors and journal editors, the imprecise use of *where* (just one illogical word) may raise more significant concerns that the writer is deficient in vocabulary and generally lacks sophistication in using language. Here's a revision (for emphasis, the changes are in bold font):

> Sleep apnea is **a disease where in which** affected individuals stop breathing for periods of 10 seconds or longer during sleep.

Another logical flaw comes in sentences that make nonsensical comparisons:

> Concerning the effects of exogenous melatonin on sleep quality, our conclusion differs from Moore and Willis (2005).

The writers literally compare their conclusion to two people. This is an apples-to-oranges sort of comparison. The following sentences more logically compare a conclusion to another conclusion.

> Concerning the effects of exogenous melatonin on sleep quality, our conclusion differs from the conclusion of Moore and Willis (2005).

> Concerning the effects of exogenous melatonin on sleep quality, our conclusion differs from that of Moore and Willis (2005).

To detect, diagnose, and solve illogical expressions and comparisons, apply the following strategies.

1. Attend to the literal meaning of words. To write logically sound sentences, you must tune into the literal meaning of words, or their primary definitions in reputable dictionaries. Try it out with the following sentence.

> Over the last two decades, college students with insomnia have exploded.

The primary dictionary definition of the verb *explode* is "to burst forth with sudden violence or noise from internal energy or pressure, or to undergo a rapid chemical or nuclear reaction with the production of noise, heat, and violent expansion of gases." By this definition, college students with insomnia obviously have not exploded. A more precise expression is called for:

> Over the last two decades, the number of college students with insomnia has increased greatly.

2. Evaluate the logic expressed in subject–predicate units. Some sentences have a mathematical structure of sorts in which subjects and their predicates form two sides of an equation, with verbs functioning as equal signs. Here's a case in point:

> Naps and caffeine ingestion are effective methods for counteracting the negative effects of sleep deprivation.

Here's the same sentence in mathematical terms:

> Naps + caffeine ingestion = effective methods for counteracting the negative effects of sleep deprivation.

The sentence makes sense because the subject and predicate are logically equal. In contrast, the following sentence does not add up.

> The three sleep deprivation conditions had significantly different scores on the vigilance test.

It's not logical to say that sleep deprivation conditions can have scores on vigilance tests—of course, conditions are not capable of taking tests. Here's a solution:

> **Participants assigned to** the three sleep deprivation conditions had significantly different scores on the vigilance test.

What is the logical flaw in the following sentence?

> Patients who received acupuncture resulted in less severe sleep apnea.

The phrase *resulted in* conveys the idea that the patients themselves were transformed into a condition of less severe sleep apnea, which does not make sense. Instead, the patients who received acupuncture *experienced* less severe sleep apnea.

3. Test the logic of your comparisons. Be aware of draft sentences that compare things, using phrases such as *more than, less than, higher than, lower than, different from, and compared to*. To test the logic, isolate the word groups that surround the comparing phrases, and then test whether the sentences make apples-to-apples comparison or apples-to-oranges comparisons. Try it with this sentence:

> The mean age of death of subjects who suffered from sleep disorders was significantly lower than the control subjects.

The sentence compares *mean age of death*, which might be a number like 81.2 years, to *control subjects*, who are people. This is an apples-to-oranges sort of comparison. The logical flaw resulted from the writer's intention to avoid using extra words, as follows.

> The mean age of death of subjects who suffered from sleep disorders was significantly lower than the mean age of death of the control subjects.

The repetitive phrase, *the mean age of death of*, adds words, but it also improves the sentence's logic, which in this case is more important. Of course, we can shorten the revised sentence as follows.

> The mean age of death of subjects who suffered from sleep disorders was significantly lower than **that** of the control subjects.

Anthropomorphism

The following sentence contains a special case of illogical expressions, called *anthropomorphism*, which means attributing human motivations, characteristics, or behaviors to nonhuman things.

> The present study asked whether sleep deprivation impairs immune function.

Read literally, the sentence is not logical because studies are not capable of interrogative thinking and communication. In other words, studies can't ask questions! To solve the problem, the writer would replace *the present study* with a logical actor, such as *we* (for a coauthored paper) or *I* (for a single-authored paper).

Anthropomorphism crops up fairly often in student papers and occasionally in published journal articles. An example is the common sentence that begins *The study found....* Read literally, the expression is not completely logical because studies are not capable of finding things. A primary definition of the verb *find* is "to come upon or discover by searching or making an effort." This definition denotes a behavior that only humans and animals can carry out. Inanimate objects cannot find things. And, in strictly logical terms, neither can studies. Novice writers may say *the study found* or *the experiment concluded* because they believe (mistakenly) that the use of first-person, the words *I* and *we*, is strictly forbidden in scientific writing. Perhaps their grade school English teachers prohibited the use of *I* to prevent students from using excessive narration in academic writing.

Or maybe these writers have been influenced by an outdated and erroneous view of science as a totally objective endeavor, one that is not influenced by human actions, values, biases, and mistakes. In reality, the style guides for all leading scientific organizations and publishers encourage authors to use first-person to attribute human actions and cognition to humans.

The following examples demonstrate revisions of problematic anthropomorphism.

> Problem: One researcher's theory (Johnson, 2002) hypothesized that chronic sleep deprivation increases the risk of heart disease by increasing levels of stress hormones.

Even though the writer cites the researcher, the sentence's subject–verb unit still conveys the irrational idea that theories are capable of formulating hypotheses.

> Solution: **Johnson (2002) hypothesized** that....

> Problem: A study conducted by Rundgren experienced a significant improvement in sleep quality after treatment with acupuncture.

Studies cannot sleep, so they cannot experience improvements in sleep quality. Also, studies cannot be treated with acupuncture.

> Solution: **Subjects in** a study conducted by Rundgren experienced a significant improvement in sleep quality **after they were treated** with acupuncture.

Detecting and diagnosing anthropomorphism requires isolating subject–verb units and determining whether the subjects are logically capable of carrying out the action of their verbs. Common cases of anthropomorphism crop up when writers mistakenly try to avoid attributing research activities and outcomes to humans. If you tend to do that—for example, if you usually write *the study concluded* rather than *the researchers concluded*—use your word processor's Find program to locate words that reflect research activities—words such as *study, results, research, conclude, data, and findings.* Then, for each instance, check whether these words are parts of logical subject–verb units.

Where do you draw the line on problematic cases of anthropomorphism? Consider, for example, expressions such as *the results indicate* and *the study showed.* Is it logical to say that results can *indicate* and that studies can *show*? The following guidelines address these sorts of questions and offer advice for avoiding anthropomorphism.

1. **When you can attribute behaviors that only humans perform to humans, always do so.** The behaviors of *concluding* and *claiming*, for example, are carried out by humans exclusively. Studies and results simply cannot do these things.

2. **If you question whether a particular behavior is carried out by humans exclusively, look up the primary definition of the behavior's verb in a reputable dictionary.** For example, the verb *indicate* means "to signify, or to serve as a sign, symptom, or token of." Because the results of a published study can signify outcomes and conclusions, it seems quite logical to write *the results indicate*.

3. **Use published articles as models.** In articles published in top-quality scientific journals, you will seldom see the logical flaw of anthropomorphism. So when you are unsure about whether your phrasing of an idea might create the problem, examine published articles for how scientists express similar ideas. For example, if you are trying to express an idea about your hypothesis, focus on how authors talk about their hypotheses in published papers. You will seldom see experienced scientific authors writing something like *this experiment hypothesized*. Instead, they write *we hypothesized*.

Dangling Modifiers

Something just isn't quite right in the following sentence's logic.

> After ingesting 200 mg of caffeine, Davis (2004) found that cognitive performance was maintained at a high level in sleep-deprived subjects.

Oddly enough, it appears that the sleep-deprived subjects maintained their cognitive performance after a researcher named Davis ingested caffeine! This is an illogical idea because, in the context of an experiment on sleep deprivation and cognitive performance, the study's subjects (not the researcher) must have ingested the caffeine. The problem, which I introduced earlier in the chapter, is a dangling modifier. The following revision clarifies who really consumed the caffeine.

> After ingesting 200 mg of caffeine, **the sleep-deprived subjects** maintained their high level of cognitive performance (Davis, 2004).

Sentences with dangling modifiers have two distinguishing features:

(1) a modifier, which is a word (or word group) that introduces, explains, or clarifies another word (or word group), such as the subject of a sentence; and

(2) an illogical relationship between the modifier and the word (or word group) that it modifies.

In the most common cases of dangling modifiers, a modifying word or phrase does not sensibly modify the subject of a sentence or an independent clause. To identify this problem in the original example, read just up to the end of its introductory modifying phrase: *After ingesting 200 mg of caffeine*. Here you may sense that the phrase is *dangling*. In other words, it's waiting for a logical word or phrase to begin completing the sentence's thought. In this example, the word or phrase should convey *who* ingested the caffeine. Because the caffeine consumer was not

Davis, which is the subject of the original sentence and its independent clause, the modifying phrase continues to dangle.

In the following examples, modifying phrases are italicized and subjects of independent clauses are underlined.

> Problem: *When comparing the sleep-deprived groups and the control group,* depression scores were highest for subjects who slept fewer than 6 hours a night.

The sentence conveys the illogical idea that the depression scores compared the two groups of study participants; the modifying word, *comparing*, dangles because the grammatical subject does not name who did the comparing.

> Solution: *When comparing the sleep-deprived groups and the control group,* **the researchers** found that depression scores were highest for subjects who slept fewer than 6 hours a night.

> Problem: Daytime secretion of interleukin-6 increases *when deprived of REM sleep.*

In this example, the dangling modifier comes at the end of the sentence, after the independent clause. The sentence illogically infers that daytime secretion of a molecule called interleukin-6 can be deprived of REM sleep.

> Solution: Daytime secretion of interleukin-6 increases *when **mice are deprived** of REM sleep.*

> Problem: *To determine the effects of caffeine on mental alertness,* the performance data were analyzed with a *t* test.

The infinitive modifier, *to determine*, dangles because the subject of the sentence, *the performance data*, did not do the determining.

> Solution: *To determine the effects of caffeine on mental alertness,* I analyzed the performance data with a *t* test.

If your drafts contain dangling modifiers, you might detect them just by reading out loud and listening closely for whether your ideas ring true. Read literally, sentences with dangling modifiers engender nonsensical and often humorous impressions (*Sleeping at least 8 hours a night, studies show that the common cold can be prevented*). Another strategy is to purposefully parse and test sentences for the problem. The following steps guide this revision process.

Step 1: Parse modifiers and independent clauses. The following sample sentence has a characteristic structure that raises suspicions about possible dangling modifiers.

> By napping during the middle of the day, decrements in cognitive performance that are caused by sleep deprivation can be avoided.

The sentence contains a modifying phrase (the entire word group before the comma) and an independent clause (the entire word group after the comma). This structure, in which the modifying phrase begins the sentence, is common; however, modifiers can also be placed in the middle or at the end of sentences. The sample sentence contains another characteristic of many dangling modifiers, which is a participle in its modifying phrase. A participle is a verb form with an *–ing or –ed* ending. Participles look like verbs but function as adjectives. The modifying phrase in our sample sentence contains the participle *napping*. The more you practice identifying and parsing sentences with these characteristic elements, the more automatically you will identify sentences that should be tested for dangling modifiers.

Step 2: Think of logical subjects for modifying words and phrases.
To apply this step to diagnosing our sample sentence, simply imagine *who* might logically carry out the action of napping during the middle of the day.

> By napping during the middle of the day, _____ (fill in the blank with who might nap).

Step 3: Test the logical subjects you would expect against the sentence's actual subject. In the previous step, you might have imagined that people in general or certain groups of people (elderly individuals, shift workers, medical residents, and so on) might nap during the day. Considering that the sentence comes from a paper on sleep deprivation in college students, you might also expect that college students would be the sentence's subject. But the actual subject is *decrements in cognitive performance*. Of course, the subject does not logically name who naps during the middle of the day. So the sentence contains a dangling modifier.

Step 4: To correct a dangling modifier, place the logical word or phrase in the sentence's subject position or in its modifying phrase. In the following solution, the logical actor, *college students*, is positioned as the subject of the independent clause.

> By napping during the middle of the day, college students can avoid decrements in cognitive performance that are associated with sleep deprivation.

In an alternative revision, the logical actor is named in the modifying phrase:

> When college students nap during the middle of the day, decrements in cognitive performance that are associated with sleep deprivation can be avoided.

(The first solution is stronger because the message is conveyed more directly, using active voice. We will discuss when to use active versus passive voice later in the chapter.)

Vagueness

Unlike anthropomorphism and dangling modifiers, which are cases of illogical ideas, vagueness describes expressions that are logical but not sufficiently clear and precise in meaning. Here's an example of a vague sentence:

> Bright light enhances alertness in sleep-deprived individuals by changing the beta rhythm of electrical activity in the brain.

The word *changing* is imprecise. How exactly does the beta rhythm of the brain's electrical activity change? Does it increase or decrease in magnitude? Does it become faster, slower, more stable, or less stable? The sentence's meaning would be clearer if the writer replaced *changing* with a word that more precisely describes *how* the beta rhythm actually changes.

Here's another case of vagueness:

> Sleep deprivation can stunt bone growth in children, but it is unknown whether this effect is caused by unfavorable influences on growth hormone.

What precisely does the phrase *unfavorable influences on growth hormone* mean? The following revision eliminates the vagueness.

> Sleep deprivation can stunt normal bone growth in children, but it is unknown whether this effect is caused by **a diminished secretion of growth hormone during slow-wave sleep.**

Once detected, vague sentences are relatively easy to revise: Just replace their unclear words and phrases with more clear and precise ones. But vagueness is especially difficult to detect independently. What you may think is perfectly precise, readers might see as hazy. So to detect cases of vagueness, read your drafts from the perspective of your audience, focusing on words and phrases that might prompt requests for clarification. In addition, seek specific feedback from reviewers on the clarity and precision of your sentences.

Unclear Pronoun Reference

Pronouns are words that stand for antecedent nouns, noun phrases, or entire ideas. Confusion results when pronouns do not refer to their antecedents clearly:

> When subjects are given caffeine and bright light, it significantly improves alertness.

To what antecedent noun or idea does the pronoun *it* refer? Does *it* refer to caffeine? Does *it* refer to bright light? Or does *it* refer to the combination of bright light and caffeine? To resolve the unclear pronoun reference, the writer replaces the ambiguous pronoun with a specific noun phrase:

> When subjects are given caffeine and bright light, **the combination of the treatments** significantly improves alertness.

A common case of unclear pronoun references involves the pronoun *this*:

> Stress-reduction techniques improved participants' ratings of sleep quality and increased their duration of REM sleep, although the techniques did not significantly influence the circadian rhythm. One reason for this is that the techniques did not influence stress hormone concentrations.

In the second sentence, does *this* refer to the beneficial effects of stress-reduction techniques on sleep quality and the increased duration of REM sleep? Does *this* refer to the nonsignificant influence of the stress-reduction techniques on the circadian rhythm? Or, does *this* refer to the previous sentence's idea that the study findings are contrasting? The meaning is unclear because *this* is not specified. The following revision adds a noun phrase that specifies the unclear pronoun by referring back to a key idea in the first sentence.

> Stress-reduction techniques improved participants' ratings of sleep quality and increased the duration of REM sleep, although the techniques did not significantly influence the circadian rhythm. One reason for **this contrast in effects** is that the techniques did not influence stress hormone concentrations.

Like nonspecific *this's*, nonspecific *which's* can also be confusing:

> Compared to the EEG recordings of young subjects, the recordings of older subjects showed greater beta activity, more K-complexes, and fewer sleep spindles, which may explain the poorer sleep quality reported by the older subjects.

To what does *which* refer: greater beta activity, more K-complexes, fewer sleep spindles, or some combination of these measures? A solution follows.

> Compared to the EEG recordings of young subjects, the recordings of older subjects showed greater beta activity, more K-complexes, and fewer sleep spindles. **These three measures may** explain the poorer sleep quality reported by the older subjects.

Writers who do not specify their *this's* and *which's* may aim to avoid wordiness. However, when a nonspecific *this* or *which* will confuse readers, the extra clarifying words are essential. Even when your nonspecific *this's* are not likely to confuse readers, you might consider specifying them. Some professors and editors view excessive instances of nonspecific *this's* as reflecting underdeveloped thinking and unsophisticated writing.

It's relatively easy to revise unclear pronouns independently. If you suspect that your drafts contain them, use your word processor's Find program to locate the pronouns that typically cause reference problems: *it, this, which, they, them, their, these,* and *those.* For each instance, identify the antecedent noun, noun phrase, or idea that you originally intended the pronoun to substitute for; this antecedent may be located in the same sentence or in a preceding sentence. Then ask whether the pronoun refers to the intended antecedent unambiguously. If readers might substitute the pronoun with a different antecedent, or if they might not clearly understand which antecedent you intended, replace the pronoun with a specific noun or noun phrase.

Illogical Tense Shifts

The tense of a verb indicates when its action occurs (in the past, present, or future); the action's state of completion; and whether the idea expressed by the action is universally true. In some cases, illogical ideas surface when verb tenses shift within and across sentences. The following example of a problematic tense shift comes from the results section of a research paper.

> The subjects who ingested caffeine experienced a significant improvement in a test for mental alertness; thus, caffeine enables the subjects to overcome the homeostatic sleep drive.

In the first independent clause, the verbs *ingested* and *experienced* are in the past tense, which is appropriate because the actions occurred at a clear-cut point in time and the reported study is long over. In the second clause, however, the shift to the present-tense verb *enables* is problematic. In context, the shift infers that even though the study was completed, the researchers are somehow still observing caffeine's effects on the subjects. The verb should be in the past tense (*enabled*), conveying that the researchers observed the result when the experiment was conducted, but not afterwards.

All tense shifts are not necessarily nonsensical. Indeed, some are essential for good logic:

> The subjects who ingested caffeine experienced a significant improvement in the test for mental alertness. Thus, caffeine enables us to overcome the homeostatic sleep drive.

Across the two sentences the shift from past to present tense is not at all problematic. In the second sentence, the writer is clearly making an inference about the study's results. The present tense appropriately reflects the writer's intention to present this inference as a general truth about caffeine that applies to everyone at all times (indicated by the word *us*) rather than only to the specific subjects of the completed study.

Successful revisions of illogical tense shifts depend on knowing the rules and uses for the following verb forms.

- Use the past tense of verbs to indicate fully completed actions and conditions; for example, past tense is appropriate for writing about the methods and results of completed studies.
- Use the present tense of verbs to indicate habitual and current actions as well as to convey truths that apply now and forever more; present tense is appropriate for stating general conclusions and interpretations of data, as you might do in discussion sections of research papers and lab reports.
- Use the future tense of verbs to indicate actions or events that will occur later; future tense is appropriate for conveying planned study methods in research proposals and grant applications (for example, *we will investigate the effects of sleep deprivation on motor performance*).

Problematic Shifts in Point of View

The following example elicits a disorienting and illogical shift in focus, or point of view.

> People who have insomnia sleep more deeply after they perform aerobic exercise. An underlying reason is that aerobic exercise increases concentrations of sleep-promoting chemicals, such as adenosine, in your brain.

The first sentence establishes an objective point of view by focusing on people who have insomnia and their response to aerobic exercise. This focus is appropriately maintained until the end of the second sentence, where the phrase *in your brain* shifts the point of view abruptly. The writer may have intended for the phrase to be interpreted in a general sense; however, an experienced scientific reader will take it literally and ask the writer, "Why are you all of a sudden talking about *my* brain?" To keep the point of view in proper objective focus, the phrase "in *your* brain" should be replaced with "in *the* brain." (In addition to its problematic shift in point of view, the preceding example breaks the scientific writing convention to avoid second-person references.)

Misplaced and Awkward Phrasing

Some cases of illogical and imprecise sentences are due to misplaced and awkward phrasing. Here's an example:

> Murray reported a significant incidence of sleep deprivation among nurses applying epidemiological statistical analyses.

Sounds like the nurses applied the epidemiological statistical analyses, which caused their sleep deprivation, doesn't it? By moving the ending modifying phrase (*applying epidemiological statistical analyses*) closer to the word it is actually intended to modify (*Murray*), the writer logically clarifies the sentence's meaning:

> **Applying epidemiological statistical analyses**, Murray reported a significant incidence of sleep deprivation among nurses.

Here's another common case of a misplaced modifier that weakens sentence logic:

> The patient only slept 5 hours last night.

Given its placement, the word *only* modifies the verb *slept*. This phrasing literally conveys that the patient did not do anything other than sleep last night. She did not eat dinner, brush her teeth, or do anything else throughout the night. She only slept. Taken literally, the sentence does not express a completely logical idea. The reason is that nighttime lasts for more than 5 hours, so the patient *must* have done something other than sleep. A slight repositioning of the word *only* makes the writer's meaning more logical:

> The patient slept **only 5 hours** last night.

To detect illogical expressions caused by misplaced and awkward phrasing, you have to read your sentences for their literal meaning. This might require parsing word groups and closely examining their logical relationships. Take the following sentence:

> Sleep apnea is characterized by pauses in breathing during sleep caused by full or partial collapse of the upper airway.

Because it contains a number of word groups that modify others, it's the sort of sentence that calls for diagnosing logical relationships. The opening clause, *Sleep apnea is characterized*, is logically followed by a string of prepositional phrases, *by pauses in breathing during sleep*. But notice the phrase *caused by full or partial collapse of the upper airway*. What does this phrase modify? On a first pass through the sentence, readers may have the nonsensical impression that *sleep* is caused by the full or partial collapse of the upper airway. This confusion is due to the placement of the modifying phrase immediately after *sleep*. More logically, the pauses in breathing that characterize sleep apnea are caused by the full or partial collapse of the upper airway. The following revision, which moves the modifying phrase closer to *pauses in breathing*, sorts out the logic.

> Sleep apnea is characterized by pauses in breathing, **which are caused by full or partial collapse of the upper airway,** during sleep.

Inappropriate Emphasis

The positioning of word groups in sentences can significantly influence the reader's sense of emphasis, logic, and meaning. To demonstrate this point, the following three sample sentences contain the same word groups but, due to their varying positions, emphasize different take-home messages.

> In the youngest subjects, muscle cramps during sleep were caused by dehydration.

This sentence's primary message concerns *what*, in the youngest subjects, caused muscle cramps during sleep: It was *dehydration* rather than some other cause.

| In the youngest subjects, dehydration caused muscle cramps during sleep.

In this sentence, the emphasis is on *when* dehydration caused muscle cramps: It was during *sleep* rather than during some other event or activity.

| Dehydration caused muscle cramps during sleep in the youngest subjects.

This sentence conveys a take-home message about *which* group of subjects experienced muscle cramps during sleep: They were the youngest subjects as opposed to older ones.

In the preceding examples, the emphasis of meaning differs because the sentences end in different word groups. Experienced readers of English look to word groups at the ends of sentences to learn what information and ideas are most important. In a way, well-structured sentences are like good jokes: The first parts provide context and background information, and the endings pack the punch line. The upshot for revising draft sentences is to check for whether word groups that you intend to emphasize, for clarity of meaning, are appropriately placed. This usually means making sure that the emphatic word groups come at the end of sentences. There are, however, exceptions to this guideline. For short sentences that convey relatively simple messages, it is usually no problem if the most emphatic information comes at the start. Moreover, it is occasionally necessary to place relatively less important information at the ends of sentences to promote paragraph coherence.

REVISING FOR STYLE AND STRUCTURE

Sentences with good style and structure are direct in expression and meaning, lively in tone, and easy to read. In this section of the chapter, we focus on solving problems that compromise sentences-level style and structure.

Weak Subjects and Verbs

If you happen to be an athlete or exercise buff, you know all about *core strength*, which refers to powerful abdominal and lower back muscles. A strong body core is essential for maintaining good posture and moving the arms and legs powerfully.

Like physically fit athletes, well-built sentences have the quality of core strength, which readers experience as the direct and forceful expression of ideas. The core muscles of sentences, so to speak, are their subjects and verbs. Here's a sentence that would benefit from a core workout:

| A decrease in cerebral metabolic rate occurs during slow-wave sleep.

Read it closely and you will sense that the sentence doesn't pack much of a punch. Indeed, its message comes across feebly. For sentences that elicit the first impression of a weak tone, it's worth isolating their core elements—that is, their subject–verb units—and testing their strength. In our sample sentence, the core is *decrease occurs*.

To continue our diagnosis, let's bring in another analogy, this one adapted from a terrific book on crafting effective sentences and paragraphs, Joseph M. Williams' *Style: Ten Lessons in Clarity and Grace*. Williams likens sentences to stories. Sentences that tell the most engaging stories have two distinguishing qualities: (a) their main characters, what the sentences are essentially about, are prominently highlighted as grammatical subjects; and (b) their main actions are directly and powerfully expressed as verbs. Applied as diagnostic criteria, these two qualities shed light on the dullness of the preceding sample sentence. The simple subject, *decrease*, does not name the sentence's main character, or the story's hero. This role should go to *cerebral metabolic rate*. Notice, however, that the writer relegates the protagonist to a supporting role by placing it in a prepositional phrase that modifies the subject.

Next, let's test our sample sentence for the strength of its verb: Does *occurs* express the most direct and powerful action? For two main reasons, the answer is *no*. First, the verb *to occur* is naturally unexciting. The occurrence of something is a fairly static condition. Second, the sentence contains another word that reflects a more lively action and that could suitably replace *occurs* as the verb: It's the word *decrease*. In the original sentence, *decrease* is a grammatical form called a nominalization, which is a verb converted into a noun. In many cases, nominalizations noticeably weaken the tone of sentences and can even impair readability.

Here, again, is the original sample sentence:

| A decrease in cerebral metabolic rate occurs during slow-wave sleep.

Based on our diagnosis, let's tell a more engaging story by replacing its weak subject and verb with stronger versions:

| Cerebral metabolic rate decreases during slow-wave sleep.

Reading this revision closely, you will be impressed by a more powerful message. This point is illustrated by a direct comparison of the two sentences' core messages.

Weak core message	Strong core message
A decrease occurs	Cerebral metabolic rate decreases

Which story would you rather read?

In the following examples, subjects are underlined and their corresponding verbs or verb phrases are italicized.

> Problem: Our <u>finding</u> that norepinephrine regulates REM sleep *is* in agreement with results from previous research.

The authors intend to tell a story about their *finding*, so the sentence's subject is strong enough. However, the verb *is* raises concern because it does not express an engaging action or condition of the subject. Notice what comes soon after the weak verb—the word *agreement* is a telltale dull nominalization. The following revision converts the nominalized verb back into a true, stronger verb.

> Solution: Our <u>finding</u> that norepinephrine regulates REM sleep **agrees** with results from previous research.

> Problem: <u>Measurements</u> of the activity of serotonergic neurons *were taken* before the subjects received treatments for restless leg syndrome.

Both elements of the core, the subject (*measurements*) and the verb phrase (*were taken*), are weak. The solution, as follows, strengthens the core and eliminates extra words.

> Solution: <u>Serotonergic neuronal activity</u> *was measured* before the subjects received treatments for restless leg syndrome.

> Problem: An <u>explanation</u> of paradoxical sleep *may have been provided* if a closer examination of the shape of EEG waveforms *had been conducted*.

Reduced to their fundamentals, the two core messages—*explanation may have been provided* and *examination had been conducted*—are plain dull. Note the sharper tone of the core messages in the following two revisions.

> Solution: <u>Paradoxical sleep</u> *may have been explained* if <u>the shape of EEG waveforms</u> *had been closely examined*.

> Solution: <u>The researchers</u> *may have explained* paradoxical sleep if <u>they</u> *had closely examined* the shape of EEG waveforms.

To walk through the steps for revising sentences with weak subjects and verbs, let's use the following example.

> The assignment of study participants to groups was implemented according to random procedures.

Step 1: Isolate the sentence's core (subject–verb) elements. Our sample sentence has one subject–verb unit. The simple subject is *assignment* and the verb phrase is *was implemented*. So the core message is *assignment was implemented*.

Step 2: Evaluate the strength of the subject. Should *assignment* get top billing as the sentence's subject? If not, who or what else should be highlighted more prominently? One could argue that the sentence is really about the study participants themselves. Even though they are named in the sentence's complete subject, the study participants take a secondary role given their position in the prepositional phrase that modifies *assignment*. So we have good reason to replace *the assignment* with *the study participants*.

Another hint of a weak subject in our sample sentence is that *assignment* is a nominalization, specifically of the perfectly lively verb *assign*. An *–ment* ending to a verb indicates a nominalization. Other common endings, as demonstrated in the following chart, are *–ion, –tion, –ence,* and *–ance.*

Verb	Nominalization
investigate	investigation
perform	performance
differ	difference
discuss	discussion
recruit	recruitment

Step 3: Evaluate the strength of the verb. Verbs should name the most revealing and lively actions that their subjects carry out. In our example, the verb phrase *was implemented* conveys what happened to *the assignment* (of study participants to groups). Certainly, the action of implementing something can be revealing. So we might not complain that the verb phrase is especially weak. But the sentence hints at a stronger verb, which is revealed by converting the nominalization *assignment* to *assign*.

Some verbs naturally tend to weaken the tone of sentences. They include *is, are, give, take, make, conduct,* and *perform*. Tune your eyes and ears, as well as your word processor's Find program, to detect and diagnose these potentially problematic verbs in your drafts.

Step 4: Replace weak subjects and verbs with stronger ones. Based on the preceding diagnosis, both the subject and verb of our sample sentence are not as strong as they could be. Here's a revision that strengthens the core and eliminates unnecessary words:

Study participants were randomly assigned to groups.

The new subject, *study participants*, names the main actor. The new verb phrase, *were randomly assigned*, focuses on the key action. Now the core message, *participants were randomly assigned*, conveys a more engaging story (recall that the original core message was *assignment was implemented*). Some might argue that the revision is still weak because it's written in the passive voice. I would urge the critics to reserve this argument until they read the very next section!

Misuses of Active Voice and Passive Voice

When a subject–verb unit directly reveals who or what is responsible for carrying out the verb's action, the sentence is written in *active* voice. Here's an example:

I Friedman identified risk factors for sleep disturbances in children.

It's active voice because the core unit (*Friedman identified*) explicitly names who did the identifying. In contrast, when the subject–verb unit does not explicitly name the actor, the sentence is in *passive* voice:

I Risk factors for sleep disturbances in children were identified.

Among students and experienced authors alike, confusion and debate abound concerning whether and when to use active voice versus passive voice in scientific writing. One view holds that all authors, regardless of their fields, should always use active voice because it communicates a more direct and emphatic tone. The other side argues that because the scientific method is objective, the language of science should be objective as well. In passive voice, the grammatical subjects of sentences are usually objects. Passive voice thus has an objective quality.

Which form of voice—active or passive—should scientific authors use? As is true for solving so many problems in the writing process, the answer is *it depends on the situation*! Writers should decide by considering a number of factors, which include clarity, precision, tone, and coherence. Earlier in the chapter, I gave an example of problematic passive voice. It was a case of ambiguity about *who* or *what* was responsible for carrying out a verb's action. The same problem characterizes the following sentence.

Because the neural modulator adenosine decreases mental alertness (Eldredge, 2004), it is concluded that an adenosine antagonist, such as caffeine, will enhance alertness and improve cognitive performance.

Notice that the second subject–verb unit, *it is concluded*, is in passive voice. Astute readers will ask *who* is responsible for the conclusion: Is it Eldredge, the sentence's author, or someone else? When passive voice creates ambiguity, a revision to active voice is warranted. To guide diagnoses and solutions for problematic uses of active voice and passive voice, Table 7.1 (pages 340–341) presents common decision-making scenarios. The table emphasizes the point that writers must rely on problem solving rather than hard-and-fast rules to determine which form is appropriate.

TABLE 7.1 When to Use Active Voice versus Passive Voice

Situation	Use active voice	Use passive voice	Examples and revisions
When those who will read and evaluate your paper either express a preference for active voice or give uncompromising instructions to use it exclusively	✓		
When the use of passive voice adds unnecessary words and weakens the tone of sentences	✓		**Problem:** Monoamine oxidase inhibitors have been known to be an effective treatment for narcolepsy. The sentence is in passive voice because the subject–verb unit doesn't name who has known that monoamine oxidase inhibitors effectively treat narcolepsy. But to understand the sentence's meaning, readers don't need that information. The verb phrase *have been known* is unnecessary, as are several other words in the original 13-word sentence. **Solution:** Monoamine oxidase inhibitors effectively treat narcolepsy. The more direct and emphatic revision, in active voice, reduces the word count to 6.
When the use of passive voice is a deceptive attempt to avoid taking responsibility for one's actions and conclusions	✓		**Problem:** In efforts to control for confounding variables that could have influenced sleep quality, exercise and dietary factors were overlooked. This is a classic case of authors acknowledging that "mistakes were made" but using passive voice to deflect personal responsibility. Skilled readers will see right through the deception. **Solution:** In efforts to control for confounding variables that could have influenced sleep quality, we overlooked exercise and dietary factors.
When the use of passive voice creates ambiguity about who or what carried out the sentence's action	✓		The previous example illustrates this case.
When those who will read and evaluate your paper either express a preference for passive voice or give uncompromising instructions to use it exclusively		✓	

When the use of active voice inappropriately indicates that the actor is more important than the object that is acted on	✓	**Problem:** Researchers use polysomnography to evaluate the severity of sleep-disordered breathing. The sentence above calls attention to *researchers* as the main subject. But one might argue that the sentence is really about *polysomnography*. The following revision highlights the true subject, creating a preferable case of passive voice. **Solution 1:** Polysomnography is used to evaluate the severity of sleep-disordered breathing. **Solution 2:** The severity of sleep-disordered breathing is evaluated by polysomnography.
When the use of active voice overloads sentences with distracting first-person references, especially in methods sections of research papers and proposals	✓	**Problem:** I recruited participants from the local community through a newspaper advertisement. I screened subjects who had been diagnosed with sleep disorders. Then I randomly assigned subjects to a treatment group and a control group. Responding to this description of study methods, readers will likely be annoyed. They already know that the author performed the methods, so there's no reason to keep repeating *I, I,* and *I*. The revision highlights the methods as the main subjects of each sentence. **Solution:** Participants were recruited from the local community through a newspaper advertisement. Subjects who had been diagnosed with sleep disorders were screened. Subjects were then assigned to a treatment group and a control group.
When active voice constructions weaken paragraph coherence	✓	**Problem:** Cessation of breathing for 10 seconds or more is called sleep apnea. Blockage of the airways, chronic respiratory disease, or dysfunction of neural pathways from the brain to the lungs commonly cause this disorder. The second sentence in the example above is in active voice. Notice that the structure weakens the coherence across the two sentences. The second sentence begins with a very long and cumbersome noun phrase, which includes all of the words leading up to the verb, *cause*. Also, the second sentence begins with new information that will be unfamiliar to readers. In the solution, the second sentence is in passive voice. **Solution:** Cessation of breathing for 10 seconds or more is called sleep apnea. This disorder is commonly caused by blockage of the airways, chronic respiratory disease, or dysfunction of neural pathways from the brain to the lungs. The revised second sentence begins with a short noun phrase, *This disorder*, which repeats a key idea, *sleep apnea*, from the previous sentence. The use of passive voice improves the coherence across the two sentences.

To demonstrate a step-by-step process for revising cases of problematic voice, let's use the following sentence.

> Sleep has been reported by Baldwin (2005) to improve immune function by enhancing lymphocyte proliferation and increasing antibody synthesis.

Step 1: Parse the sentence's core (subject–verb) elements. The sample sentence has only one subject–verb unit. The subject is *sleep*, and the verb phrase is *has been reported*.

Step 2: For each core element, identify whether the subject directly names the actor of the verb. This step is for determining whether core elements are constructed by active or passive voice. Our example sentence is written in passive voice because the subject–verb unit, *sleep has been reported*, does not name the person(s) who actually reported the effects of sleep on immune function. This is true even though the actor, Baldwin, is named in the prepositional phrase that follows the subject–verb unit.

Step 3: For each core element, decide whether the voice construction (active or passive) is problematic or preferred. This decision depends on the factors presented in Table 7.1. Suppose that (for some odd reason) a professor instructed the author of our sample sentence to use passive voice exclusively. In this case, passive voice is preferred (that is, unless the author wants to argue that the instruction is far too restricting). If no instructions were given, the author should consider the common situation in which passive voice can add unnecessary words and weaken the tone of sentences. In our sample sentence, the tone is indeed weak because the verb phrase *has been reported* does not express the main action of the subject, *sleep*. In addition, the verb phrase is redundant because readers will clearly understand that the researcher named Baldwin reported the finding about sleep and immune function.

Step 4: If necessary, convert active voice to passive voice, or vice versa. Some sentences in active voice would actually be stronger if they were converted to passive voice. An example, which occasionally crops up in methods sections of research papers and proposals, is a paragraph in which numerous consecutive sentences contain first-person subjects (*I* and *we*). This automatically creates active voice and can result in excessively *I*-heavy or *we*-heavy prose: *I recruited...I measured...I analyzed.* A sound revision is to convert active voice to passive voice by turning the sentences' subjects into objects rather than actors (see the example in Table 7.1). Another solution is to use passive voice for *most* of the sentences of methods sections but to occasionally intersperse a sentence in active voice. This solution adds variety to sentence structure.

In contrast to the preceding case, some sentences would be strongest in active voice. This is true for our demonstration sentence, which we might revise as follows.

> Sleep improves immune function by enhancing lymphocyte proliferation and increasing antibody synthesis (Baldwin, 2005).

In active voice, the core element (*sleep improves*) is more emphatic, and the sentence conveys the same message in fewer words. An alternative revision, with two subject-verb units in active voice, highlights the scientist's contribution:

> Baldwin (2005) reported that sleep improves immune function by enhancing lymphocyte proliferation and increasing antibody synthesis.

Wordiness

Wordy sentences contain more words than are necessary to convey essential messages. In addition to weakening writing style, wordiness can impair readability and comprehension. It's especially important to eliminate unnecessary words in scientific papers because their maximum length is usually restricted. However, this aspect of revision can be truly challenging when writing projects demand explanations of complex concepts and extensively supported arguments.

Revising to eliminate wordiness is a bit like playing golf: Low score wins—that is, as long as reducing the word count does not alter or confuse a sentence's original meaning. Take a swing at revising this sentence:

> In the study conducted by Garcia and Newman (2003) it was found that women spend a relatively long time in slow-wave sleep.

The phrase *Garcia and Newman (2003)* obviously refers to the authors' study, so it's redundant to include *in the study conducted by*. In addition, the construction *it was* has no direct meaning and serves only to delay the researchers' findings. Eliminating the redundant and meaningless words, we have the following revision.

> Garcia and Newman (2003) found that women spend a relatively long time in slow-wave sleep.

It's seven words (or approximately 32%) shorter than the original, makes the same point more directly, and is easier to read.

Even for the most experienced writers, wordiness comes with the territory of drafting. It's a perfectly good problem to overlook until the revision process. But it's also an especially difficult flaw to detect independently. You really have to know what, specifically, to look and listen for in your draft's sentences.

1. Look and listen for long sentences. Try reading your draft out loud, tuning your ears to sentences that seem to go on forever. Whereas long sentences are not automatically wordy, they should still be flagged for evaluations, applying the following guidelines.

2. Look for redundant words, phrases, and ideas. The following 27-word sentence demonstrates wordiness characterized by redundancy.

> What is meant by the term paradoxical sleep refers to the stage of sleep during which the brain's neural activity is similar to that of waking conditions.

The phrase *what is meant by the term* replicates the idea in the phrase *refers to*. In addition, the idea in the phrase *of sleep* can be easily inferred from the word *stage*. The following revision cuts eight words, or approximately 30% of the original sentence.

> ~~What is meant by the term~~ Paradoxical sleep refers to the stage ~~of sleep~~ during which the brain's neural activity is similar to that of waking conditions.

Here's another sentence with redundant ideas:

> One possible reason for the conflicting findings between studies may be because scientists have used various different definitions of normal sleep.

The 21-word sentence has three cases of needless repetition: (a) *may be* replicates *possible*; (b) *because* replicates *reason*; and (c) *different* replicates *various*. So only one of the words or phrases in each set is necessary. In addition, the sentence has a weak verb, *used*, which adds to the wordiness. The revision that follows is 16 words, approximately 24% shorter than the original.

> One ~~possible~~ reason for the conflicting findings between studies ~~may be because~~ is that scientists have ~~used various different definitions of~~ defined normal sleep differently.

3. Look for words and phrases that readers can infer from other sentence elements.

> Normal sleep has been reported to enhance the primary response to antigens in animals that are exposed to viruses (Martens, 2002).

In this example, the verb phrase *has been reported* is not necessary because readers will infer that Martens reported the findings. The following revision is approximately 21% shorter than the original.

> Normal sleep ~~has been reported to enhance~~ enhances the primary response to antigens in animals that are exposed to viruses (Martens, 2002).

4. Look for words and phrases that have little, if any, meaning.

> Problem (16 words): It is this author's contention that REM sleep is essential for neural development in young animals.
>
> Solution 1 (13 words, ~19% shorter than the original): ~~It is this author's contention that~~ I contend that REM sleep is essential for neural development in young animals.
>
> Solution 2 (10 words, ~37% shorter than the original): ~~It is this author's contention that~~ REM sleep is essential for neural development in young animals.

Solution 1 eliminates the meaningless words *it is* and expresses the claim more directly, using first-person. Solution 2 would be appropriate for conveying established knowledge rather than stating an original claim.

5. Look for overstated phrases that are common in scientific writing. Scientists have the infamous reputation for using a stock set of wordy phrases, examples of which are presented in Table 7.2 (pages 346–347). If you recognize any of these phrases in your writing, replace them with their concise revisions.

Unnecessary Jargon

Among nonscientists the #1 complaint about scientific writing is that it is unreadable due to excessive jargon. In response, many scientists contend that jargon is absolutely essential for their effective communication. The confusion centers on the different ways that jargon is defined. We will rely on the primary definition of jargon as the technical terminology or the characteristic idiom of a specialized activity or group. Accordingly, jargon is generally not problematic in scientific papers intended for scientific audiences. In fact, highly technical and specialized language defines and facilitates communication in the scientific community. But scientists and science students sometimes use jargon when it is not warranted—that is, they use inflated, pretentious, and uncommon language to represent things and ideas that can be expressed in plain English. So instead of jargon *per se*, the justly criticized flaw in scientific writing is *unnecessary* jargon. The following example may exaggerate the problem somewhat, but it's really not too far from the overblown sentences that you may occasionally come across in published scientific papers.

> Individuals afflicted by the persistent condition of anti-somnolence are advised to conform to parameters that incorporate the application of periodic and systematic physical exertion and the attenuation of disturbances in the homeostasis of mental states.

For no good reason the writer has turned ideas that can be expressed perfectly well in plain English into mumbo jumbo. The phrase *persistent condition of*

TABLE 7.2 Examples of Wordy Phrases and Concise Revisions

Wordy	Concise
a considerable number of	many
a great majority of	most
-accounted for by the fact that -due to the fact that -for the reason that -in view of the fact that -on the grounds that	because
a plethora of	many
a total of 64 subjects	64 subjects
along the lines of	like
are of the same opinion	agree
as a means of	to
as of the current date	now, today
as to whether	whether
at the present time	now
because of the fact that	because
blue in color	blue
by means of	By
causal factor	cause
despite the fact that	although
diminutive in size	small
during the time that	while
end result	result
fewer in number	fewer
first of all	first
for the purpose of	for
give an account of	explain
give rise to	cause
in the absence of	without

TABLE 7.2 *continued*

it goes without saying that	(omit)
it has long been known that	(could omit)
it is evident that	(omit)
it is of interest to note that	note that (or omit)
in the course of	during
in the event that	if
made note of	noted
made reference to	referred to
of the opinion that	think that
on a daily basis	daily
on the basis of	by
performed a study of	studied
place a major emphasis on	emphasize
prior to	before
referred to as	called
so as to	to
subsequent to	after
take into consideration	consider
the reason why is because	the reason is that
through the use of	by
whether or not	whether

anti-somnolence really means *lack of sufficient sleep*. Instead of saying *regular exercise*, the writer puzzles us with *the application of periodic and systematic physical exertion*. Instead of *stress*, the writer creates the eyesore *disturbances in the homeostasis of mental states*. In plain English the sentence says, *People who cannot sleep should exercise regularly and reduce their stress*. This is how it should have been written in the first place!

In addition to impairing readability, unnecessary jargon is a major problem because it reflects negatively on the writer's credibility. Professors and editors view needlessly overblown language as the writer's hot air and perhaps even as deliberate attempts at putting up smokescreens. Some novice writers mistakenly believe that they will impress readers by using inflated language. But more often than not, unnecessary jargon signals that writers do not really understand their subjects.

To revise for the flaw at hand, turn on your *jargon detector* and identify uncommon words and phrases in your drafts. Suppose that you are writing about how researchers measure brain activity during distinct stages of sleep. Your draft might include terms such as *electroencephalography, beta rhythm, K complexes,* and *reticular activating system.* Of course, these words and phrases represent the specialized jargon that aids communication in the field of sleep physiology. For example, to describe the method for measuring the brain's electrical activity during sleep, you will find no better word than *electroencephalography.* If your readers need you to define and describe *electroencephalography,* do so the first time you present the word and its appropriate abbreviation, *EEG.* Thereafter, don't waste space by trying to translate EEG into plain English. The point is that EEG is good jargon. And so is any other word that passes the following tests.

- Scientists in the field agree that it is the most appropriate and precise word to use.
- Translating the word into plain English would reduce its precision, create confusion, and waste space.
- The word would never be used outside of communicating about the scientific topic at hand.

While you are identifying the good jargon in your drafts, also look for cases of *tuxedo-at-the-beach language.* Wearing a tuxedo to the beach, you would obviously be dressed unsuitably for the occasion. Carried over to scientific writing, the analogy applies to embellished words and phrases that needlessly dress up language. Some common examples of such unnecessary jargon in scientific communication, along with their plain-English revisions, are presented in Table 7.3.

Excessive Separation of Subjects and Verbs

Setting out to read any given sentence, we are especially keen to learn two things: *what the sentence is about* and *what will happen* (to what it is about). In other words, we purposefully attend to the core messages of sentences, which are conveyed by their subjects and verbs. When these elements are obscured or difficult to connect, readability and comprehension suffer, as demonstrated in the following example.

> Our results revealed in the present study of a progressive decline in alertness and cognitive performance in subjects who were deprived of sleep, indicated in the data showing that vigilance deteriorated over the 48 hours of testing agree with those from previous studies.

To grasp its message, you probably had to read the sentence more than once. When a sentence requires several readings to comprehend, it's worth checking

TABLE 7.3 **Examples of Unnecessary Jargon and
Plain-English Revisions**

Unnecessary jargon	Plain-English revision
ameliorate	improve
apprise	inform
causal factor	cause
commence	begin
component	part
dialogue (verb)	talk
effectuate	cause
elucidate	explain
employ	use
endeavor	try
eventuate	happen
evidenced	showed
fabricate	make
finalize	finish, end
impact (verb)	affect, influence
implement (verb)	apply
initiate	start
quantitate (verb)	count
terminate	end
transpire	occur
utilize, utilization	use

the location of its subject–verb units. Often the problem is an excessive separation of subjects and their verbs. In the preceding example, the main subject (*results*) is easy enough to locate because it is positioned toward the sentence's beginning, where we normally expect to find grammatical subjects. But the verb that complements *results* is not so easy to find. Immediately following the subject, the word *revealed* looks like it might be the verb, leading us to think that we are going to learn what the results actually showed. But in the context of the sentence, *revealed* is not a verb at all. Instead, it's a participle that modifies the subject. Seventeen words after the subject comes a verb phrase, *were deprived*; it, however, is the verb that complements the preceding pronoun *who*. Twenty-one words after the subject, the word *indicated* (another participle that modifies the subject but does not serve as its verb) tricks us again. We finally locate the true verb, which is the word *agree*, a whopping 35 words after the subject *results*. The sentence's core message turns out to be *results agree*. But it is truly difficult to identify the core message because the subject and verb are so far separated. The following two-sentence revision solves the problem.

> We found that alertness and cognitive performance progressively declined in sleep-deprived subjects, as indicated in the data showing that vigilance deteriorated over the 48 hours of testing. Our results agree with those from previous studies.

The revision has four subject–verb units with no words or only one word separating the core elements:

- we found (0 words between the subject and verb);
- alertness and cognitive performance progressively declined (1 word between the subject and verb);
- vigilance deteriorated (0 words between the subject and verb); and
- results agree (0 words between the subject and verb).

One way to detect excessive subject–verb separation is to read your sentences out loud. In the middle of a sentence, listen for whether your pace slows. And pay attention to whether you have to keep returning to the start of the sentence to make sense of it. These are signs of long stretches between subjects and verbs. You can also parse simple subjects and their verbs, counting the number of separating words. As a general guideline, when more than six or seven words come between a subject and its verb, consider rearranging word groups to bring the core elements closer together. In some cases, as just demonstrated, a good solution is to form two sentences. This involves moving subordinate information—that is, relatively less important ideas that come in long phrases that separate subjects and verbs— into introductory elements or newly formed sentences.

Long Noun Trains

Scientific writing commonly requires using nouns as adjectives that modify other nouns. In the following examples, nouns that serve as adjectives are italicized: *sleep* deprivation, *electroencephalography* signals, *brain* waves, and *heart* rate. These phrases are perfectly fine. A problem may result, however, when more than two or three noun-adjectives are stacked one after the next, forming a long noun train: *caffeine alertness effect*, *wakefulness extension condition*, *sleep loss behavior results*, and *daytime baseline test interval*. The problem is demonstrated in the following sentence.

> The *insomnia drug effect data analysis* indicated that sleep quality depended on the drug's dose.

No fewer than four nouns—*insomnia, drug, effect,* and *data*—are stacked to form a compound adjective that modifies the fifth noun in a row, the simple subject *analysis.* It's difficult to get through the sentence in one smooth pass. The reason is that, to understand the idea that the writer intends to convey in the long noun train, we have to sort out what is modifying what. A solution is to break up the long noun train by adding prepositions that indicate the logical relationships between its nouns. Compare the following revision to the original sentence.

> The data analysis for the effect of the insomnia drug indicated that sleep quality depended on the drug's dose.

The revision has (a) shorter noun trains, consisting of only two words (**data analysis** and **insomnia drug**); and (b) added prepositions that indicate the relationship between nouns (data analysis **for** the effect **of** the insomnia drug).

Any phrase with more than two noun-adjectives in a row is worth diagnosing for the problem at hand. A four- or five-word noun train is almost always problematic, as in the following example.

> A *beta-wave electroencephalography amplitude measurement reduction* was observed in subjects who took sleeping pills.

The following revision applies the strategy of rephrasing the sentence, breaking up the noun train by adding prepositions.

> A reduced amplitude for the measurement of beta-wave electroencephalography was observed in subjects who took sleeping pills.

It's certainly an improvement over the original sentence; however, the resolution of the long noun train reveals new flaws that warrant iterative revision. Consider, for example, the new noun phrase *A reduced amplitude for the measurement of beta-wave electroencephalography.* The simple subject, *amplitude,* is separated from its verb phrase, *was observed,* by six words. This is getting close to excessive separation of the subject and its verb. In addition, the revised sentence contains a weak core unit: *amplitude was observed.* The verb, *observed,* does not reflect the main action or condition of the subject, *amplitude.* The following revision strengthens the core and brings the subject and verb closer together.

> As indicated in the electroencephalography measurement, beta-wave amplitude was reduced in subjects who took sleeping pills.

A noteworthy feature of this solution is the added introductory phrase (everything up to the comma), which includes subordinate information from the original lengthy noun phrase.

Lack of Parallel Structure

Chapter 6 presented a paragraph-level problem in which successive sentences that serve a similar function lack parallel structure. The same flaw occurs at the sentence level when naturally coordinated ideas are not expressed in consistent grammatical form. Here's an example:

> Sleep deprivation impairs academic performance, increases risks of upper respiratory tract infections, and personal and social activities are negatively impacted.

The sentence immediately elicits the vague sense of a broken pattern. Examining the sentence more closely, we hone in on the word group, *personal and social activities are negatively impacted*. To diagnose the problem, notice that the sample sentence presents a list of things that sleep deprivation does. The list's first two items indicate that sleep deprivation

- impairs academic performance, and
- increases risks of upper respiratory tract infections.

These items have the same grammatical structure, comprising a verb (*impairs, increases*) followed by a direct object (*academic performance, risks of upper respiratory tract infections*). Now examine the grammatical structure of the list's third item:

- personal and social activities are negatively impacted.

A compound noun (*personal and social activities*) is followed by a verb phrase (*can be negatively impacted*). So this item is not parallel, or coordinated, with the first two items in the list. This is a problem because all three items are naturally coordinated in function, or in the role that they are meant to play in the sentence. That is, the items are meant to convey consequences of sleep deprivation. The following revision makes the sentence parallel and thereby improves its readability.

> Sleep deprivation impairs academic performance, increases risks of upper respiratory tract infection, and **negatively impacts personal and social activities**.

In the previous example, the coordinated items make up a list. The guideline for maintaining parallel structure also applies to comparing items. The following sentence lacks a parallel comparison.

> The severity of insomnia is greater in people who work at night than in a person who works during the day.

The problem is that the severity of insomnia is compared in *people* (a plural form) versus *a person* (a singular form). A solution, as follows, coordinates the items being compared.

> The severity of insomnia is greater in people who work at night than in people who work during the day.

Often, a lack of parallel structure is easier to detect when you hear it versus when it is read. If you did not catch the lack of parallel structure in my last sentence, go back and read it out loud. The clause *when you hear it* is in active voice, but the clause *when it is read* is in passive voice. The sentence would be easier to follow with both clauses in the same voice form. Accordingly, here's one solution in parallel active voice: Often, a lack of parallel structure is easier to detect when you hear it versus when you read it. And here's a more economical solution: A lack of parallel structure is easier to detect when heard versus read.

A lack of parallelism can crop up in sentences that list or compare things. Of course, list items are separated by commas, and the last two items are joined by the coordinating conjunction *and*. Comparisons rely on words like *as, than, versus,* and *compared to*. Being mindful of lists and comparisons in your drafts, test their items and ideas for whether they have the same grammatical structure. If they don't, you will improve the readability of your sentences by restructuring them in parallel form.

REVISING BASIC GRAMMAR ERRORS

Especially for native speakers of English, sentence-level revision does not require much recalling and applying basic rules of grammar. Native speakers follow these rules instinctively. Take a sentence in which the subject is the noun *pattern* and its verb is *to disrupt*. If English is your first language, or if it's a second language that you learned in childhood, you won't think twice about the correct verb form. In other words, you won't stop to recall the rule that third-person singular subjects have verbs with –s or –es endings. You would just naturally use the correct verb form, *disrupts*, because it sounds right. Consider, however, a common case of subject–verb disagreement in scientific papers:

> The pattern by which college students deprive themselves of sleep during the week in order to study and participate in social events and then catch up by sleeping an excessive number of hours on the weekends disrupt the normal circadian rhythm.

The simple subject, *pattern*, does not agree in number with its verb, *disrupt*. The error is attributable to a flaw that we have already discussed: excessive separation of subjects and their verbs. The writer simply lost track of the main subject and mistakenly chose to make the closest noun, *weekends*, agree in number with the verb. So the problem was not caused by insufficient knowledge about the rules of grammar. Instead, the writer just had a lapse of memory, which can happen when subjects are placed too far from their verbs.

By no means am I dismissing the importance of knowing the rules of grammar for successful scientific writing. Even experienced scientific authors should be concerned about matters of grammar because long-standing rules of Standard English may change. Our language is always evolving under societal influences and by our efforts to communicate more effectively. For the most accurate and up-to-date scientific information, we seek the latest work of expert researchers; the same approach applies to learning about the leading ideas and practices in English grammar. The experts, called grammarians, serve on editorial boards and "usage panels" for journals, publication manuals and style guides, and dictionaries. In educational resources on English usage, grammarians express their views on whether long-standing conventions are worth maintaining or abandoning for more functional alternatives. Several reputable books on contemporary English usage are available for free on the *Bartleby.com Great Books Online* Web site (http://www.bartleby.com/usage/). The books feature current views on English usage, ones that might contrast lessons you learned in grade school. For example, most leading grammarians find no problem at all with prepositions placed at the end of sentences. In addition, they contend that it is acceptable to intentionally split infinitives, as I have just done. To locate other helpful online resources for learning or remembering rules of English grammar, do an Internet search for university-based online writing centers. One of the best is *The Purdue Online Writing Lab* (http://owl.english.purdue.edu/). The upcoming sections highlight a few cases of grammatical errors to be aware of in revising scientific papers.

Sentence Fragments

Like a complete sentence, a sentence fragment begins with a capital letter and has ending punctuation such as a period or a question mark. Unlike a sentence, a fragment cannot stand alone and express a complete thought. An example is the italicized word group as follows.

> The survey indicated that the subjects slept between 5.7 and 8.3 hours per night. *Showing large individual differences in the need for sleep.*

This is a sentence fragment because it lacks a subject–verb unit and does not express a finished idea. As a phrase, the italicized word group is a common type of sentence fragment. Suspecting that their drafts might contain sentence fragments in the form of phrases, writers should read slowly and assess each sentence for whether it conveys a complete, stand-alone message. Another approach involves parsing sentences for their subject–verb units. Although the italicized phrase in the preceding example contains three nouns (*differences*, *need*, and *sleep*), none is a subject. To be a subject, a noun must have a complementary verb that indicates its action or state. The italicized word group in our sample fragment does not contain such a verb.

One way to correct phrase-fragments is to attach them with a comma to neighboring sentences:

> The survey indicated that the subjects slept between 5.7 and 8.3 hours per night, **showing large individual differences in the need for sleep**.

Another solution is to convert the phrase into a complete sentence:

> The survey indicated that the subjects slept between 5.7 and 8.3 hours per night. **This result shows large individual differences in the need for sleep.**

Another common type of fragment is a dependent clause, or a word group that contains a subject and a predicate but does not express a complete thought. Here's an example:

> Scientists know much about the need for sleep. *Whereas they know little about the functions of sleep.*

The italicized word group looks like a sentence. It has a subject (*they*), a verb (*know*), and a direct object (*little about the functions of sleep*). But it's not a sentence because it lacks an independent clause and, therefore, does not express a complete idea. The subordinating conjunction *whereas* sets up the requirement for an independent clause to be included within the sentence. The following revision uses a comma to attach the dependent clause to the preceding sentence, which then becomes an independent clause.

> Scientists know much about the need for sleep, **whereas** they know little about the functions of sleep.

To revise fragments that are dependent clauses, you have to know about the roles of subordinating conjunctions, which are words such as *although, after, because, before,* and *whereas* (see **Conjunctions** in Appendix B).

Subject–Verb Disagreement

As demonstrated many times so far in this chapter, the fundamentals of well-crafted sentences are their subjects and verbs. Consider some of the major sentence-level flaws that we have covered: illogical expressions and comparisons, anthropomorphism, dangling modifiers, and problematic cases of active voice and passive voice. Revising for all of these problems requires isolating subjects and verbs and then running diagnostic tests on them. This fundamental approach also applies to cases of subject–verb disagreement. As I discussed above, this is usually not a major problem for native speakers of English. But a few special instances of subject–verb disagreement, described as follows, are worth attending to in scientific writing.

1. Excessive separation of subjects and verbs. Examples of this glitch and suggestions for resolving it were presented earlier in the chapter.

2. Cases involving *data*. The word *data* is the plural form of *datum*. So, in scientific papers, the word *data* is usually accompanied by a plural verb. Many professors and editors will thus make the following correction.

> The data ~~indicates~~ indicate that infants sleep between 14 to 16 hours per day on average.

However, leading grammarians would not automatically correct the singular verb. They would argue that *data* can be interpreted as a singular mass of information rather than as all of the separate individual values, or datum, that make up a data set. Grammarians thus advise writers to use a singular verb when they intend *data* to be interpreted as a mass noun (for example, see The Columbia Guide to Standard American English at http://www.bartleby.com/68/19/1619.html). But note that most scientists still use a plural verb for the noun *data* in all cases.

3. Cases involving compound subjects. Verbs that have two or more subjects (compound subjects) take the plural form. The following correction applies this rule.

> The amplitude and frequency of the EEG signal **indicates** the stage of sleep.

The plural form, *indicate*, is required because the sentence has a compound subject composed of two nouns: the amplitude (of the EEG signal) and the frequency (of the EEG signal). The original flaw, using the singular verb *indicates*, may have occurred because the writer mistakenly viewed *the EEG signal* as the subject. Or the writer mistakenly viewed *amplitude and frequency* as a singular noun composed of grouped entities (like *peanut butter and jelly* or *the long and short of it*).

4. Cases involving inverted sentences. In an inverted sentence, the subject appears after the verb. This construction sometimes results in subject–verb disagreement. Here's a correct revision of the problem:

> Of particular significance in patients with sleep apnea ~~is~~ are the high incidence of stroke and the condition of bradycardia.

In its inverted position, the sentence's true subject is *the high incidence of stroke and the condition of bradycardia*. This compound subject requires the plural verb *are*. Originally using the singular verb *is*, the writer mistakenly viewed *significance* as the subject. The problem is revealed by reversing the order so that the subject comes before the verb, as in the following case of subject–verb disagreement.

> The high incidence of stroke and the condition of bradycardia is of particular significance in patients with sleep apnea.

5. Cases involving *et al.* The Latin word *et alii,* abbreviated as *et al.,* means "and others." Because it is a plural noun, *et al.* requires a plural verb, as demonstrated below.

> Barringer et al. ~~argues~~ **argue** that the main function of sleep in animals is to conserve energy and to thereby ensure survival.

Noun–Pronoun Disagreement

A long-standing convention in academic writing is that pronouns should agree with their antecedents in number. The following sentence breaks the convention.

> After each participant was informed about the purpose of the experiment, they underwent a medical examination.

The noun *participant* is singular, but the pronoun *they* is plural. One solution is to change *they* to *he.* However, this revision would not make sense if the participants included women. In addition, some readers view the generic male form, the pronoun *he,* as sexist language. The best solution is to make both the noun and pronoun plural, as follows.

> After **the participants** were informed about the purpose of the experiment, **they** underwent a medical examination.

The pronouns *who* and *whom* refer to people, whereas the pronouns *that* and *which* generally refer to objects and animals. The following sentences demonstrate revisions that match nouns and their pronouns accordingly.

> There are few scientists ~~that~~ who would argue that sleep deprivation does not negatively impact fine motor performance.

> Ten subjects were selected for the experiment, seven of ~~which~~ whom were men.

If you suspect that your drafts might contain instances of noun–pronoun disagreement, isolate the pronouns that are commonly involved: *they, who, whom, which,* and *that.* Then, linking the pronouns to their antecedents, test for agreement by applying the rules presented above. (There are a number of other rules for noun–pronoun agreement, which you can learn about in basic grammar books.)

REVISING FOR WORD CHOICE

Our divide-and-conquer approach to revision has progressed from larger to smaller units of discourse. Beginning with revisions of global structure in Chapter 5, we have gradually worked our way to sentences. Here we narrow our focus to individual words. Actually, you have already seen instances in which a single poorly chosen word can weaken writing quality. Take the problem of anthropomorphism in this sentence: *The study hypothesized that sleep is necessary for consolidating memory.* One word, *hypothesized*, does not fit logically in the sentence because studies are not capable of hypothesizing. At the paragraph level, an inappropriately used transitional word can lead readers down a confusing path. And at the content level, a single misused word can hinder readers' understanding of a definition, an explanation, or even an entire argument. The best writers know that every word matters, so they pay careful attention to word choice when revising.

Some cases of inappropriate word choice are attributable to the writer's lacking knowledge of a scientific field's specialized vocabulary. Here's a case in point:

> Adenosine is a hormone that plays a role in regulating sleep and wakefulness.

The inappropriate word is *hormone*. Scientists generally do not define the regulatory molecule adenosine as a hormone. Instead, they define it as a neurotransmitter or neuromodulator. To recognize the inappropriate word choice, you would need to know the vocabulary of neuroscience. The take-home messages are to be mindful of specialized scientific terms in your drafts and to question whether you are using them correctly. When you are uncertain, refer to a discipline-specific scientific dictionary (see Table 1.6).

Other cases of inappropriate word choice involve common words that come in potentially confusing groups. Examples are *affect* and *effect, compose* and *comprise,* and *than* and *then.* This section focuses on these more general matters of word choice.

Affect, Effect

Misuses of *affect* versus *effect* stem from overlooking the fact that each word has two different definitions, which depend on its part of speech. Both *affect* and *effect* can be used as nouns or verbs. A common flaw, demonstrated as follows, is using the noun *affect* incorrectly.

> Our study addresses the affect of sleep on immune competence and recovery from infectious disease.

In this example, the word *affect* is a noun that serves as the object of the verb *addresses.* The dictionary definition of the noun *affect* is "a feeling or emotion." So the sentence literally conveys the idea that the study addresses the feelings or emotions of sleep. Of course, this is not logical because sleep does not have feelings or emotions! The solution is to replace *affect* with the noun *effect,* which means "a result, or something brought about by a cause."

Here, *effect* is used incorrectly as a verb:

> Caffeine can significantly effect mood.

The verb *effect* means "to create, or to bring into existence." So the sentence above literally conveys that caffeine brings mood into existence. This is not logical because everyone already has some sort of existing mood, whether it's a good one, a lousy one, or something in between. Rather than bringing mood into existence, caffeine *influences* a mood that already exists. A verb that means "to influence" is *affect*. Here's a sentence that uses *affect* correctly as a verb and as a noun:

Ⅰ   Like affection, caffeine positively affects my affect.

If you are uncertain about whether you are using *affect* and *effect* correctly, use your word processor's Find program to locate each instance of the words in your draft. Then determine whether you intend for the word to serve as a noun or a verb in the sentence. In place of the word *affect* or *effect*, substitute its definition:

- Affect (noun)—a feeling or emotion.
- Affect (verb)—to have an influence on; to bring about a change in something.
- Effect (noun)—a result or outcome; something brought about by a cause.
- Effect (verb)—to bring into existence; to produce as a result.

If the sentence does not make sense, substitute the correct word.

Consider that you may never need to use *effect* as a verb. The verb *cause* usually conveys the ideas of bringing something into existence and producing an outcome. In addition, unless you are writing about the psychological experiences of feelings and emotions, you will never need to use *affect* as a noun.

As, Since, Because

To indicate cause-effect relationships and explanatory ideas, the word *as* is not as accurate as *because*. The primary meaning of *because* is "for the reason that." If, for example, a writer intends to explain why some doctors do not prescribe sleeping pills, the following revisions would be appropriate.

Ⅰ   Some doctors do not prescribe sleeping pills, [as] [since] because patients can become addicted to these drugs.

The primary definitions of *as* are "to the same degree" and "at the same time that." In the preceding example, these definitions do not convey the writer's precise meaning. *Because* would also be more appropriate than *since* in the sample sentence. The primary meaning of *since* refers to relationships between events in time: "from then until now or between then and now."

Amount, Number

The word *amount* refers to quantities of objects, phenomena, and concepts that are measured and described as whole, or continuous, units. So it's correct to write about the "amount of rapid eye movement (REM) activity." Here, *amount* refers to the phenomenon of REM activity, which can be measured as an overall unit (e.g., by the total area under the curve of an EEG signal). Suppose, however, that you were describing the individual back-and-forth movements of the eyes during REM sleep. These movements are counted in discrete units. In this case, reference to "the *amount* of rapid eye movements" would be problematic. For describing measurements of countable individual objects, phenomena, and concepts, the appropriate word is *number*, as in "the number of rapid eye movements per minute."

Compose, Comprise

Compose means "to make up the constituent parts of." In the following sentence, *compose* is used correctly.

| Six women and four men composed the treatment group.

Comprise means "to include, or to contain the parts that make up the whole." As follows, *comprise* is used correctly.

| The treatment group comprised six women and four men.

To determine whether you are using these words correctly, apply this conventional rule: The whole comprises the parts, and the parts compose the whole. Accordingly, an entity cannot be *comprised of* things, but it can be *composed of* things. The following revision is thus correct.

| The treatment group was ~~comprised of~~ composed of six women and four men.

Gender, Sex

When you are writing about research involving women (or girls) and men (or boys), the word *sex* refers to biological distinctions, or to factors underlying sex-specific characteristics. In contrast, the word *gender* refers to societal and cultural distinctions. In the following sentence, *gender* is used incorrectly.

| During sleep, men and women experience distinct fluctuations in the release of gender-specific hormones.

Because hormones distinguish women and men on a biological basis, rather than a social or cultural basis, the sentence should be revised to replace *gender-specific hormones* with *sex-specific hormones*.

Decisions about when to use *gender* versus *sex* can be tricky, and sometimes the implications can be a bit risqué. Suppose that an author is reporting the results of a cross-cultural study on differences in sleep habits between women and men; the study did not concern biological differences. In the following sentence, the author misuses the word *sex*.

| Across many cultural groups, sleep quality is not influenced by sex.

As you might imagine, readers may easily misinterpret the sentence's meaning. A more appropriate revision would replace *sex* with *gender*. Better still, to avoid all confusion, the writer might revise the sentence as follows.

| Across many cultural groups, sleep quality does not differ between men and women.

Less, Few, Fewer

The word *less* should be used to compare variables that cannot be counted in discrete units. So the phrases *less time*, *less blood*, and *less sleep* are correct. In contrast, *few* and *fewer* apply to comparing discrete, countable variables: *fewer milliseconds*, *few blood cells*, and *fewer people falling asleep*. In scientific papers, the most common flaw involving these words is corrected in the following sentence.

| Subjects who slept at least 8 hours a night made significantly ~~less~~ **fewer** errors on the vigilance task.

Study, Experiment

These words are often used interchangeably with no problem. But their precise meanings differ. As a noun, *study* refers to all approaches to conducting research. The noun *experiment* refers to a specific type of study in which researchers manipulate treatments and conditions directly. If a study involves observing outcomes without systematically manipulating treatments and conditions, it is not an experiment. Here's a revision that illustrates the distinction:

| In ~~an experiment~~ a study on sleep habits in infants, researchers asked parents to record the number of nightly hours that their babies slept.

Because the researchers did not intervene to influence the infants' sleep habits, the study was not an experiment.

That, Which

These words are pronouns that, among other uses, serve to introduce and form dependent clauses (which are also called subordinate clauses). The traditional rule is to use *which* in dependent clauses that are nonrestrictive. These are clauses that (a) are not intended to limit, or narrowly define, the meaning of the antecedent noun that *which* refers to; and (b) are not essential for communicating the sentence's main message. The following example uses *which* correctly.

> Nelson's hypothesis, which was published in 2004, centers on the negative effects of sleep deprivation on short-term memory.

The sentence contains an independent clause and a dependent clause.

- Independent clause: Nelson's hypothesis centers on the negative effects of sleep deprivation on short-term memory
- Dependent clause: which was published in 2004

The dependent clause is nonrestrictive because it does not limit Nelson's hypothesis to the one published in 2004. In other words, the use of *which* indicates that the author did not intend to distinguish Nelson's hypothesis from ones published in *other* years. In addition, the dependent clause is not essential for conveying the sentence's main message. To understand that the hypothesis involves the negative effects of sleep deprivation on short-term memory, you do not need to know that it was published in 2004. The nonrestrictive clause simply provides extra information.

The other part of the traditional rule is to use *that* for restrictive clauses. These are clauses that (a) are intended to limit the meaning of the antecedent noun and (b) are essential for communicating the sentence's message. Here's a correct example:

> The hypothesis that Nelson has advanced involves the negative effects of sleep deprivation on short-term memory.

The pronoun *that* limits the meaning of its antecedent, *hypothesis*, to the hypothesis that one particular scientist, named Nelson, advanced. The dependent clause introduced by *that* is essential for conveying an important aspect of the sentence's meaning.

According to the traditional rule, the use of *which* in the following sentence is incorrect.

> The hypothesis which Nelson has advanced involves the negative effects of sleep deprivation on short-term memory.

It's worth noting, however, that grammarians do not unanimously agree on the rule that *which* must always be reserved for nonrestrictive clauses. Some experts would say that the sentence above is perfectly fine because its meaning is clear. At least until the experts agree, you might as well follow the traditional rule, using *which* for nonrestrictive clauses and *that* for restrictive clauses.

Than, Then

In making comparisons, writers sometimes inadvertently use the adverb *then* (which denotes an action occurring at some point in time) instead of the conjunction *than*. The sentence below shows the problem and a correct revision.

> The time to fatigue was longer in the treatment group ~~then~~ than in the control group.

If you know that you tend to confuse *then* and *than*, use your word processor's Find program to mark instances of these words. Then (not than!), run the following checks.

- *For than*, make sure that you are making comparisons.
- *For then*, make sure that you are denoting an action occurring at a point in time.

Who, Whom

The pronoun *who* substitutes for the subject of a sentence or clause, as correctly demonstrated in the following example.

> Cynthia Yancy, who studies the effects of menopause on sleep, is recognized as a leading researcher in the field.

Who functions as the pronoun that refers to Cynthia Yancy and as the subject of the verb *studies*.

The pronoun *whom* functions as the object of a verb or preposition.

> Cynthia Yancy, whom many scientists recognize as a leading researcher, studies the effects of menopause on sleep behavior.

In this example, *whom* is used correctly as the object of the verb *recognize*. The clause *whom many scientists recognize as a leading researcher* conveys that many scientists recognize Cynthia Yancey.

To work through the steps for revising problems involving *who* and *whom*, we will use the following sentence.

> Catastrophic accidents have involved transportation workers who supervisors reported as being sleep deprived.

First, let's isolate the clause in which *who* or *whom* appears. In this case, the clause is *who supervisors reported as being sleep deprived*. Next we need to determine whether *who* or *whom* is (a) the subject of a verb or (b) the object of a verb or preposition. In our sample clause, *who* does not function as the subject of a verb. The only verb in the clause is *reported*, and its subject is *supervisors*. The situation would be different if the sentence read this way: *Catastrophic accidents have involved transportation workers who were sleep deprived.* From this sentence, we would isolate the clause *who were sleep deprived.* In this correct case, *who* is the subject of the verb *were.* In our original sentence, however, *who* is the object of the verb *reported.* The supervisors reported the transportation workers. The next diagnostic step is to apply the rules for using *who* and *whom.* Doing so, we determine that *who* is used incorrectly in our sample sentence. The correct word to indicate the object of a verb or preposition is *whom:*

> Catastrophic accidents have involved transportation workers **whom** supervisors reported as being sleep deprived.

REVISING PUNCTUATION AND MECHANICS

This section addresses two general types of problems: (1) missing and misplaced punctuation, such as commas, semicolons, colons, and apostrophes; and (2) errors involving capitalization. Compared to revising for matters of logic, structure, style, and grammar, the revision processes for punctuation and mechanics are usually more straightforward. This aspect of sentence-level revision generally relies more heavily on knowing and applying definite rules to detect, diagnose, and solve problems.

Problems Involving Commas

Commas separate parts of sentences to enhance clarity and readability. Revisions of common problems involving commas are described and demonstrated as follows.

Omitting a Necessary Comma between an Introductory Word Group and an Independent Clause
A bit of confusion arises in the following sentence.

> When the study participants awakened the researchers administered a questionnaire to assess sleep quality.

Midway through the sentence, you may have stopped to ponder why the researchers were sleeping and had to be awakened by the study's participants! At first glance, *the researchers* seems to be the object of the verb *awakened.* But the writer really did not intend to say that the study participants had to wake up sleeping researchers. The confusion is resolved in the following revision, in which the introductory dependent clause is appropriately separated from the sentence's independent clause by a comma.

> When the study participants awakened**,** the researchers administered a questionnaire to assess sleep quality.

A comma is optional when the introductory word group is short and does not run confusingly into the subject of the independent clause:

> In this study we sought to determine the effects of moderate sleep loss on short-term memory.

Omitting a Necessary Comma to Set Apart a Nonrestrictive Clause

> Research indicates that individuals with type 2 diabetes are likely to suffer from sleep disorders which exacerbate insulin resistance.

The preceding sentence has two major parts: an independent clause (the word group leading up to the pronoun *which*) and a dependent clause (the word group that begins with *which* and completes the sentence). The dependent clause describes something about the sleep disorders that individuals with type 2 diabetes are likely to suffer from. Notice, however, that the description can be interpreted in two ways. One meaning is that individuals with type 2 diabetes are likely to suffer from a restricted class of sleep disorders, specifically those that exacerbate insulin resistance. In an alternative interpretation, the dependent clause does not restrict the meaning of *sleep disorders*; instead, it provides supplemental information about them. The first interpretation is based on reading the dependent clause as a restrictive clause. If this is how the writer intended the clause, an appropriate revision would be to replace *which* with *that* (see the explanation of when to use *that* versus *which* on page 361). But if the writer intended the dependent clause to be nonrestrictive, a comma should be added to offset the independent clause from the dependent clause:

> Research indicates that individuals with type 2 diabetes are likely to suffer from sleep disorders, which exacerbate insulin resistance.

Omitting the Second Comma in a Pair That Sets off a Parenthetical Element

A parenthetical element, as demonstrated by the word group enclosed by this sentence's first and second commas, is a word, phrase, or clause that provides supplemental or clarifying information. A pair of commas should be used to set off a parenthetical element from the main parts of its sentence. As demonstrated in the following sentence, it's an error to omit the second comma in the pair.

> The circadian rhythm, which is the body's internal clock that operates on an approximately periodic 24-hour cycle is disrupted in individuals who have sleep disorders.

The first comma effectively signals the start of a parenthetical element that provides supplemental information about the circadian rhythm. But it's not immediately obvious where the parenthetical element ends and the sentence's main idea picks up again. A comma placed after *cycle* and before the main verb, *is*, would make this necessary distinction.

Omitting a Necessary Comma between Independent Clauses That Are Joined by a Coordinating Conjunction

> Stage 1 sleep is characterized by normal muscle tone and vivid dreams do not tend to occur during this stage.

Midway through this example, readers might be confused by the role of the phrase *vivid dreams*. It could easily be read as a second characteristic of stage 1 sleep, which the writer is adding to the first characteristic, *normal muscle tone*. But the writer actually intended for *vivid dreams* to be the subject of a new independent clause. The coordinating conjunction *and* joins the sentence's two independent clauses. For the sake of clarity, and as a general rule, a comma should be placed before the coordinating conjunction that joins two independent clauses:

> Stage 1 sleep is characterized by normal muscle tone**,** and vivid dreams do not tend to occur during this stage.

A comma is not necessary, however, when the two independent clauses are short and its omission does not confuse readers.

Comma Splices

A comma splice is a type of run-on sentence in which a comma separates two independent clauses (or two complete sentences) that are not joined by a coordinating conjunction. Here's an example:

> Many investigations have focused on the effects of sleep loss on cognitive performance, the findings from these studies conflict.

One solution is to place the appropriate coordinating conjunction, in this case *but*, immediately after the comma and before the second independent clause. Alternative solutions are to replace the comma with a semicolon or to use a period to form two separate sentences.

Check the following sentence for whether it contains a comma splice.

> Periodic limb movements during sleep are common in individuals who suffer from sleep apnea, however the physiological basis for this association is unclear.

The sentence contains two independent clauses that are separated by a comma and joined by the word *however*. The comma may appear to be correct because the sentence is structured similarly to one in which a coordinating conjunction joins two independent clauses. But the word *however* is not a coordinating conjunction. Instead, it's a conjunctive adverb. Examples of conjunctive adverbs are *consequently, however, moreover, then, therefore,* and *thus.* When a conjunctive adverb joins two independent clauses, it should not be preceded by a comma. So our example sentence has a comma splice. One correct revision is to replace the comma with a period and to begin a new sentence with *However.* Another solution, demonstrated as follows, is to replace the comma with a semicolon.

> Periodic limb movements during sleep are common in individuals who suffer from sleep apnea; however, the physiological basis for this association is unclear.

Note, however, that in the current sentence, a semicolon does not precede *however.* This is because the *however* in the previous sentence does not function to join two independent clauses.

Commas Separating Two-Part Compound Predicates

We know that commas are correctly used to separate two independent clauses that are joined by coordinating conjunctions. But this is *not* what the comma is doing in the following sentence.

> The brain's arousal system can be activated by sensory input, and inhibited by adenosine accumulation.

The coordinating conjunction *and* does not serve to join two independent clauses. Instead, it joins two parts of the sentence's predicate, which is everything after the complete subject, *The brain's arousal system.* The first part of the compound predicate is *can be activated by sensory input*; the second part is [*can be*] *inhibited by adenosine accumulation.* The comma before *and* should be deleted because it breaks the flow of the sentence. In addition, it breaks the rule that commas should not be used to separate compound predicates that have two parts. It's a different story if the compound predicate has more than two parts:

> The brain's arousal system can be activated by sensory input, inhibited by adenosine accumulation, and stimulated by cholinergic drugs.

Because the predicate contains a list of more than two things that can happen to the brain's arousal system, the commas are used correctly.

Commas Separating Verbs from Their Subjects

In the sentence below, the strikeout reflects the rule that commas should not be used to separate verbs from their subjects.

> People who claim that herbal teas are effective for treating insomnia and other sleep disorders; are basing their arguments on anecdotal evidence.

Commas after the Words *Including* and *Such as*

Commas should not be used after *including* or *such as* when these words introduce a list:

> Disruptions of the circadian rhythm can lead to several negative outcomes including, decreased alertness, impaired performance, and negative moods.

The comma immediately following *including* should be deleted.

Commas after *Although* and Other Subordinating Conjunctions

When a subordinating conjunction introduces a dependent clause, the conjunction should not be followed by a comma. A common error is placing a comma after the subordinating conjunction *although*:

> Although, our results indicate that sleep restriction improves mood, other researchers have reported conflicting findings.

The comma after *Although* should be deleted.

Problems Involving Semicolons

One of two primary purposes for semicolons in academic writing is to join independent clauses that are closely related in meaning. The following sentence correctly reflects this function.

> Compared to the control mice, the sleep-deprived mice did not differ in lymphocyte responses to various antigens; this finding does not support the hypothesis that sleep deprivation impairs immune function.

The semicolon above signals readers that the ideas in the two independent clauses express a unitary message. Because the semicolon indicates the relationship between the two independent clauses, a coordinating conjunction is not necessary.

In the following sentence the semicolons correctly serve their second primary purpose, which is to separate listed items that are punctuated internally with commas.

> During paradoxical sleep we experience increases in heart rate, respiration rate, and ocular movements; a predominance of high-frequency, low-amplitude brain waves; and temporary paralysis of the skeletal muscles.

Problematic uses of semicolons are presented as follows.

Semicolons That Connect an Independent Clause to a Dependent Clause or a Phrase

When a semicolon is used to connect two closely related ideas, the rule is that they must be expressed in independent clauses. Here's a sentence that breaks this rule:

> Sleep-related breathing disorders such as snoring can cause hypertension; possibly by increasing sympathetic neural drive.

To diagnose the problem, determine whether the word groups on both sides of the semicolon are independent clauses. If they are, they can stand alone and express a complete thought. The first word group, everything that comes before the semicolon, passes the test. The second word group, everything thereafter, does not express a complete thought because it is a phrase rather than an independent clause. So the semicolon is used incorrectly. One solution is to replace the semicolon with a comma. An alternative solution, demonstrated as follows, converts the second word group into an independent clause.

> Sleep-related breathing disorders such as snoring can cause hypertension; a possible mechanism is an increase in sympathetic neural drive.

Using a Semicolon to Introduce a List

Semicolons should not be used to introduce lists, as in the following problematic sentence.

> Previous studies on the effects of sleep on immune function have been flawed for the following reasons; inconsistent methods for depriving animals of sleep, imprecise measures of antibody activity, and poor controls for stress responses that are not due to sleep deprivation.

The correct punctuation for introducing lists is a colon.

Overuse of Semicolons

Some writing experts warn against using semicolons because they can make prose look and sound choppy. Semicolons do serve useful purposes, as illustrated by the preceding correct examples. But their overuse, especially in many successive sentences of a paragraph, is problematic.

Problems Involving Colons

The primary uses of colons in scientific writing are the following: (a) to introduce long lists with items described by long word groups, (b) to alert readers to an upcoming emphatic idea, (c) to alert readers that an upcoming idea summarizes or provides an example of the immediately preceding idea, and (d) to indicate ratios (e.g., a 3:1 ratio of female to male subjects). The first colon in the previous sentence is used correctly to introduce a long list. The next example effectively demonstrates how a colon can focus readers' attention on an emphatic idea.

> Whereas many college students think that they concentrate better and learn more by studying all night before exams, the research indicates otherwise: Sleep deprivation negatively affects performance of tasks that involve vigilance, concentration, and memory.

When a colon is used to introduce a list or to focus readers' attention on an emphatic or exemplary idea, the word group that precedes the colon should be a complete sentence or an independent clause that can stand alone in meaning. The word group following the colon may be a phrase, a list, or a complete sentence. According to APA publication guidelines (American Psychological Association, 2001), when the word group following a colon is a complete sentence, the sentence should begin with a capital letter. Other scientific publication organizations give writers the option of capitalizing the first word, as long as they are consistent throughout the manuscript.

Colons That Separate Verbs from Their Complements or Objects

In academic writing, a colon should not be used to separate a verb from its complement or object. Here's a flawed case:

> Two types of memory affected by sleep loss are: declarative and procedural memory.

In the correct revision that follows, the phrase *affected by sleep loss* now complements the verb *are* and forms a complete sentence before the colon.

> Two types of memory are affected by sleep loss: declarative and procedural memory.

If the writer is not trying to deliberately draw extra attention to the two types of memory, the sentence should be rewritten without a colon:

> Two types of memory affected by sleep loss are declarative and procedural memory.

Colons That Separate Prepositions from Their Objects

The following example is problematic because the colon interrupts an independent clause, separating a preposition (*of*) from its compound object (*attention, depression, anger, and vigor*).

> Our questionnaire included measures of: attention, depression, anger, and vigor.

Here are two good solutions:

> Our questionnaire included measures of the following: attention, depression, anger, and vigor.

> Our questionnaire included measures of attention, depression, anger, and vigor.

Problems Involving Apostrophes

In informal writing, one use of apostrophes is to form contractions:

| In scientific writing, authors *don't* use contractions.

The convention in scientific writing is to spell out words rather than to use apostrophes to indicate omissions of letters. The main use of apostrophes in scientific writing is to indicate ownership or origin:

| Professor Marcum's hypothesis contradicts evidence showing that cortisol levels fall during sleep.

The apostrophe *s* in the preceding sentence indicates that the hypothesis belonged to, or originated with, Professor Marcum.

The main rules for using apostrophes to indicate the possessive case of nouns are as follows.

- For singular and plural nouns that do not end in *s*, add an apostrophe and an *s*: *the author's argument, the children's behaviors.*
- For singular nouns that end in *s*, rules vary by publication organization. So it's best to check the style guide and author-instructions for the scientific discipline in which you are writing. Some style guides direct authors to add an apostrophe only, as in *Edwards' findings.* Others give the instruction to add an apostrophe and an *s*, as in *Edwards's findings.* A common guideline is to omit the added *s* for singular nouns that end with an *s* and make an "iz" or "uz" sound. As an example, consider the difference in reading *Bridges' results* compared to the awkward *Bridges's results.*
- For plural nouns that end in *s*, add an apostrophe only: *the scientists' conclusions were identical.*

Omitting a Necessary Apostrophe to Indicate the Possessive Case
The following sentence omits a necessary apostrophe to indicate ownership.

| The subjects recorded medical histories indicated whether they had taken medications for insomnia.

The recorded medical histories belong to the subjects, so the noun *subjects* is possessive; because it is a plural noun, an apostrophe should be added to its end (*subjects'*). If you are uncertain whether a noun is possessive, use it in a phrase with the word *of.* Consider the phrase *the peaks of the circadian rhythm.* It indicates that the peaks belong to the circadian rhythm. So the phrase could be written *the circadian rhythm's peaks.* Note, however, that some style guides advise against using apostrophes to indicate the possessive case for inanimate objects. So *the peaks of the circadian rhythm* may be preferable to *the circadian rhythm's peaks.*

Omitting a Necessary Apostrophe to Indicate the Possessive Case of a Gerund

Compare the meanings of the following two sentences.

> I take strong exception to Professor Booth claiming that everyone needs more than 8 hours of nightly sleep.

> I take strong exception to Professor Booth's claiming that everyone needs more than 8 hours of nightly sleep.

In the first sentence, the writer strongly objects to Professor Booth. In the second sentence, the writer strongly objects to the professor's action of claiming (that everyone needs more than 8 hours of nightly sleep). If the writer did not intend a personal attack on Professor Booth, the second sentence and its apostrophe are correct. The word *claiming* is a special type of noun called a gerund, which is formed by adding *-ing* to a verb. When a gerund is preceded by a possessive noun, the noun requires an apostrophe.

Using an Apostrophe for a Noun That Is Not Possessive

Here's an incorrect use of an apostrophe:

> The subjects' recorded their medical histories.

The word or phrase immediately following a possessive noun is what the noun owns. In the preceding example, the subjects do not own *recorded*. In addition, *recorded* is not part of a noun phrase indicating something that the subjects could actually own. Instead, *recorded* is the sentence's verb. So the apostrophe should be deleted.

Misplacing an Apostrophe in a Plural Possessive Noun

In the following example, the apostrophe is misplaced.

> The subject's medical histories indicated their use of medications for insomnia.

The correct revision is to place the apostrophe after the *s*. The reason is that the possessive noun *subjects* is plural (we know this because the pronoun *their* is plural).

Misused Apostrophes for the Possessive Pronoun *Its*

An apostrophe should never be used to indicate possessive case for the pronoun *its*. In fact, the only correct use of *it's* is as the contraction of *it is*. The strikeout in the following sentence indicates the correct revision.

> Although the study did not reveal significant effects of bright light on the circadian rhythm, it's findings showed a positive effect of caffeine.

Without the revision, the sentence literally reads *it is findings showed*.

Problems Involving Hyphens

In scientific writing, hyphens serve the following primary purposes.

- Forming compound nouns (*follow-up, cross-reaction, self-regulation*).
- Linking two or more words that form compound adjectives (*high-intensity stimulus, large-bowel blockage, short-term memory*).
- Forming fractions and compound numbers that are spelled out (*one-third, ninety-four*).
- Forming compound adjectives that begin with numbers (*32-degree difference in angle, 8-year-old child*).
- Connecting prefixes to root words (*post-1990 studies, self-report questionnaire*).
- Separating elements of chemical and biological compounds (*acetyl-CoA, 2,3-DPG*).
- Indicating the words *to* or *through* in a series of numbers, conditions, or pages (*individuals between 25-40 years of age, see pages 102-105*).

When you are not certain whether two or more nonscientific words should be hyphenated to form a compound word, consult a reputable dictionary. For clarification on whether to hyphenate scientific word groups, consult a style guide for the discipline in which you are writing.

Some common problems and solutions involving hyphens are presented as follows.

Omitting a Necessary Hyphen in a Compound Adjective

The classic, although somewhat gross, example of this case highlights the distinction in meaning between a *large bowel movement* and a *large-bowel movement* (in which *large-bowel* refers to part of the gastrointestinal tract). Here's another case in which an omitted hyphen can significantly influence meaning:

In a study on sleep deprivation and accidents in shift workers, Drexler investigated the causes of two car collisions.

Readers might very well ask whether Drexler investigated the causes of (a) two separate collisions that might have involved multiple cars or (b) more than one collision that all involved just two cars. If the writer intended the latter meaning, the phrase should be hyphenated as *two-car collisions*.

Hyphenating a Compound Adjective Placed after the Noun That It Modifies

A hyphen is used to form a compound adjective *before* the noun it modifies. However, by rule, a compound adjective should not be hyphenated when it comes *after* the noun:

The effects of sleep loss on higher cortical functioning are well-established.

This example breaks the rule, so the hyphen should be deleted.

Hyphenating the Combination of an Adverb Ending in *-ly* and an Adjective

The hyphen in the following sentence should be deleted.

> The Profile of Mood States (POMS) is a frequently-used test for studying the effects of sleep disorders on mood.

A word group with an adverb ending in *-ly* followed by an adjective is not technically a compound adjective. So the rule for this construction is to omit the hyphen.

Problems Involving Quotation Marks

Scientific authors rarely quote published literature, preferring instead to summarize and paraphrase. However, as discussed in Chapter 4, direct quotations are appropriate for selected purposes in scientific writing. The rules for punctuation within quoted material vary across subdisciplines of the life sciences. For example, in papers that follow APA guidelines, periods and commas are always placed inside of quotation marks. In papers that follow CSE guidelines (Council of Science Editors, 2006), periods and commas are placed inside of quotations marks only if the punctuation was part of the text that the writer is quoting.

Problems Involving Capitalization

Some of the rules for capitalization in writing are straightforward and universally accepted. They include the rules to capitalize the first word of a sentence, people's names, and places on geographic maps. Other cases of capitalization may raise questions and cause problems. Here's a pop quiz:

> Dr. Sandra Richardson, an [assistant professor or Assistant Professor?], teaches a course on [sleep physiology or Sleep Physiology?] in the [department of integrative biology or Department of Integrative Biology?] at a major research [university or University?].

To know whether to capitalize the words in question, you must know the general rule that proper nouns are capitalized but common nouns are not. Proper nouns designate *specific* persons, places, or things—that is, nouns referring to one-and-only cases. Common nouns designate *general* persons, places, or things. Table 7.4 highlights the distinction between proper and common nouns.

Applying the general rules for capitalization, we have the answer key to our pop quiz:

> Dr. Sandra Richardson, an assistant professor, conducts research on sleep physiology in the Department of Integrative Biology at a major research university.

TABLE 7.4 **Guidelines for Capitalization**

Proper nouns (specific designation)	Common nouns (general designation)
Assistant Professor Sandra Richardson	an assistant professor
Sleep Physiology 3000	a junior-level course in sleep physiology
The Department of Integrative Biology	a biology department
The University of California at Berkeley	a major research university

Compared to the rules for capitalization in general writing, the rules in scientific writing are more numerous and complex. In the life sciences, rules for capitalization vary by journals and publication organizations. So when you are uncertain about which words to capitalize in your scientific papers, refer to the author-instruction documents and styles guides for the subdiscipline in which you are writing.

REVISING FOR BIASED AND INADVERTENTLY OFFENSIVE LANGUAGE

Many life science studies are conducted on people who are categorized by their ages, gender, disease conditions, and ethnic and racial backgrounds. When describing people by these categories, scientists try to use the most precise and concise language. This is true for the following sentence.

> The results indicated that the type 2 diabetics had significantly lower scores for sleep quality than the normal subjects.

Look closely at how the writer describes the two groups that were studied: *the type 2 diabetics* and *the normal subjects*. The descriptions are indeed concise, which is usually a virtue in scientific writing. But the descriptions are problematic for a few important reasons. First, the phrase *the type 2 diabetics* defines the study participants entirely by their disease. The phrase conveys the idea that they are not people who *happen to have* type 2 diabetes—instead, they *are* type 2 diabetics. The description thus depersonalizes the individuals. Second, the phrase *the normal subjects*, referring to members of the control group (who did not have type 2 diabetes), indirectly implies that people with type 2 diabetes are not normal. Some readers may view these observations as trivial matters of political correctness. But most people who are suffering from a disease do not want to be defined exclusively by it. They might be offended by someone who stereotypes them, even inadvertently, in this way. For these reasons, the sample sentence can be revised to eliminate the biased and inadvertently offensive language. Here's a solution:

> The results indicated that **the participants diagnosed with type 2 diabetes** had significantly lower scores for sleep quality than **the participants in the control group**.

Although extra words have been added, making the sentence less concise than the original, they are essential words for solving the problem of biased and inadvertently offensive language. Of course, successive sentences might very well be cumbersome if the writer continues to use the unbiased, fleshed-out descriptions. One solution for the successive descriptions is to refer to the study groups in more general terms. For example, *the participants diagnosed with type 2 diabetes* can later be called *the experimental group*.

Sexist Language

Earlier in the chapter, I raised the concern about using the male pronoun *he* generically—that is, to refer to men as well as women. Here's a case:

> If a researcher wants to control for placebo effects in studies on acupuncture and sleep disorders, he should include a group of individuals who receive sham acupuncture.

Many readers would consider the sentence's generic use of *he* as sexist language because it does not recognize that women also conduct research on acupuncture and sleep disorders. One solution is to use pronouns for both men and women:

> If a researcher wants to control for placebo effects in studies on acupuncture and sleep disorders, **he or she** should...

But this solution is clumsy, especially if it is overused. A better revision is to make the noun and the pronoun plural:

> If **researchers** want to control for placebo effects in studies on acupuncture and sleep disorders, **they** should...

The sentence can also be revised to eliminate the problem altogether:

> To control for placebo effects in studies on acupuncture and sleep disorders, researchers should include a control group that receives sham acupuncture.

Another case of sexist language is using *man* to refer to all people. The following example shows a good solution to this problem.

| Why ~~man requires~~ people require sleep is a mystery to researchers.

Age-Biased Language

The main instance of age-biased language in scientific writing involves descriptions of older adults. Many people who are in their 60s, 70s, 80s, and even 90s do not define themselves primarily as *old*. The following sentence corrects this potentially offensive description.

| The study involved health outcomes associated with insomnia in ~~old~~ people~~.~~ **between 75 and 86 years of age**.

Gerontologists, scientists who study aging, often describe their subjects as *older individuals* rather than *old individuals* or *the elderly*. The most unbiased and precise solution is to refer to groups by their specific age or age ranges, as in the preceding example.

Biased Language Involving Ethnic and Racial Groups

The term *race* generally refers to genetic factors that underlie common physical characteristics, such as skin color, in groups of people. *Ethnicity* is most often used to distinguish groups of people who share common cultural practices. In one case of inadvertently offensive language, authors describe racial and ethnic groups with terms that the groups would not use to describe themselves. Examples are using the outdated terms *Oriental* and *Afro-American* to describe people from Asia and Americans of African descent. Another problem is labeling a group of people with terms that are inaccurate or too broad. For example, writing about the sleep habits of people in Mexico, you should not use the term *Hispanic* to describe the population. Along with people from Mexico, those from Puerto Rico, Cuba, and South American countries may accurately be described as *Hispanic*, which means "of or relating to a Spanish-speaking people or culture." But the sleep habits of Mexicans may be different from those of Cubans, Puerto Ricans, and other Spanish-speaking cultures. So the specific population should be described as *Mexicans* or as *people who live in Mexico*.

A vital principle guides conscientious and unbiased scientific authors in writing about people who can be categorized by disease, disability, gender, age, and racial and ethnic background: Consider the feelings and preferences of the people who are grouped in these ways. How do they describe themselves? What labels do they think are offensive? It's certainly worth learning more about the best practices for avoiding biased and culturally insensitive language. Publication guidelines for many scientific organizations, including the American Psychological Association and the Council for Science Editors, offer extensive discussions and advice in this area.

SUMMING UP AND STEPPING AHEAD

Well-crafted sentences make positive first impressions in scientific papers. Such sentences are characterized by their strong logic and clarity of expression, by their sound structure and style, by their correctness in grammar and punctuation, and by their appropriate and unbiased use of language. Of course, many sentences that make the most favorable first impressions on their readers were not crafted perfectly on first attempts by their writers. The take-home message at the core of this chapter is that sentence-level revision is a vital part of the scientific writing process.

This chapter marks the end of our process-based approach to learning about scientific writing. A central theme of this approach—conveyed in previous chapters that focused on planning, organizing, drafting, and revising content—is that the process is ideally guided by rhetorical goals. Next, we take a closer look at the conventional rhetorical goals for the primary types of scientific papers.

CHAPTER 8

Rhetorical Goals for Scientific Papers

Exercise Science

These days, nearly everybody views exercise as an essential part of a healthy lifestyle. Many of us put the principle into practice daily without second thought. Not so long ago, however, most people who exercised regularly were serious athletes in training for competition. Beginning in the late-1970s in America, folks with absolutely no aspirations of winning Olympic medals started walking, jogging, lifting weights, and playing sports mainly for health. Partly in response to this fitness boom, the last decades of the twentieth century witnessed a research boom in a discipline that is now called *exercise science*. Over time, this burgeoning field has branched into numerous subdisciplines with specialized agendas. In exercise epidemiology, for example, researchers compare the risks of developing diseases in large populations of physically active versus sedentary people. In the relatively new area of exercise immunology, investigations focus on the influence of physical activity on the body's capacity to defend against pathogens. In exercise psychology, a primary aim is to understand the cognitive and behavioral factors that explain why some people successfully adhere to their exercise programs while others unfortunately drop out before gaining benefits. A number of other subdisciplines, including exercise physiology, emphasize the study of human performance and athletic skill.

Like researchers in all life science disciplines, exercise scientists communicate their discoveries through various types of documents, the primary ones being research papers, review papers, and research proposals. At their best, these papers demonstrate a truly distinguishing hallmark of successful writing: the use of rhetorical goals to generate and shape content; to organize information, ideas, and arguments; and to refine language and style. Set in the context of exercise science, this chapter presents conventional rhetorical goals for the primary types of scientific papers.

INTRODUCTION

As defined in Chapter 2 and demonstrated throughout the book, rhetorical goals are statements of intention that guide writers in performing key planning, drafting, and revising activities. Earlier I likened rhetorical goals to power tools for crafting successful content in scientific papers. To extend the analogy, we might imagine that accomplished writers possess *mental toolboxes* fully stocked with specialized rhetorical goals. The experts select and skillfully apply the best goals to meet situational demands. Think of this chapter as an extensive catalog of these power tools, which you can use to stock your mental toolbox and guide your writing process for the best outcomes.

The chapter is organized by the following major sections of the primary types (and subtypes) of scientific papers.

- The introduction section to all of the primary paper types: IMRAD-structured papers that report original studies (including research papers, lab reports, theses, and dissertations); review papers (including informative reviews, conceptual reviews, and position papers); and research proposals and grant applications.
- The methods sections of IMRAD-structured papers and research proposals.
- The results section of IMRAD-structured papers.
- The discussion section of IMRAD-structured papers.
- The body section of review papers.
- The conclusion section of review papers.

For each of these major sections, the chapter presents their key rhetorical goals. Each goal is accompanied by a set of strategies (from Chapter 2, recall that a strategy is a specific way to achieve a generally stated rhetorical goal). If rhetorical goals are akin to power tools, you might think of strategies as their specialized accessories—like drill bits or saw blades.

For any given writing project, you will not need all of the goals and strategies cataloged in the chapter. Ideal selections will depend on the type of paper you are writing, your assignment instructions, discipline-specific guidelines, concerns of audience, and other situational factors. To better appreciate this point, consider a conventional rhetorical goal that generally applies to the introduction to all scientific papers: *Convince the audience that your research issue is important and truly worth resolving.* For a paper you are planning, suppose that your intended readers already believe that its central issue is *tremendously* meaningful. Through audience analysis you realize that nothing you could say would strengthen their preconceptions. In this case you can leave the rhetorical goal in your mental toolbox. Why waste space writing about something that your audience already understands perfectly well, strongly accepts, and will bring to mind naturally? Along the way I will offer more advice about how to choose rhetorical goals and strategies to meet situational demands.

In Chapter 2, I pointed out that well-constructed statements of rhetorical goals may include audience-affecting cues, or phrases that focus our attention on how we want readers to respond to our ideas and arguments. Sometimes our intentions for readers are obvious, so explicit audience-affecting cues need not be included in goal statements. However, it's always extremely helpful to reflect on how rhetorical goals serve concerns of audience. This chapter's descriptions of conventional rhetorical goals and strategies take an audience-centered perspective.

To demonstrate the effective use of selected rhetorical goals and strategies, the chapter presents models of writing from published journal articles and student papers. I chose the models for their strong content; they are not necessarily flawless in features at the paragraph and sentence levels. Rather than showing entire papers, I have selected sections and, in some cases, individual paragraphs and sentences that highlight especially effective applications of goals and strategies. To obtain the entire published article from which sample text has been excerpted, refer to the article's citation information in the book's reference list.

RHETORICAL GOALS FOR INTRODUCTION SECTIONS

In all types of scientific papers, introduction sections can be especially challenging to write. Just deciding what to say in a paper's first sentence can sometimes seem like a tall mountain to climb. So where do you even begin? How much background information and literature review should you present? What approaches will you take to defining precisely what your paper is about and to convincing readers that the topic is meaningful? When assignment instructions and discipline-specific conventions restrict the overall length of your paper, how will you manage to pack everything that seems essential into a relatively short introduction? The best answers depend on a sound understanding of the rhetorical goals for the introduction sections of scientific papers. The following set of goals, along with their corresponding strategies, applies to introducing all of the primary paper types. An overview of the goals and strategies is presented in Figure 8.1 (pages 382–383).

Rhetorical Goal 1: Present Your Research Issue and Explain Its Unresolved Status

Every well-conceived research and writing project is inspired by an unresolved issue—that is, one or more unanswered questions or unsolved problems. Even purely informative review papers are based on questions and problems that reflect knowledge gaps in their research fields. The ultimate objective for all scientific papers is to communicate meaningful resolutions to their research issues. If readers do not clearly grasp the central issue that inspired your paper, they will miss its purpose and point entirely. For this reason, the most fundamental rhetorical goal for the introduction section is to present your research issue and explain its unresolved status.

To demonstrate this goal, as well as several others for introduction sections, we will examine a published research paper written by physiologist Jay Campisi and

Rhetorical Goal 1: Present your research issue and explain its unresolved status. (381)

Strategies
1.1: Identify your research issue directly. (383)
1.2: Provide the essential background information that readers will need to understand your research issue. (384)
1.3: Reveal the knowledge gaps that underlie your research issue. (386)
1.4: Summarize studies on your research issue that have produced contrasting outcomes. (387)
1.5: Summarize contrasting concepts and theories that underlie your research issue. (387)
1.6: Explain why your research issue has not been sufficiently resolved. (388)

Rhetorical Goal 2: Convince readers that your research issue is truly important and therefore worth resolving. (388)

Strategies
2.1: Describe the problems in nature and society that underscore the importance of your research issue. (389)
2.2: Describe the basic scientific problems that underscore the importance of your research issue. (390)
2.3: Reflect on the potential benefits that you envisioned as the consequences of resolving your research issue. (390)
2.4: Discuss the negative implications of failing to resolve your research issue. (390)

Rhetorical Goal 3: State your hypotheses and explain their rationale. (390)

Strategies
3.1: State your hypotheses directly and in detail. (391)
3.2: Provide the essential background information that readers will need to understand your hypothesis. (392)
3.3: Explain the rationale for your hypotheses. (392)
3.4: Convince readers that even though your hypotheses are plausible, they still need(ed) testing. (392)

Rhetorical Goal 4: Introduce the novel and unique features of your research and writing project. (393)

Strategies
4.1: Introduce your study's innovative methods. (393)
4.2: Introduce your paper's innovative conceptual approaches. (395)

FIGURE 8.1 Rhetorical goals and strategies for introduction sections. The pages on which the goals and strategies are introduced are shown in parentheses.

> ## Rhetorical Goal 5: Present the specific purposes of your research and writing project. (395)
>
> Strategies
> 5.1: State your purposes directly and in detail. (396)
> 5.2: Provide an overview of how you achieved, or plan to achieve, your main purposes. (396)
>
> ## Rhetorical Goal 6: Present your claims. (396)
>
> Strategies
> 6.1: State your claim directly. (396)
> 6.2: Elaborate on essential aspects of your claim and establish its boundaries. (397)

FIGURE 8.1 *continued.*

his colleagues (Campisi et al., 2003). The paper reports a study on exercise and a phenomenon called the *stress response*. This refers to the body's reactions to psychological and physical stimuli that organisms perceive and experience as harmful and that disrupt normal functioning. Campisi et al. induced the stress response in rats by administering brief electrical shocks to their tails. One group of rats exercised regularly; they lived in cages equipped with freely revolving running wheels. Without any coercion, rats naturally take to the wheels and typically run several miles per week. A second group of rats, serving as control subjects, remained sedentary; they lived in cages with locked running wheels. The study was based on an interesting observation from previous research and common experience: Regular exercise seems to buffer, or reduce, the harmful effects of stress on our bodies and minds. Campisi et al. sought to determine exactly *how* exercise engenders this protective influence. In other words, they aimed to identify the underlying mechanisms.

Strategy 1.1: Identify Your Research Issue Directly

Not very far into any well-written introduction section, a single sentence (or a set of related sentences) directly points out the paper's central issue. In Campisi et al.'s introduction, this sentence comes in the first paragraph, which is presented in Figure 8.2 (page 384).

The paragraph's first two sentences provide background information about how exercise protects against stress-related disorders and diseases. The third sentence reveals the study's motivating issue. Notice that the authors logically phrase it as a problem statement. Another common approach is to convey research issues as direct questions: *What are the mechanisms that underlie the stress-buffering effect of physical activity?*

Physically active organisms are protected from many of the damaging effects of stressor exposure. For example, it has been previously reported that physical activity can prevent the negative effect of stress on behavioral depression (9, 35), anxiety (11), immune function (12, 15, 23), and splenic apoptosis (2). **The mechanisms for the stress-buffering effect of physical activity remain unclear.** One feature of the stress response that could be altered by physical activity status and can contribute to stress resistance that has not yet been thoroughly investigated is the cellular stress response.

FIGURE 8.2 A directly expressed research issue (bolded for emphasis). From a published research paper authored by Campisi et al. (2003). Reproduced with permission from the American Physiological Society.

From the very start of introduction sections, more than anything else, experienced scientific readers want to know the questions and problems that papers are intended to answer and solve. So a useful guideline is to identify your research issue within the first or second paragraph of the introduction. As illustrated in the preceding example, only a single sentence may be necessary to satisfy readers' initial curiosity. The sentence is truly essential because it serves as the cornerstone for the entire paper. However, more than a single sentence is necessary to present unresolved research issues clearly, completely, and convincingly. The additional content should answer questions that readers will naturally raise about succinctly stated research issues. Consider, for example, questions that we might ask after learning that the mechanisms underlying the stress-buffering effects of exercise are unclear: What, *exactly*, is unclear about the mechanisms? What are some viable candidate mechanisms, and how do they work? What have researchers learned about the candidate mechanisms through previous studies? Which specific mechanisms did Campisi et al.'s study focus on? Why haven't researchers been able to elucidate the mechanisms definitively? These questions reflect the reader's need to understand the unresolved status of research issues. The remaining strategies for Rhetorical Goal 1 address this need.

Strategy 1.2: Provide the Essential Background Information That Readers Will Need to Understand Your Research Issue
Suppose that Campisi et al.'s introduction began as follows.

In prehistoric times, human beings were naturally physically active. Evidence suggests that hunter-gatherers ran for hours every day, tracking down animals for food. Millions of years ago, daily life involved regular exercise simply for survival. Stone Age

humans had no television, video games, cars, and other technologies of entertainment and convenience, the very technologies that
have contributed to the sedentary existence of modern man....
*[imagine that the next several paragraphs continue to detail the
history of physical activity. Then, at the introduction's end, the
authors finally pinpoint their issue involving the stress-reducing
effects of exercise!]*

Sooner or later in this rambling introduction, you might throw up your arms
and cry out to the authors, "Why are you presenting all of this background information? How does it actually relate to what you studied and wrote about?" The
problem is that the background information is far too general, serving no other
use than to inform. A more appropriate purpose for presenting background
information is to clarify a research issue. This strategy entails defining key terms,
explaining complex concepts, and summarizing relevant findings from previous
research. To decide on the necessary details and how extensively to develop them,
you must consider what your readers already know and need to know to deeply
understand your paper's central issue.

Campisi et al.'s introduction section nicely reflects the strategy to provide
appropriate background information. The model content begins with the last sentence of the first paragraph (Figure 8.2). The sentence conveys the idea that habitual physical activity might protect against harmful effects of stress by influencing
something called *the cellular stress response*. Campisi et al. give background information about this phenomenon in the second paragraph of their introduction,
which is presented in Figure 8.3. Before reading the paragraph, note that the
researchers focused on a specific component of the body's cells, called *heat-shock
proteins*, or *HSPs*. As Campisi et al. explain, our cells synthesize HSPs in response
to stressful conditions. The various forms of these molecules, including HSP70
and HSP72, protect our cells against stress-related damage and disease.

The generation of the stress response is under tight regulatory control and is
designed to facilitate an organism's chance of survival during challenge. This
is accomplished at both the system (e.g., muscular, cardiovascular, respiratory,
and neuroendocrine) and single-cellular levels. One way cells can resist damage

FIGURE 8.3 The second paragraph of Campisi et al.'s introduction section. Most
of the paragraph contains background information that clarifies
the study's motivating research issue (Strategy 1.2) as well as the
authors' hypothesis (Strategy 3.2). The last sentence presents a
general hypothesis statement (Strategy 3.1). Reproduced with
permission from the American Physiological Society.

and/or death induced by stress is to synthesize a highly conserved set of intracellular proteins, termed heatshock proteins (HSPs) (25). A variety of stressors are known to induce HSPs, and these proteins confer protection and/or tolerance against stress-induced oxidative, heat, and cytokine insult (24, 36). The 70-kDa HSP (HSP70) family of proteins includes a constitutive 73-kDa protein (HSC73) and a highly stress-inducible 72-kDa protein (HSP72) (37). HSP72 is essential or cellular recovery after stress as well as survival and maintenance of normal cellular function. Furthermore, HSP72 prevents protein aggregation and also refolds damaged proteins. Importantly, expression of high levels of HSPs has been associated with an increased ability of cells to withstand challenges that would otherwise lead to cell injury and/or death (36). It would be reasonable to speculate, therefore, that one potential adaptation produced by physical activity that could contribute to stress resistance would be improvements in the cellular HSP response.

FIGURE 8.3 *continued.*

Strategy 1.3: Reveal the Knowledge Gaps That Underlie Your Research Issue

Every research issue is inspired by knowledge gaps, which are uncertainties and inconsistencies in the data, ideas, concepts, and theories that define an area of study. A good way to reveal knowledge gaps is to first explain what is known about an issue and then to elaborate on what is unknown. This pattern of ideas is evident in the third paragraph of Campisi et al.'s introduction, reproduced in Figure 8.4.

In fact, there is some evidence to support this idea. Fehrenbach et al. (13) demonstrated that in vitro heat shock of human peripheral blood leukocytes significantly stimulated HSP27 and HSP70 mRNA, with physically active individuals exhibiting the greatest increases. Additionally, Gonazalez et al. (17) demonstrated that HSP72 was expressed at higher levels in the skeletal muscles of

FIGURE 8.4 The third paragraph of Campisi et al.'s introduction section. Serving several rhetorical goals and strategies, the paragraph (a) presents the knowledge gaps that inspired the reported study (Strategy 1.3); (b) points to methodological limitations of previous studies (Strategy 1.6); (c) provides supporting rationale for the authors' hypothesis (Strategy 3.3); and (d) argues that, despite supporting evidence, the hypothesis still needed to be tested (Strategy 3.4). Reproduced with permission from the American Physiological Society.

treadmill-trained rats compared with sedentary animals after an exhaustive exercise challenge. Although these studies lend support to the hypothesis that habitual physical activity might better prepare an organism to deal with subsequent stressor exposure at the cellular level, they are limited for a variety of reasons. First, HSPs are ubiquitously induced after stress (29, 36, 39, 45), yet these studies only investigated peripheral blood lymphocytes and skeletal muscle. Furthermore, whole organism stressors affect multiple tissues (36), and physical activity selectively changes features of the stress response such that some systems demonstrate adaptations while others do not (9, 10, 15, 27, 35, 47); thus it is important to examine more than a single tissue or system during an experiment. Last, in light of the facts that previous preconditioning (i.e., exposure to stress) results in a larger HSP response to a subsequent stressor (25) and that treadmill training has been reported to be chronically stressful to rats (34), the results from the previously mentioned study using treadmill training are difficult to interpret.

FIGURE 8.4 *continued.*

The paragraph's first two sentences summarize discoveries from previous studies on exercise and HSP activity. We learn that, in response to stress, the cells of physically active individuals synthesize large amounts of HSP—that is, compared to the cells of sedentary individuals. Then, beginning in the third sentence, the authors directly identify the knowledge gap that motivated their study: Previous research has focused on HSP activity in only a limited subset of cells and tissues (specifically, in a type of immune cell called blood lymphocytes and in skeletal muscle). It is unknown, however, whether physical activity promotes protective HSP activity in *multiple* cells and tissues throughout the body. By revealing this knowledge gap, the first five sentences of the model paragraph in Figure 8.4 bring Campisi et al.'s specific research issue into sharp focus.

Strategy 1.4: Summarize Studies on Your Research Issue That Have Produced Contrasting Outcomes

Some research issues are unresolved mainly due to contrasting outcomes in previous studies. To present this sort of issue, authors summarize key details about the conflicting studies' methods, results, and conclusions. Note that this strategy is *not* intended to produce a general, informative review of literature. Instead, the content serves specifically to convince readers that a paper's research issue has not been resolved, at least up until the author's contribution.

Strategy 1.5: Summarize Contrasting Concepts and Theories That Underlie Your Research Issue

Like the previous strategy, this one applies to research issues that are characterized by disagreements in the scientific literature. In this case, the different sides of unresolved issues are defined by their contrasting concepts and theories. For example, some exercise physiologists argue that stretching prevents musculoskeletal injuries by enabling the muscles to withstand great amounts of contractile force. A directly contrasting view is that stretching can actually weaken the muscles and thereby cause injuries. If you were writing a position paper on this issue, you might devote part of your introduction to summarizing the contrasting arguments about the underlying mechanisms by which stretching influences injury risk. By doing so, you will convince readers that your paper is based on a truly unresolved issue.

Strategy 1.6: Explain Why Your Research Issue Has Not Been Sufficiently Resolved

When an author clearly demonstrates that a research issue is unresolved, readers will naturally ask *why*. That is, they will wonder about the barriers that have kept scientists from reaching resolutions. The most common barriers are methodological shortcomings of previous studies. For example, prior to Campisi et al.'s study, many researchers used forced treadmill running, rather than voluntary wheel running, to investigate the effects of exercise on the cellular stress response in rodents. The problem is that forced treadmill running causes stress in animals. This, of course, is a confounding effect in studies intended to determine whether exercise buffers the negative effects of stress. In the last sentence of their introduction's third paragraph (Figure 8.4, pages 386–387), Campisi et al. address this methodological shortcoming of previous studies. The sentence reinforces the authors' explanation of the unresolved nature of their research issue.

Rhetorical Goal 2: Convince Readers That Your Research Issue Is Truly Important and Therefore Worth Resolving

A well-presented research issue naturally inspires questions about its practical importance in nature, society, and science:

- What specific problems or negative conditions associated with the issue make it important?
- Who or what is influenced by these problems or negative conditions?
- What benefits might be realized through the ultimate resolution of the issue?
- What are the implications of failing to resolve the issue?

Responses to these questions sum to answer a big-picture question that conscientious readers always have for authors: *Why should we care about your work and your efforts to communicate it?* If, in the introduction section, you can convince your audience that your paper addresses a truly meaningful issue, they will be encouraged to keep reading. They will be eager to learn about your discoveries, arguments, and resolutions.

As I discussed earlier, the importance of a research issue is best judged by a paper's intended readers. Consider this point and its implications for audience analysis when you are developing and shaping ideas to accomplish the rhetorical goal at hand.

Strategy 2.1: Describe the Problems in Nature and Society That Underscore the Importance of Your Research Issue

Many research and writing projects in the life sciences are inspired by pressing problems in nature and society. By describing these real-world problems and their implications, authors convey the importance of their research issues. An excellent model of content for achieving this strategy is presented in Figure 8.5, which is the entire introduction section from a published review paper on relationships between asthma and physical activity levels in children and adolescents (Williams et al., 2008). The paper's lead author is Dr. Brian Williams of the University of Dundee in Scotland.

Williams et al. very effectively describe the existing problems associated with asthma among youth, detailing their implications for individuals who suffer from the disease as well as for specialists who are concerned with public health

Asthma in children and young people is a major health problem [1–4]. For reasons that are not fully understood the incidence and prevalence of asthma have increased considerably over the last thirty years, particularly for children and young people [5,6]. Despite a reduction in incidence since it peaked in 1993 the UK asthma rate continues to be six times higher among children now than it was in 1976 [3,7]. The prevalence rate in the 2–15 years age group is reportedly one-fifth of the population (20.7%) and when compared with 56 other countries the UK has one of the highest prevalence rates of wheeze and asthma in 13–14 year old children [5]. The burden on health services from asthma is extensive and increasing [5]. At primary care level asthma accounts for 1 in 5 of all child GP consultations [8], while within secondary care in England and Wales in 1999, there were over 30,000 asthma-related hospital admissions and 25 child deaths [7]. Asthma is thus a common condition affecting a substantial and increasing proportion of children and young people [3] and as a result provides a challenge in management for the individual sufferer and for health services as a whole.

In addition to the direct effect the condition has on the wellbeing of an individual, it is increasingly recognised that the health of sufferers can be further compromised because of the impact that asthma may have on their participation in physical activity [6,9]. This paper presents a narrative synthesis based on a scoping review of the literature in order to provide an overview of evidence and argument in this area, and therefore inform decisions about the future direction of empirical studies and/or more specifically focussed systematic reviews [10,11]. It reviews and discusses recent literature outlining the growing problem of physical inactivity among young people with asthma and explores the psychosocial dimensions that may explain inactivity levels and potentially relevant interventions and strategies, and the principles that should underpin them.

FIGURE 8.5 Exemplary content for convincing readers that a research issue is important. From a published review paper authored by Williams et al. (2008). Reproduced with permission.

management. The author documents the problems by presenting statistics on the increasing prevalence of asthma among youth in the United Kingdom, on the large number of associated hospital admissions, and on the number of deaths attributable to the disease. The content is especially fitting for the readership of the journal in which Williams et al.'s article is published—it's a journal called *BMC Family Practice*, which serves primary health care providers and public health officials. In addition to helping readers understand the negative implications of asthma, Williams et al. directly address the importance of the core issue that inspired their review paper: This is the problem that asthma may limit youth from participating in health-promoting physical activity. Notice that most of the content of Williams et al.'s introduction section is devoted to convincing readers that the paper's central research issue is important and worth resolving. This is certainly appropriate for issues that have extraordinary implications for nature and society.

Strategy 2.2: Describe the Basic Scientific Problems That Underscore the Importance of Your Research Issue

Some papers primarily address *basic* research issues, which do not have immediate applications in nature and society. To convey the importance of these issues, authors describe the problems that have restricted the advancement of knowledge in their fields. The problems may include knowledge gaps, methodological limitations, conflicting study outcomes, and debatable concepts and theories. Note that this strategy overlaps several strategies that we have discussed for presenting research issues (Rhetorical Goal 1).

Strategy 2.3: Reflect on the Potential Benefits That You Envisioned as the Consequences of Resolving Your Research Issue

When you originally conceived your research and writing project, what positive outcomes did you imagine would result from resolving its motivating issue and communicating the resolution? In what specific ways did you envision that nature, society, and science would benefit? Direct answers to these questions contribute to building strong arguments for the importance of research issues. This strategy is demonstrated in the last sentence of Figure 8.5, where Williams et al. comment that their review paper on asthma and physical activity participation in youth "explores the psychosocial dimensions that may explain inactivity levels and potentially relevant interventions and strategies, and the principles that should underpin them."

Strategy 2.4: Discuss the Negative Implications of Failing to Resolve Your Research Issue

A forward-thinking argument for the importance of a research issue includes ideas about what might happen if it is *not* resolved.

Rhetorical Goal 3: State Your Hypotheses and Explain Their Rationale

As defined in Chapter 1, a hypothesis is a statement that conveys a researcher's expectations or tentative explanations for a study's outcomes. Well-conceived hypotheses are extremely valuable for guiding the design and execution of progressive research. So it makes sense that they are presented in the introduction sections to many IMRAD-structured papers that report original studies as well as in research proposals and grant applications. Hypotheses are truly warranted, however, only when they meet certain criteria. Above all, a hypothesis is worth presenting if it can be justified with plausible scientific knowledge, evidence, and reasoning. When they are based purely on guesswork, hypotheses serve no use and can even be misleading. A paradoxical quality of a strong hypothesis is that it is falsifiable. In other words, it can be rejected or confirmed through research. In addition, a hypothesis is warranted if it is testable against alternative and competing hypotheses. If you cannot offer plausible alternatives to your hypothesis, readers might think that it is biased and therefore not worth presenting and testing at all.

Strategy 3.1: State Your Hypotheses Directly and in Detail

Stating a hypothesis is usually a straightforward matter, as demonstrated in the following examples.

> **Hypothesis statement for a grant application:** The proposed study will test the hypothesis that exercise enhances cognitive function in Alzheimer's disease patients by inhibiting the neurodegenerative effects caused by proinflammatory cytokines.

> **A set of related hypotheses for a research paper:** This study was based on the following hypotheses concerning physiological measures of sea-level endurance performance in athletes subjected to high-altitude training: (a) maximal oxygen consumption will increase significantly; (b) the expected increases in maximal oxygen consumption will be strongly correlated with altitude-induced increases in hemoglobin and hematocrit levels; and (c) lactic acid concentrations, measured during a submaximal endurance test, will decrease significantly.

Campisi et al.'s introduction contains two versions of the authors' hypothesis. A general statement comes in the last sentence of the second paragraph:

> It would be reasonable to speculate, therefore, that one potential adaptation produced by physical activity that could contribute to stress resistance would be improvements in the cellular HSP response.

A more specific version ends the entire introduction section:

> We hypothesize that habitually physically active rats will have both greater and faster HSP72 responses to stress, suggesting that physical activity may contribute to stress resistance at the cellular level.

In the last example, notice the directness and detail. The hypothesis statement conveys information about the specific experimental treatment (physical activity in rats) and the precise outcomes that the researchers anticipated (greater and faster HSP72 responses to stress).

Strategy 3.2: Provide the Essential Background Information That Readers Will Need to Understand Your Hypothesis

This strategy, which is similar to the one for presenting background information about research issues (Strategy 1.2), applies when terms and concepts that are integral to a hypothesis statement might be unfamiliar or confusing to your audience. In Campisi et al.'s introduction, the same paragraph that presents background information to clarify the research issue also clarifies the authors' hypothesis (Figure 8.3, pages 385–386).

Strategy 3.3: Explain the Rationale for Your Hypotheses

On its own, a hypothesis statement may be viewed by experienced readers as shallow conjecture. To convince readers that your hypothesis truly has merit, you must present its underlying rationale. This refers to the scientific knowledge, evidence, and reasoning that make the hypothesis plausible. One approach to developing the rationale for a hypothesis is to summarize the results of previous studies that offer provisional support for it. Recall that Campisi et al.'s third paragraph presents evidence from several studies in which physically active individuals exhibited large stress-induced HSP responses in a limited set of cells and tissues (Figure 8.4, pages 386–387). This evidence provides tentative support for Campisi et al.'s hypothesis, which applies to HSP responses in a wider range of cells and tissues throughout the body. Another approach to explaining the rationale for a hypothesis is to present its conceptual and theoretical foundations. In Campisi et al.'s introduction, the second paragraph (Figure 8.3) serves the dual strategies of providing general background information about the authors' hypothesis and developing the specific conceptual rationale for it.

Strategy 3.4: Convince Readers That Even Though Your Hypotheses Are Plausible, They Still Need(ed) Testing

The rationale for a hypothesis should encourage readers to consider its viability but, at the same time, question it. To better appreciate this point, imagine a grant application in which the authors go completely overboard with the previous strategy. Their rationale is so convincing that it leaves no doubt that the proposed study's hypothesis is valid. Here's how the grant's peer reviewers will likely respond to the authors: "Because the supporting evidence and reasoning for your hypothesis are apparently undeniable, we are not convinced that your study is necessary. Why should we recommend funding a study intended to test a hypothesis that has apparently already been confirmed?" To avoid this response, the authors must actually raise some doubts about the viability of their hypothesis. Effective approaches are to identify gaps and conflicts in the supporting evidence and reasoning for a hypothesis, explain the rationale for viable alternative hypotheses, and point out methodological shortcomings to previous studies that produced provisional support.

In the third paragraph of their introduction, after summarizing the studies that offer provisional support for their hypothesis, Campisi et al. raise the following pivotal idea:

> Although these studies lend support to the hypothesis that habitual physical activity might better prepare an organism to deal with subsequent stressor exposure at the cellular level, they are limited for a variety of reasons.

To argue that their hypothesis is not a foregone conclusion, Campisi et al. explain the shortcomings of the previous studies that focused on HSP responses in a limited number of cells and tissues. In addition, the authors raise doubts about their hypothesis by pointing out that HSP responses might *not* be enhanced throughout the body:

> ...physical activity selectively changes features of the stress response such that some systems demonstrate adaptations while others do not (9, 10, 15, 27, 35, 47); thus it is important to examine more than a single tissue or system during an experiment.

In response to the uncertainty that Campisi et al. skillfully raise, readers will be convinced that the authors' hypothesis truly needed to be tested.

Rhetorical Goal 4: Introduce the Novel and Unique Features of Your Research and Writing Project

In one way or another, the first three rhetorical goals serve to introduce the central problems that motivate research and writing projects. This next goal is for introducing solutions. It is a useful tool for answering questions that experienced readers normally raise in response to effectively presented research problems: *What is new and different about the author's approaches to solving them? Why should we read the author's paper instead of, or in addition to, others on the research issue?* The strategies for answering these questions involve introducing the novel and unique methodological and conceptual solutions that a research and writing project offers.

Strategy 4.1: Introduce Your Study's Innovative Methods

This strategy applies to introducing studies that feature new and improved research methods, such as the following.

- Stronger experimental designs than have been employed previously.
- The selection of subjects with characteristics that have not been previously investigated.
- A broader range of variables than scientists have studied in the past.
- Applications of pioneering instruments and technologies for data collection and analysis.
- Creative new approaches for controlling extraneous variables.

This strategy is *not* for presenting methodological details (their proper place is in the methods sections of IMRAD-structured papers and research proposals). Instead, it is for highlighting the most innovative methods that distinguish the author's problem-solving approach.

For a demonstration of the strategy, let's recall two of the major problems that motivated Campisi et al.'s study. First, whereas previous studies had confirmed that physically active individuals demonstrate enhanced HSP responses to stress, the studies were limited by the restricted types of cells and tissues examined. Second, the previous studies on rats were limited because the mode of exercise training involved forced treadmill running. When rats are forced to run on treadmills, rather than allowed to run at will on freely revolving wheels, they actually experience stress. This is a confounding variable in research intended to determine the mechanisms by which exercise reduces stress effects. Campisi et al. highlight their study's innovative solutions to these problems by (a) describing the extensive range of cells and tissues that they examined and (b) explaining how their choice of voluntary freewheel running accounted for the confounding effects of treadmill running. The model content, presented in Figure 8.6, comes in the last paragraph of Campisi et al.'s introduction.

Therefore, the purpose of the current study was to determine whether prior voluntary freewheel running facilitates the stress-induced induction of HSP72 in central (brain), peripheral, and immune tissues. Voluntary freewheel running was chosen because this type of physical activity does not activate the stress response (33, 37). The following tissues were examined: brain (hypothalamus, hippocampus, dorsal vagal complex, prefrontal cortex), peripheral tissues (pituitary, adrenal, heart, liver, triceps), and immune tissues (spleen and mesenteric lymph nodes). These tissues were chosen to extend the previous observations and because they are known to be stress responsive. The triceps muscle was selected because previous studies have indicated that this particular muscle shows physiological adaptations to freewheel running (41), and it is unknown if the HSP72 response to stress is modulated by voluntary freewheel running. In addition to examining multiple stress-responsive tissues, the generality of the effect across stressors was tested by assessing the effect of exposure to two different stressors, i.e., inescapable tail-shock stress (IS) and treadmill running to exhaustion [exhaustive exercise stress (EXS)]. Finally, to begin to investigate how physical activity changes HSP72 responses, examination of the rate of HSP72 induction after stress was tested in lymphocytes from physically active and sedentary rats. We hypothesize that habitually physically active rats will have both greater and faster HSP72 responses to stress, suggesting that physical activity may contribute to stress resistance at the cellular level.

FIGURE 8.6 The fourth, and last, paragraph of Campisi et al.'s introduction section. The first sentence presents the purpose of the study (Strategy 5.1). The last sentence presents the authors' specific hypothesis (Strategy 3.1). The rest of the paragraph provides an introductory overview of Campisi et al.'s innovative research methods (Strategy 4.1) and their approach to achieving the overall purpose of their study (Strategy 5.2).

Strategy 4.2: Introduce Your Paper's Innovative Conceptual Approaches
This strategy applies to papers with novel and unique approaches to framing research issues, to applying scientific knowledge that advances new concepts, and to making arguments. Take the example of a position paper that a student named Joyce is writing about whether exercise prevents colon cancer. To present her research issue, Joyce summarized the contrasting outcomes of published studies that have addressed it. Then she discussed the importance of resolving the issue. In the last paragraph of her introduction, presented in Figure 8.7, Joyce states her claim and summarizes her innovative conceptual approach to supporting her argument.

In this position paper I will argue that regular aerobic exercise can reduce the risk of developing colon cancer. As a whole the previous literature on this issue is highly contradictory. However, a closer analysis of published studies indicates a consistent pattern in the findings. This analysis, which is the basis of my argument, reveals a classical dose-response effect of exercise. Studies in which subjects exercised at low intensities over short time periods have generally produced results indicating no effects of exercise on colon cancer risk. In contrast, most studies based on high-intensity, long-duration exercise programs have produced results indicating significant preventive effects. To support my argument in favor of regular aerobic exercise as a preventive measure against colon cancer, I will demonstrate that this dose-response effect is evident in the results of numerous published studies.

FIGURE 8.7 An effective example of Strategy 4.2 in action.

Rhetorical Goal 5: Present the Specific Purposes of Your Research and Writing Project

Purpose statements convey authors' specific aims for resolving research issues and testing hypotheses. Ideally placed toward the end of the introduction, an effectively written purpose statement sharply focuses the reader's attention on the paper's direction. After stating their purposes, experienced authors often provide an overview of their approaches to achieving them.

Strategy 5.1: State Your Purposes Directly and in Detail
In most cases, purpose statements can be expressed sufficiently in a sentence or two. An example is the sentence that begins the last paragraph of Campisi et al.'s introduction:

> Therefore, the purpose of the current study was to determine whether prior voluntary freewheel running facilitates the stress-induced induction of HSP72 in central (brain), peripheral, and immune tissues.

A more elaborate example, presented as follows, conveys a set of related purposes for a research proposal to study whether pre-exercise stretching influences injury risk.

> The specific aims of the proposed study are to
>
> (1) confirm the results of previous studies which have revealed that pre-exercise stretching reduces injury risk in lower-limb muscle and connective tissue;
> (2) determine the effects of post-exercise stretching on injury risk; and
> (3) investigate whether the effects of pre- and post-exercise stretching on injury risk differ across the sexes.

Strategy 5.2: Provide an Overview of How You Achieved, or Plan to Achieve, Your Main Purposes

If you are writing an IMRAD-structured paper or a research proposal, this strategy dovetails with Strategy 4.1, which is for presenting a brief overview of a study's key methods. Both strategies are effectively achieved in the last paragraph of Campisi et al.'s introduction (Figure 8.6, page 394). If you are writing a review paper, Strategy 5.2 involves providing advance information about the content and structure of sections that follow the introduction, to reveal how they serve your explicitly stated purposes.

Rhetorical Goal 6: Present Your Claims

As defined in Chapter 3, claims in scientific arguments are assertions that authors want readers to accept. Some types of review papers, including position papers and conceptual reviews, are essentially based on claims. Their presentation, ideally toward the end of the introduction section, sets up the argument's supporting evidence and reasoning, which follow in the paper's body.

Strategy 6.1: State Your Claim Directly

Consider a hypothetical position paper intended to argue for appropriate exercise guidelines for preventing the age-related bone disease, osteoporosis. The author's directly stated claim, which begins the introduction's final paragraph, is as follows.

> To prevent osteoporosis in later life, young women should adopt a regular program of high-intensity weightlifting exercise.

The sentence conveys important details about the author's claim without overwhelming readers.

Strategy 6.2: Elaborate on Essential Aspects of Your Claim and Establish Its Boundaries

This strategy applies when succinctly stated claims demand elaboration for clarity and detail. In response to the preceding example claim, for instance, readers might raise the following questions: At what specific ages should young women begin regular exercise for preventing osteoporosis in later life? What, precisely, defines *regular* exercise? And what does *high-intensity weightlifting exercise* actually entail? To elaborate on the claim, the author might round out the paragraph as demonstrated in Figure 8.8.

> To prevent osteoporosis in later life, young women should adopt a regular program of high-intensity resistance exercise. More specifically this guideline prescribes that, beginning in their early 20s, women will gain optimal benefits by performing weight-lifting exercises at least three times per week. The specific exercises should emphasize the large muscle groups in the legs and back. To induce a sufficiently intense training effect, the amount of weight lifted should be great enough to completely fatigue the muscles within 8 to 10 repetitions.

FIGURE 8.8 Demonstrating Strategy 6.2, content that elaborates on essential aspects of a claim for a position paper.

Note how the author's elaboration of the claim establishes the boundaries within which it is valid and generalizable. This feature serves to effectively focus the argument.

RHETORICAL GOALS FOR METHODS SECTIONS

Methods sections describe and justify the designs, procedures, and analyses of research projects. They are integral to IMRAD-structured papers that report completed original studies as well as to research proposals and grant applications. Whether these documents are ultimately successful depends partly on how well their methods sections are written. Consider, for example, that peer scientists may use the descriptions of methods in published research papers to replicate reported studies; this is a common practice for verifying exciting new findings and conclusions. If the methods of an original study are ineffectively described, the scientists carrying out replication studies may obtain misleading results and then reach flawed conclusions. So a poorly written methods section can ultimately undermine progress in resolving its paper's research issue.

Another crucial point about methods sections is that they contribute appreciably to building strong arguments in scientific papers. As discussed in Chapter 3, an argument's validity depends on the quality of the procedures used to obtain its supporting data. So when scientists report original research in IMRAD-structured papers, the strength of their conclusions depends partly on how effectively they describe and justify their study methods. In research proposals and grant applications, methods sections are essential to convincing reviewers to approve and fund studies. The methods section is a fitting place for scientists to argue for their innovative procedural and analytical solutions, for their advanced technical knowledge, and for their specialized ability to execute proposed studies successfully.

The methods sections of IMRAD-structured papers, research proposals, and grant applications share a common set of rhetorical goals and strategies, which are outlined in Figure 8.9. For demonstrations of their applications, we will

---

Rhetorical Goal 7: Describe the methods that you used, or plan to use, in carrying out your study. (399)

<u>Strategies</u>

7.1: Describe your study's setting, participants (subjects), and materials. (400)

7.2: Describe your methods for recruiting and screening subjects. (400)

7.3: Summarize your study's overall design and general procedures. (400)

7.4: Describe your methods for assigning subjects to groups, conditions, or treatments. (403)

7.5: Describe your methods for administering treatments and measuring independent variables. (403)

7.6: Describe your methods for measuring and analyzing dependent variables (study outcomes). (405)

7.7: Describe your methods for avoiding potential confounding effects of extraneous variables. (407)

7.8: Describe your study's statistical analyses. (407)

Rhetorical Goal 8: Justify your use of selected study methods. (407)

<u>Strategies</u>

8.1: Explain the relevance of selected study methods. (408)

8.2: Justify the validity, precision, and reliability of selected instruments and techniques for collecting data and measuring outcomes. (409)

8.3: Argue for the superiority of your study's methods over viable alternatives. (410)

FIGURE 8.9 Rhetorical goals and strategies for the methods sections of IMRAD-structured papers, research proposals, and grant applications. The pages on which the goals and strategies are introduced are shown in parentheses.

examine the methods section of a published research paper on physical activity and chronic fatigue syndrome (CFS). This is a disorder characterized by persistent incapacitating fatigue, muscle and joint pain, cognitive impairments (such as memory loss), and general symptoms of sickness. The paper was authored by Christopher Black, Patrick O'Connor, and Kevin McCully, all of the University of Georgia (Black et al., 2005). These exercise scientists conducted a 6-week study to determine the effects of increased physical activity on mood, pain symptoms, and feelings of fatigue in individuals with CFS.

In the excerpts from Black et al.'s methods section, all of the verbs describing the study's procedures and analyses are in the past tense. Here's an example (with bold font added for emphasis):

> Control participants **were chosen** to be similar in age, height, and weight to the CFS patients and **were defined** as sedentary by self-report of one bout of regular exercise per week or less.

If Black et al. had written a research proposal or a grant application, verbs describing the planned study's methods would be in the future tense:

> Control participants **will be chosen** to be similar in age, height, and weight to the CFS patients and **will be defined** as sedentary by self-report of one bout of regular exercise per week or less.

Presented next are two rhetorical goals for methods sections. One is for describing methods, and the other is for justifying them.

Rhetorical Goal 7: Describe the Methods That You Used, or Plan to Use, in Carrying Out Your Study

Methods sections mainly communicate straightforward descriptions of research procedures and analyses. In a logical and commonly used outline, successive subsections are devoted to describing the following information about a study's methods.

1. Setting, participants, and materials.
2. Methods for recruiting and screening subjects.
3. Overall design and general procedures.
4. Methods for assigning subjects to groups, conditions, and treatments.
5. Methods for administering treatments and measuring independent variables.
6. Methods for measuring and analyzing dependent variables.
7. Approaches to avoiding confounding effects of extraneous variables.
8. Statistical analyses.

The following strategies reflect this organizing theme.

Strategy 7.1: Describe Your Study's Setting, Participants (Subjects), and Materials

In applying this strategy, you should describe only the relevant characteristics of your study's setting, subjects, and materials. A given characteristic is relevant if it was (or will be) essential to resolving your research issue and testing your hypotheses. If your study involves human or animal subjects, you may need to describe them by their age, sex, genetic markers, anthropometric measures (such as height, weight, and body composition), health status, behaviors, and living conditions. If you are conducting research on biological materials such as cells and molecules, descriptions should include characteristics of the organisms from which the materials were taken, their chemical composition, and their size and weight.

Strategy 7.2: Describe Your Methods for Recruiting and Screening Subjects

How did, or will, you find subjects for your study? For instance, if you are proposing a new study, will you obtain a random sample from the entire population of interest? Or will you recruit volunteers from your local community by posting advertisements about your study? Once you have identified candidate subjects, by what criteria will you screen and eliminate those who lack relevant characteristics or who have extraneous characteristics? The answers to these questions are fundamental to convincing readers that your study's sample represents its targeted population and lacks biasing factors. Presented in Figure 8.10, the first subsection of Black et al.'s methods section is aimed at Strategies 7.1 and 7.2.

In addition to providing straightforward details about how many participants were initially recruited and how they were recruited, the subsection effectively describes the methods by which participants were screened and eventually selected for the study. Among other key characteristics, the authors describe the participants' initial physical activity levels, psychological health status, and medication use. In the complete article, the participants' ages and physical characteristics are included in a table (not shown here). An especially important bit of information about the subjects is that "a physician's diagnosis of CFS was required for inclusion." One suggestion for enhancing this description is to add details about the qualifying features of the diagnosis (e.g., how long individuals had symptoms of CFS, the number of hours that individuals slept daily, and specifics about medication frequency and dosage).

Strategy 7.3: Summarize Your Study's Overall Design and General Procedures

Well-written methods sections provide precise details about every step in executing research. Sometimes, however, the details are so specific and numerous that they overwhelm readers and cause them to lose sight of the bigger picture of a study's design. To prevent this occurrence, skilled authors preface methodological

Methods

Participants

All experimental procedures were approved by the Institutional Review Board at the University of Georgia, and informed consent was obtained from each participant. All participants were recruited from the general community and either responded to a newspaper ad, responded to a flyer placed around campus, or were referred to the study by their physician. Seventeen CFS and twenty-one controls responded and were screened as possible study participants. A physician's diagnosis of CFS was required for inclusion. Additionally, CFS patients were required to confirm a self-report of decreased physical activity compared to pre-CFS levels, and self-reported inability to sustain high levels of physical activity without a subsequent exacerbation of CFS symptoms for study inclusion. CFS participants with a self-reported history of depression or other psychiatric illness were excluded. Control participants were chosen to be similar in age, height, and weight to the CFS patients and were defined as sedentary by self-report of one bout of regular exercise per week or less. The most sedentary participants (those with desk jobs, etc.) were given first priority for study inclusion. Control participants were also apparently healthy and reported no illnesses or disease conditions. Medications were monitored in all participants. The CFS participants were found to be taking many medications, both prescription and over-the-counter. Analgesics such as Vioxx, Celebrex, Advil, and Aleve were common.

FIGURE 8.10 Content demonstrating strategies for describing study settings, participant characteristics, and methods for recruiting and screening participants (Strategies 7.1 and 7.2). From a published research paper authored by Black et al. (2005). Reproduced with permission.

details with summaries of their studies' overall designs and general procedures. A good example of this strategy in action comes in the second subsection of Black et al.'s methods section, presented in Figure 8.11.

Study Design

(1) Initially, participants received instructions for wearing the activity monitors and for completing a daily activity log. **(2)** A "pre" score from the 30-item Profile of Mood States short form questionnaire (POMS) was obtained (The Educational and Industrial Testing Service, San Diego, CA). **(3)** Participants proceeded

FIGURE 8.11 From Black et al.'s methods section, a subsection devoted to summarizing a study's overall design (Strategy 7.3). Reproduced with permission.

to wear the monitors for two weeks during which time they were instructed to maintain normal daily activity. **(4)** After the two weeks, data were collected from the activity monitors, and the monitors were reset. **(5)** The participants then wore the monitors for an additional four weeks. **(6)** CFS patients were asked to increase their daily physical activity (30% above baseline) during this four week period by walking a prescribed amount each day in order to approximate the daily physical activity of a healthy sedentary person. **(7)** This was based upon averaging the findings of others that suggested CFS patients had activity levels that were 15% to 40% reduced from healthy but sedentary individuals [3,4,6]. **(8)** CFS patients were given neutral instructions as to whether or not increasing their daily physical activity would alter their mood and fatigue symptoms. **(9)** Control participants maintained their normal activity for a six week period. **(10)** Daily activity logs were completed each day. **(11)** Participants recorded all daily activities, time spent in each activity, as well as time periods when the monitor was not worn (e.g., bathing). **(12)** Participants also completed a series of questions documenting their daily mood, perceptions of fatigue and muscle pain intensity, and the duration of time fatigue and muscle pain were experienced each day.

FIGURE 8.11 *continued.* (The sentences are numbered for reference to them in the ensuing discussion of the model.)

The content is organized by a chronological theme, which is especially fitting for the strategy at hand. In the first two sentences, we learn what happened at the beginning of the study, when subjects provided information about their daily activity, filled out a questionnaire for assessing their mood states, and received instructions concerning the *accelerometers*, or activity monitors, that they wore throughout the study. Sentence 3 provides a brief overview of the 2-week baseline period during which the subjects' normal activities were monitored. Sentences 5 and 6 summarize the main events of the 4-week period during which participants increased their physical activity above baseline levels. Toward the end of the paragraph, sentence 9 quickly covers the procedures followed by the control group (participants who did not have CFS) over the entire 6-week period. And sentences 10–12 review methods for collecting data on key study outcomes.

Black et al.'s chronological summary of their overall study design enables us to see the forest (of their methodology) without getting lost in the trees. In sentence 6, for example, we learn that the CFS group was instructed to increase their daily physical activity by 30% over baseline levels. However, in this early subsection, the authors do not bog us down with details about how they calculated the higher activity levels. As you will see, the details come in the next subsection of Black et al.'s methods section.

Black et al.'s subsection devoted to presenting a bird's-eye view of their study is effectively headed *Study Design*. Other common headings for such subsections are *Experimental Design* and *General Procedures*. An especially reader-friendly technique for summarizing general procedures is to accompany the text with a flow chart or a timeline that depicts key events in the overall scheme. A sample timeline graphic is presented in Figure 8.12. The graphic comes from a published research paper on the combined effects of resistance exercise, which is weightlifting, and a supplemental hormone called melatonin on the body's production of various hormones that promote muscle growth (Nassar et al., 2007).

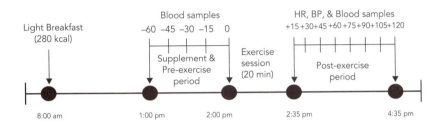

FIGURE 8.12 A graphic overview of a study's general procedures from a published research paper authored by Nassar et al. (2007). Adapted with permission.

From the timeline, readers can quickly grasp the experimental procedures, in which subjects received melatonin supplements 5 hours after a light breakfast, had blood samples taken every 15 minutes during a 60-minute pre-exercise period, performed exercise for 20 minutes, and then had more blood samples taken, along with measures of heart rate (HR) and blood pressure (BP), every 15 minutes for a 2-hour post-exercise period.

Strategy 7.4: Describe Your Methods for Assigning Subjects to Groups, Conditions, or Treatments

This strategy is for informing readers about whether study participants were (or will be) self-selected for pre-existing characteristics and conditions or whether they were (or will be) randomly assigned to groups, conditions, or treatments. In addition, the strategy may entail describing procedures for matching subjects in treatment and control groups in order to avoid bias. The content may be placed in a subsection devoted to describing participant characteristics, in a subsection that summarizes a study's overall design, or in its own subsection.

Strategy 7.5: Describe Your Methods for Administering Treatments and Measuring Independent Variables

As defined in Chapter 1, independent variables are the factors that scientists intentionally manipulate or observe to determine influences on targeted study outcomes. All methods sections should describe their studies' independent variables in sufficient detail. For studies with experimental designs, this requires explaining what the treatments, or experimental manipulations, were (or will be) and how they were (or will be) administered. For studies with observational designs, the explanations should address the methods for measuring the key independent variables that define study groups.

In Black et al.'s study on physical activity and CFS, the main independent variable was the amount of physical activity that subjects performed. The study had both observational and experimental phases. Over the initial 2-week baseline period, participants' physical activity levels were observed without direct manipulation. CFS and control subjects were fitted with accelerometer devices that measured bodily movements throughout the day and even while subjects slept. The accelerometers stored the movement data, which the researchers entered into a computer program that yielded measures called *counts*, which corresponded to individual movements. A count would be assigned, for example, to a swinging leg movement during walking, a reaching of the hand to pick up a toothbrush, and a turning of the torso during sleep. Black et al.'s 2-week baseline observational study enabled the researchers to document differences in daily physical activity between the study's CFS patients and the control group composed of healthy sedentary individuals.

In the ensuing 4-week experimental phase of their study, Black et al. manipulated the participants' physical activity levels by instructing the CFS patients how to increase their daily physical activity. The subsection devoted to describing the researchers' observations and manipulations of physical activity is reproduced in Figure 8.13.

Descriptions of methods for administering and measuring independent variables should provide sufficient information to enable peer researchers to replicate the procedures exactly. Imagine, for example, that you are planning a study on physical activity and CFS. To measure your subjects' daily movements objectively, and to compare the data directly to the outcomes of Black et al.'s study, you would need to use the same brand of accelerometers. What was the brand? Who makes the accelerometers? Following convention for describing study equipment and instruments, Black et al. answer these questions in their subsection devoted to describing their study's physical activity measurements. With the parenthetical information at the end of the first sentence in Figure 8.13, you could easily use the Internet to track down contact information for the company that made Black et al.'s accelerometers. The first paragraph of the subsection also answers other important questions that readers will ask about the activity-measuring devices: How were they fitted to the participants' bodies? How much of the day were they worn? How were the data extracted from them?

In the second paragraph of Figure 8.13, Black et al. provide essential details about the experimental treatment—specifically, information about how they went about increasing the activity levels of the CFS patients. The authors explain that, using each subject's baseline accelerometer data, they calculated a 30% increase in activity counts. Then, they related the additional counts to numbers of walking steps. Subjects were instructed to increase their daily activity by walking an individually specified number of steps, using pedometers (devices that count steps)

Objective Measurement of Physical Activity

Daily physical activity was assessed by a CSA accelerometer (Computer Science Associates Inc., Fort Walton Beach, FL). To ensure accurate measurements, the procedure recommended by CSA was followed. The monitors were positioned over the subject's anterior superior iliac spine with the belt fitting snuggly so as to limit extraneous monitor movement. Participants were asked to wear the monitors at all times of the day, including during sleep. Two minute epochs were used. Data were retrieved from the monitors using a specially designed docking module that input data into a computer.

Recommended percent increases in daily activity were calculated based on each subject's average daily counts during their two-week baseline activity period. Counts are arbitrary units assigned to movements detected by the accelerometer. Counts are assigned based upon the magnitude of a change in velocity during a given time period. The number of counts needed to raise daily activity approximately 30% was calculated. Participants were then asked to walk on a treadmill at what they considered to be a comfortable walking pace. Walking speed was recorded, and used to calculate the recommended daily walking time. The approximate number of counts per minute for various walking speeds was assessed prior to the onset of the study (unpublished observations). Additionally, a pedometer was also used to aid participants in achieving the desired daily activity increase. Participants were given an approximate number of steps to take each day during their walk based upon their prescribed walking pace and time. Steps per minute for various walking speeds were assessed in a similar manner as counts per minute prior to the study (unpublished observations).

FIGURE 8.13 From Black et al.'s methods section, a subsection devoted to describing the administration and measurement of the study's independent variable (Strategy 7.5). Reproduced with permission.

for verification. Black et al. provide quite adequate details about their methods for administering and measuring their study's independent variable, especially when one considers typical length restrictions on published research papers. If the authors had unlimited space, one suggestion for revising the content in Figure 8.13 would be to present a sample calculation of the required increase in daily walking steps for one subject.

Strategy 7.6: Describe Your Methods for Measuring and Analyzing Dependent Variables (Study Outcomes)

Dependent variables are the main outcomes of interest in research. In well-structured methods sections, authors devote subsections to describing the instruments and procedures they used (or plan to use) to collect and analyze data for their studies' dependent variables. In excellent applications of Strategy 7.6, authors provide sufficient details to leave readers with no doubt about how a study's data were (or will be) collected and analyzed. The details ensure that peer researchers can replicate procedures. In addition, they ultimately enable readers to clearly understand study results and interpretations of them in the two major sections that follow the methods section (the results and discussion sections).

There were four dependent variables in Black et al.'s study on physical activity and CFS: daily physical activity, mood, fatigue, and muscle pain. Figure 8.14 reproduces most of the subsection in which the authors describe their measurements of these outcomes.

Self-report of Daily Activity and Feelings

Data concerning daily activity, mood, fatigue, and muscle pain were obtained from each subject via daily self-report. Participants were asked to report all daily activities and time spent engaged in each. Time periods when the activity monitor was not worn were also reported. Participants ranked, using a 10 cm (0 to 100 mm) visual analog scale, their general overall daily mood for that day (with 0 being their best possible overall mood and 100 being their worst possible overall mood), the intensity of their fatigue that day (with 0 being no fatigue and 100 being the highest intensity fatigue imaginable), and the intensity of their muscle pain that day (with 0 being no pain and 100 being the worst imaginable pain). A similar visual analog scale has been used previously in CFS patients to rate daily fatigue [13]. Additionally, participants reported the amount of time each day they experienced fatigue as well as muscle pain. Once each week participants completed a Profile of Mood States (POMS) short form questionnaire consisting of 30 questions (The Educational and Industrial Testing Service, San Diego, CA) in which participants reported how they had been feeling during the prior seven days. These forms were scored for both fatigue and vigor ratings.

FIGURE 8.14 From Black et al.'s methods section, a subsection devoted to describing measurements of the study's dependent variables (Strategy 7.6). Reproduced with permission.

Notice the details concerning the procedures for measuring subjects' self-reported mood, fatigue, and muscle pain. The researchers used a simple but effective instrument called a visual analog scale (VAS), which is essentially a ruler presented on a sheet of paper or computer screen. Beginning four sentences into Figure 8.14, Black et al.'s straightforward description of the VAS is sufficiently detailed for readers to replicate its application and understand the resulting data. Moreover, following a useful convention, in the subsection's fifth sentence the authors refer readers to a published study in which a VAS was used to measure self-reported daily fatigue in CFS patients. Black et al.'s reference to this study, which is citation 13 in their reference list, lets readers know where they can find additional details about the VAS method in the published literature. Figure 8.14 is another example of well-detailed descriptions of methods, despite significant length restrictions for published research papers. Perhaps the authors could have provided other useful details. Responding to the last sentence of the subsection, for instance, readers who are not familiar with the Profile of Mood States short-form questionnaire might ask which 30 questions the study's participants answered. If Black et al. had unlimited space, they might have included the questions in a figure or an appendix.

Strategy 7.7: Describe Your Methods for Avoiding Potential Confounding Effects of Extraneous Variables

As discussed in Chapter 3, factors other than a study's targeted independent variables can potentially influence outcomes of interest. When these extraneous variables are not eliminated or otherwise controlled, they can lead to flawed results and invalid conclusions. Procedures for avoiding confounding effects include administering placebo controls and matching subject groups evenly for extraneous variables. By describing these procedures in detail, authors build convincing arguments for their studies' methodological strengths. Especially in research proposals and grant applications, it's worth highlighting these arguments by dedicating focused subsections to achieving this strategy. The subsections might be headed something like *Controls for Extraneous Variables*.

Strategy 7.8: Describe Your Study's Statistical Analyses

Advanced projects—including Honors theses, graduate dissertations, and research papers that will be submitted for publication—usually require statistical analyses of data. In describing these analyses, authors include information about the specific statistical tests they used (or plan to use), the procedures for conducting the tests, and the computer software involved. If a power analysis was (or will be) used to determine the appropriate number of subjects for a study, the details should be included. In addition, if criterion P values were (or will be) used to guide hypothesis testing, they should be reported.

Rhetorical Goal 8: Justify Your Use of Selected Study Methods

Sometimes the straightforward description of a procedure or analysis prompts readers to ask for more: How did (or will) the method serve the study's overall purpose? How can we be certain that the method was (or will be) accurate and dependable? Is the method notably superior to viable alternatives? These questions call for authors to justify their study procedures and analyses, adding an often-essential element of argument to methods sections.

Strategy 8.1: Explain the Relevance of Selected Study Methods

In a typical methods section, the relevance of many study procedures and analyses is often obvious. From their descriptions alone, readers will clearly grasp how the methods contributed to (or will contribute to) resolving a study's motivating research issue and testing its hypotheses. Consider a research proposal for a study on exercise and weight control—the author need not *justify* or *argue for* plans to recruit overweight subjects and to measure their body fat composition. Sometimes, however, a straightforward description of a study method does not immediately convey its relevance and convince the audience of its necessity. Strategy 8.1 applies in this case. Often, only a sentence is necessary to explain the relevance of a procedure or an analysis. This is true in Figure 8.15, which comes from the methods section of Campisi et al.'s research paper on exercise and the stress response.

Freewheel running. Rats were individually caged in Nalgene Plexiglas cages (45 × 25.2 × 14.697 cm) with a stainless steel open running wheel attached (1.081 wheel circumference). Physically active rats (Active) had a mobile running wheel and ran for 8 wk before stressor exposure. Sedentary controls (Sed) were housed in the same environment except that the running wheel was locked and remained immobile. **Voluntary freewheel running was the chosen modality because, in contrast to forced treadmill training, voluntary freewheel running does not produce negative adaptations that are indicative of chronic stress in rats (34, 35).** The caging environment meets National Institutes of Health floor space standards for a single rodent. Total daily running distances were monitored hourly by computer with the Vital-View Automated Data-Acquisition System (Bend, OR). Rats were weighed weekly.

FIGURE 8.15 A single-sentence justification of a study method (bolded for emphasis), demonstrating Strategies 8.1 and 8.3 (Campisi et al., 2003). Reproduced with permission from the American Physiological Society.

In the noteworthy sentence (highlighted with bold font), the authors justify their use of voluntary freewheel running as the exercise training method for the study's animal subjects. Recall that the introduction section to Campisi et al.'s research paper also justified freewheel running as a superior method compared to forced treadmill exercise. However, it's such an important justification, one that reflects a major strength of the study, that the authors have good reason to repeat it in their methods section.

More elaborate justifications for the relevance of study methods are sometimes warranted, especially in research proposals and grant applications. The following example (Figure 8.16) justifies the use of a control group in a student's proposal to study exercise as a potential treatment for depression.

A control group consisting of 15 subjects will meet three times per week in a local fitness club; however, these individuals will not perform any exercise. Instead, the meetings will provide opportunities for social interaction. The rationale for including this control group is based on the potential for social factors to introduce confounding effects in studies on exercise and depression. Many previous studies have shown that physically active individuals have a relatively low risk for developing this disease (5, 8, 9, 11, 12). However, the studies have not been designed to determine whether a direct cause-effect relationship exists. One theory is that exercise directly influences concentrations of serotonin and norepinephrine in the brain, thereby enhancing mood states and protecting against symptoms of depression (3, 6). A contrasting theory is that the beneficial effects of exercise are essentially social in nature, given that many people exercise with friends and family (2, 4). This psychosocial theory holds that the lower incidence of depression in regular exercisers is attributable to the good feelings promoted by interacting with people. By including a control group that interacts socially but remains sedentary, I will be able to determine the extent to which the protective effects of exercise against depression are biological versus psychosocial in nature.

FIGURE 8.16 Applying Strategy 8.1, a well-developed justification of a study method.

The rationale for including the control group effectively contributes to the author's overall argument for the necessity of her proposed study.

Strategy 8.2: Justify the Validity, Precision, and Reliability of Selected Instruments and Techniques for Collecting Data and Measuring Outcomes

Consider a hypothetical research paper reporting a study to determine the effects of aerobic exercise on self-esteem. Let's say that the subjects were asked to respond to a relatively new questionnaire called the *Self-Esteem Diagnostic Inventory*, or SDI. In their methods section, the researchers described the questionnaire in great detail, providing representative questions from its various subscales. Suppose, however, that peers who specialize in research on self-esteem would be likely to raise concerns about the questionnaire. They might, for example, ask about its validity: How do we know that the instrument truly measures self-esteem rather than other psychological traits? Readers might also question the instrument's precision: Can the SDI distinguish between fine differences in self-esteem across individuals? Another question concerns the questionnaire's reliability: Does the SDI yield consistent, or reproducible, outcomes in repeated applications? Anticipating these sorts of questions, the authors should justify the instrument's validity, precision, and reliability. If convincing arguments have already been made in the published literature, the authors can refer readers to the sources. If supporting research is lacking, the justification can be based on sound scientific knowledge and reasoning.

Strategy 8.3: Argue for the Superiority of Your Study's Methods over Viable Alternatives

Across various studies on any given research issue, different methods may be used for common purposes. For instance, two studies on the same issue may vary in their approaches to recruiting subjects, in the instruments and technologies used for data collection, and in the types of statistical tests performed. When more than one viable approach exists for executing some aspect of a study, informed readers will ask for justification of the one that the scientists used (or plan to use). Anticipating this response, skilled authors argue for the superiority of their chosen methods. These arguments are usually backed by one or more of the following lines of support.

- Discussions of flaws in the alternative methods, detailing how they led to problematic results and conclusions in previous studies.
- Presentations of research findings that demonstrate greater validity, precision, and reliability in chosen methods compared to the alternatives.
- Explanations of how chosen methods are ideally suited for resolving research issues and for testing hypotheses.

The content for Strategy 8.3 has a logical place in methods sections because it immediately addresses readers' concerns about viable alternative approaches. As presented earlier, Figure 8.15 is a case in point. In IMRAD-structured papers, however, authors often delay arguments for the methodological strengths of their studies until the discussion section. This tactic is sensible when the arguments are extensive and their presentation in the methods section would be disruptive, breaking the flow of step-by-step procedural descriptions.

RHETORICAL GOALS FOR RESULTS SECTIONS

At the core of all IMRAD-structured papers, results sections communicate the essential outcomes of research, or the data that scientists use to reach conclusions that contribute to resolving their studies' motivating issues. The data presented in well-written results sections thus directly address the research questions, problems, and hypotheses that authors first raise in their papers' introduction sections. Take the example of Black et al.'s research paper on physical activity and CFS. In their introduction, the authors presented their research issue, which can be paraphrased by the following two questions.

1. To what extent do physical activity levels differ between CFS patients and healthy individuals?

2. What are the effects of increasing daily physical activity on mood, fatigue, and muscle pain in CFS patients?

The bulk of Black et al.'s results section is reproduced in Figure 8.17. After an introductory subsection in which the authors present data to characterize the final participants of the study, the last two subsections are organized by its primary dependent variables, which correspond to the two preceding research questions. Specifically, the subsections headed *Daily Physical Activity* and *Self-Report Mood/Feeling Ratings* present the essential results that Black et al. used to answer their first and second research questions, respectively. Black et al.'s results section

Results

Participants

No participants (CFS or controls) reported adverse events associated with the increased activity program. Data were obtained from six CFS patients as well as seven sedentary control participants. Additionally, two other CFS patients began the testing protocol but were removed from the study at their own request. One was removed on a doctor's recommendation due to an injury (unrelated to the study) and the second was removed due to a change in residence. The physical characteristics of all participants are presented in Table 1. Mean age, height, and weight were not different between the CFS patients and the sedentary controls.

Daily Physical Activity

Individual as well as mean group daily activity counts are presented in Table 2. All 24-hour periods in which the monitor was not worn for at least 23 hours were removed from the analysis. No trends were observed in daily activity counts in either subject group across all monitoring periods. Day-to-day counts were also relatively stable within a given activity period. Based on this, activity levels are presented as average daily counts during a given activity period. During baseline activity, CFS participants demonstrated 39% lower daily activity counts compared to controls ($P = 0.017$). All six of the CFS participants were successful in increasing their daily physical activity. Their daily activity counts increased 28%, on average, during the four-week training period ($P < 0.001$). However, it should be noted that 4 of the 6 CFS participants did not reach the prescribed 30% increase in daily activity. Interestingly, following their activity increase the CFS participants had activity levels that were still 24% reduced from those of the control group ($P = 0.08$).

FIGURE 8.17 Exemplary subsections of Black et al.'s results section. Reproduced with permission.

Self-Report Mood/Feeling Ratings

Figures 1, 2, and 3 contain daily ratings of overall mood, fatigue intensity, and muscle pain intensity averaged over two-week periods. Days where missing data were found (i.e. ratings were not completed) were checked against the activity monitor data to ascertain if the participants simply forgot to fill out the form or if some problem was present that could prevent them from filling out the ratings. Out of 692 possible days, missing data were found for only 19 days. Activity monitor data appeared normal on all 19 of these days. This suggests that the missing data were likely the consequence of participants forgetting to fill out the form rather than any adverse medical event.

Figure 1 demonstrates a significant group-by-time interaction between the CFS and control participants ($P = 0.016$, $Eta^2 = 0.311$) in overall mood. CFS participants reported a worsening of overall mood over time compared to controls. Neither an interaction nor a time main effect was observed in ratings of fatigue intensity. A significant group difference was observed between the CFS and control participants (Fig. 2, $P < 0.001$, $Eta^2 = 0.892$). Although not statistically significant, ratings of fatigue intensity in the CFS group did increase from 58.2 ± 8.5 to 67.0 ± 17.5 (indicating a worsening of symptoms). A significant group-by-time interaction ($P = 0.03$, $Eta^2 = 0.295$) was seen between the CFS and control participants in their ratings of muscle pain intensity (Fig. 3). As their daily activity was increased, the CFS participants reported higher intensity muscle pain compared to controls.

The amount of time spent with fatigue each day as well as the amount of time spent with muscle pain each day was also reported by both groups. During baseline activity, the CFS participants reported experiencing a significantly greater amount of time spent with fatigue per day compared to the control participants (930 ± 397 min/day. vs. 43 ± 73 min/day; $P < 0.001$). Additionally, the CFS participants reported experiencing a significantly greater amount of time spent with muscle pain each day (552 ± 505 min/day vs. 9 ± 22 min/day; $P = 0.011$). A significant time main effect was found for time spent with fatigue each day ($P = 0.047$, $Eta^2 = 0.243$). A significant difference was found between baseline activity, 451 ± 528 min/day, compared to the final two weeks of increased activity, 521 ± 566 min/day ($P = 0.048$, $Eta^2 = 0.287$). The CFS participants also demonstrated a non-significant increase in time spent with muscle pain each day during baseline activity, the first two weeks of increased activity, and the final two weeks of increased activity (554 ± 507 min/day vs. 642 ± 546 min/day vs. 713 ± 557 min/day). Control participants demonstrated no change over time in time spent each day with fatigue or muscle pain.

FIGURE 8.17 *continued.*

reflects an ideal general structure in which each subsection presents results for a key variable (or a set of related variables) and maps directly to a motivating question, problem, or hypothesis.

As outlined in Figure 8.18, our approach to generating the content of results sections is based on just one rhetorical goal and a set of accompanying strategies, which are applied recursively to construct each subsection.

Rhetorical Goal 9: Present the results that are essential to reaching and supporting your conclusions. (413)

Strategies

9.1: Explain any preliminary observations that led you to revise planned methods for analyzing your study's data. (413)

9.2: Explain how selected results served your study's overall purposes. (414)

9.3: Report your study's essential outcomes. (415)

9.4: Guide readers to and through complex graphics. (417)

9.5: If necessary, acknowledge problems that may have compromised your results. (418)

FIGURE 8.18 The main rhetorical goal and accompanying strategies for results sections. The pages on which the goal and its strategies are introduced are shown in parentheses.

Rhetorical Goal 9: Present the Results That Are Essential to Reaching and Supporting Your Conclusions

A scientific conclusion is an answer to a question, a solution to a problem, or a resolution to a hypothesis. The overall rhetorical goal for results sections is to present the relevant data that authors need to reach valid conclusions and, ultimately, to develop convincing arguments for them. The following strategies serve this aim.

Strategy 9.1: Explain Any Preliminary Observations That Led You to Revise Planned Methods for Analyzing Your Study's Data

After preliminary observations of their study results, researchers sometimes adapt their original plans for analyzing data. Take a study to determine the effects of a carbohydrate-rich diet on endurance performance. In the methods section of the paper reporting the study, suppose that the authors described and justified plans to analyze results for female and male subjects separately. However, in a preliminary analysis, the researchers found no difference in the diet's effects on endurance performance in females versus males. So the researchers decided to revise their planned analysis by collapsing the data across the sexes. Under the circumstances the decision was sound because the pooled data increased the sample size and power for statistical analyses of differences between the study's experimental and control groups. Applying Strategy 9.1, the authors reported their preliminary observations, along with their adapted approach to analyzing the data.

Strategy 9.1 is demonstrated in the following excerpt from the *Daily Physical Activity* subsection of Black et al.'s results section.

> All 24-hour periods in which the monitor was not worn for at least 23 hours were removed from the analysis....Day-to-day counts were also relatively stable within a given activity period. Based on this, activity levels are presented as average daily counts during a given activity period.

In addition to justifying the elimination of selected physical activity data from their analyses, the authors explain that the counts recorded by the monitors (the accelerometers that the study participants wore) were consistent from day to day across designated monitoring periods. Accordingly, Black et al. found no reason to analyze the counts by separate days of the week or monitoring periods. This preliminary analysis justifies the authors' decision to present their results as "average daily counts." For the study's 2-week baseline phase and 4-week experimental phase, respectively, all data values from daily monitoring periods were collapsed to calculate one mean daily activity count for the CFS group and the healthy control group, respectively. The complete data set for all participants, as it appeared in Table 2 of Black et al.'s results section, is presented in Figure 8.19.

Strategy 9.2: Explain How Selected Results Served Your Study's Overall Purposes

This optional strategy is useful when readers might not immediately grasp how selected results actually contribute to resolving a study's motivating issue and testing its hypotheses. Suppose, for example, that Black et al. were concerned that their audience might not clearly understand the relevance of the results for muscle pain, as reported in the subsection headed *Self-Report Mood/Feeling Ratings* in Figure 8.17. The authors might have anticipated this question from readers: *Why did you need to measure subjects' ratings of muscle pain in order to determine the effects of increased physical activity on CFS?* The answer is that muscle pain is a common symptom of CFS. So if Black et al. had indeed been concerned that their audience might not get the connection, they might have prefaced their presentation of the muscle pain results with a brief explanation of the relevance. In reality, the authors did not need to take this approach because, in their introduction section, they clearly established that muscle pain is a symptom of CFS. In general, Strategy 9.2 also is unnecessary when authors have clearly established the relevance of key outcome measures in their methods sections.

TABLE 2: Average daily activity counts for CFS and control participants. Data are mean ± SD.

Subject	Average Daily Activity (Counts × 10³)		%Difference
CFS	Baseline	Increase	
1	88.3 ± 24.1	143.4 ± 42.5	+62.3
2	126.7 ± 19.5	179.1 ± 44.9	+41.4
3	199.8 ± 38.7	226.7 ± 57.5	+13.5
4	167.4 ± 32.8	197.4 ± 32.5	+17.9
5	234.2 ± 37.2	263.8 ± 31.2	+12.6
6	158.3 ± 32.9	193.5 ± 33.0	+22.2
Mean	**162.5 ± 51.7**	**200.3 ± 41.2**	
Control	Baseline 1	Baseline 2	
1	415.1 ± 79.4	360.7 ± 67.3	−13.1
2	284.1 ± 87.1	289.5 ± 81.3	+1.9
3	150.6 ± 61.6	146.2 ± 48.7	−3.0
4	254.9 ± 60.8	281.9 ± 76.5	+10.6
5	223.7 ± 73.5	190.9 ± 68.8	−14.7
6	263.7 ± 60.7	233.9 ± 58.1	−11.3
7	278.1 ± 106.2	274.0 ± 113.9	−1.5
Mean	**267.2 ± 79.5**	**253.9 ± 70.0**	

FIGURE 8.19 From Black et al.'s results section, a table presenting individual data and group means for physical activity levels in chronic fatigue syndrome (CFS) patients and control participants over the study's baseline and experimental phases. Physical activity levels are expressed as daily accelerometer counts × 10³. Reproduced with permission.

Strategy 9.3: Report Your Study's Essential Outcomes

This strategy generates the core content of results sections. For every essential analysis of a study variable or relationship between variables, the strategy entails presenting the following information.

(1) A general statement, or a take-home message, that conveys the main result.
(2) The actual data value(s) for the main result.
(3) The outcomes of statistical tests associated with the main result.

These three elements are contained in the following straightforward sentence from the *Daily Physical Activity* subsection of Black et al.'s results section.

> During baseline activity, CFS participants demonstrated 39% lower activity counts compared to controls ($P = 0.017$).

Addressing Black et al.'s research question about differences in physical activity between CFS patients and healthy individuals, the sentence gives (a) the take-home message that the CFS patients are less active; (b) the supporting data, in this case a percentage difference between study groups; and (c) the key statistical outcome, which is the significant P value.

For another demonstration of Strategy 9.3, Figure 8.20 presents an excerpt of text and an associated line graph that report Black et al.'s results for the effects of increased physical activity on overall mood ratings. The text and figure caption directly convey the take-home message that (contrary to what we might have expected) over a 4-week period, increased physical activity worsened mood states in CFS patients. This interpretation is supported by two key statistics: a significant P value and an Eta^2 value which, on a scale of 0 to 1, indicates the strength of association between variables. A noteworthy feature of Black et al.'s line graph is the detailed figure caption, which effectively explains what the data points represent and the outcome of their statistical analysis.

Figure 1 demonstrates a significant group-by-time interaction between the CFS and control participants ($P = 0.016$, $Eta2 = 0.311$) in overall mood. CFS participants reported a worsening of overall mood over time compared to controls.

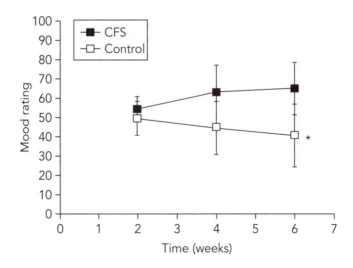

Figure I

Overall mood ratings. Each time point represents an average of the scores from the previous two weeks. "0" represents the best possible mood and "100" represents the worst mood imaginable. For CFS participants, the two week time point is from baseline activity and the four and six week time ponts are from increased activity. For control participants, all time points are from baseline activity. * Significant group × time interaction ($P = 0.016$). Values are mean ± SD.

FIGURE 8.20 From Black et al.'s results section, text and an associated line graph that effectively demonstrate the reporting of a study's essential outcomes (Strategy 9.3).

It's usually simple task to report a study's main results, although challenges occasionally crop up. One involves deciding whether to present data in the form of text, graphics (tables or figures), or both. Guidelines for making this decision were presented in Chapter 4. When presenting lots of complex data as text, authors must figure out how to best structure them in individual sentences. From Black et al.'s results section, here's an effective technique in which the data and statistical outcomes are offset in parentheses:

> During baseline activity, the CFS participants reported experiencing a significantly greater amount of time spent with fatigue per day compared to the control participants (930 ± 397 min/day. vs. 43 ± 73 min/day; $P < 0.001$).

Another challenge for writing the results section is determining which data, from all that you have collected, to present. Should you provide the raw data for your study's individual subjects? Would it be more appropriate to calculate and compare group means and measures of variability, such as standard deviations? Is it necessary to present statistical correlations between variables? The answers depend on the questions you sought to answer, the hypotheses you tested, your study's design and sample size, and the nature of your results. For example, take another look at Table 2 from Black et al.'s results section (Figure 8.19, page 415). The table presents the data for physical activity counts for each subject who completed the study. The approach of reporting individual data values makes good sense because the study included relatively few subjects; so the details are by no means overwhelming for readers. In addition, the physical activity counts varied considerably across subjects, especially in the CFS group. This sizeable interindividual variability, which Black et al. convey directly in their table, is an especially important outcome of the study.

Strategy 9.4: Guide Readers to and through Complex Graphics

As effectively demonstrated by the figure caption and supporting text in Figure 8.20, this strategy applies when study data are presented in a table or figure rather than, or in addition to, in the text. The objectives are to direct readers to the graphic, to help them navigate and isolate its key features, and to explain how the graphical information and data support the main messages about a study's results.

Strategy 9.5: If Necessary, Acknowledge Problems That May Have Compromised Your Results

Among other common glitches in scientific research, subjects may drop out midway through a study, instruments may fail at critical points, or an experiment might be shortened because its initial results indicated potentially harmful effects of a treatment. When methodological and logistical problems interfere with planned data collection and analyses, authors should explain the problematic events. If a sufficient explanation requires no more than a few sentences, it will fit nicely in the results section. If, however, the problems are fairly extensive, their explanations should be delayed until the discussion section.

RHETORICAL GOALS FOR DISCUSSION SECTIONS

The last section of IMRAD-structured papers, the discussion section, serves the overall goal of answering the questions, solving the problems, and resolving the hypotheses that motivate original studies. As outlined in Figure 8.21, the section

---

Rhetorical Goal 10: Briefly reintroduce the defining features of your study. (420)

Rhetorical Goal 11: State your conclusions and support them with your study's results. (420)

Strategies
11.1: State your conclusions directly. (421)
11.2: Support your conclusions by presenting your study's most relevant results and by developing any necessary warrants. (422)
11.3: Acknowledge and respond to viable alternative conclusions. (423)
11.4: Support your conclusions by arguing for your study's methodological strengths. (423)
11.5: Highlight unanticipated study outcomes that are especially interesting and important. (423)

Rhetorical Goal 12: Relate your study's outcomes to those from previous studies on your research issue. (424)

Strategies
12.1: Synthesize the outcomes of previous studies that support your results and conclusions. (425)
12.2: Synthesize the outcomes of previous studies that contrast your results and conclusions. (426)

FIGURE 8.21 Rhetorical goals and strategies for the discussion section of IMRAD-structured papers. The pages on which the goals and strategies are introduced are shown in parentheses.

12.3: Argue for the greater validity of your results and conclusions (when such an argument is warranted). (426)

Rhetorical Goal 13: Discuss the mechanisms that underlie your study's main results, and argue for the most plausible underlying mechanisms (when such an argument is warranted). (427)

Strategies

13.1: Address all viable explanatory mechanisms. (428)

13.2: In sufficient depth, explain *how* plausible mechanisms may have influenced your study's results. (428)

13.3: Argue for the mechanisms that *best* explain your study's results. (429)

Rhetorical Goal 14: Acknowledge significant methodological shortcomings to your study, and explain how they may have influenced its outcomes. (430)

Strategies

14.1: Acknowledge only those methodological shortcomings that may have influenced your study's outcomes significantly. (431)

14.2: Explain *how* significant methodological shortcomings might have influenced your study's outcomes. (431)

14.3: Speculate on how your study might have turned out if its methodological shortcomings had been avoided. (432)

14.4: Convince readers that, despite your study's methodological shortcomings, its outcomes are still valid. (432)

Rhetorical Goal 15: Discuss the practical implications and applications of your study's results. (433)

Strategies

15.1: Discuss the practical value of your study's outcomes to society and nature. (433)

15.2: Discuss the practical value of your study's outcomes to the scientific community. (434)

Rhetorical Goal 16: Propose future studies on your research issue. (435)

Strategies

16.1: Explain the problems that underlie your suggestions for future research. (435)

16.2: Present your hypotheses for proposed new studies. (436)

16.3: Describe and justify the key methods for your proposed future studies. (436)

16.4: Speculate on potential beneficial outcomes of your proposed research. (436)

FIGURE 8.21 *continued.*

is based on a diverse set of rhetorical goals that demand the most advanced critical thinking skills in scientific writing. Presented as follows, the goals and their associated strategies engage authors in interpreting data, synthesizing research outcomes, evaluating study methods, constructing convincing arguments, and proposing creative new ideas for advancing scientific knowledge.

Rhetorical Goal 10: Briefly Reintroduce the Defining Features of Your Study

Picture in your mind a long IMRAD-structured paper with especially complex methods and results sections. As readers work their way to the paper's discussion section, they may lose their focus on the reported study's defining features. Presented way back in the introduction section, these features may include information about the study's motivating research issue, the authors' hypotheses, and an overview of novel experimental methods. In an audience-friendly overture, authors begin the discussion sections of such IMRAD-structured papers by reorienting readers to essential study-defining information and ideas. The rhetorical goal at hand, by far the most straightforward of all rhetorical goals for discussion sections, can usually be achieved sufficiently in only a paragraph, as in Figure 8.22.

> The purpose of this study was to determine the effects of resistance exercise on balance and locomotor function in 80- to 90-year-old men. Previous studies on resistance training in this population have not addressed such functional outcomes, which are critical for maintaining independence in daily living. We tested the hypothesis that training-induced increases in muscle mass and strength, which have been observed in previous studies on resistance exercise in older adults,[7–9] would be associated with improvements in functional outcomes, specifically improved postural stability and coordination in a walking task.

FIGURE 8.22 A demonstration of Rhetorical Goal 10, which involves reorienting readers to the defining features of a reported study's motivating issue. The example begins the discussion section of a research paper on resistance training and functional outcomes in older men.

Rhetorical Goal 11: State Your Conclusions and Support Them with Your Study's Results

In any report of original research, ultimately the most pivotal content comprises the author's conclusions and arguments for them. This content logically comes in the discussion section of IMRAD-structured papers.

Strategy 11.1: State Your Conclusions Directly

Each question, problem, or hypothesis that an author raises in the introduction section of an IMRAD-structure paper should have a corresponding conclusion in the discussion section. Take the hypothesis that Jay Campisi and coworkers raised in the introduction to their research paper:

> We hypothesize that habitually physically active rats will have both greater and faster HSP72 responses to stress, suggesting that physical activity may contribute to stress resistance at the cellular level.

Now have a look at Campisi et al.'s overall conclusion, which they effectively present in the first paragraph of their discussion section (Figure 8.23). In the paragraph's second sentence, the authors directly state their conclusion, which is that

Discussion

The purpose of the present investigation was to test the hypothesis that habitually physically active rats have both greater and faster HSP72 responses to stress, suggesting that physical activity may contribute to stress resistance at the cellular level. The results of the current studies support this hypothesis. Physically active rats exposed to either IS or EXS stress responded with a greater induction of HSP72 in response to stress than did sedentary rats. Physically active stressed rats had increased HSP72 in nearly every tissue tested. In contrast, sedentary stressed rats had increased HSP72 only in pituitary, adrenal, liver, and spleen, and the increase was smaller than in the physically active-stressed rats. Moreover, physically active rats exposed to IS stress induced HSP72 in lymphocytes at a faster rate than did sedentary rats. Thus the impact of stress on HSP72 induction was potentiated in physically active rats. This study is the first comprehensive examination of the effects of voluntary freewheel running on stress-induced HSP72 expression in multiple stress-responsive tissues. The results are consistent with previous work conducted on skeletal muscle (17) and peripheral blood leukocytes (13) demonstrating that physical activity status can modulate the HSP response to stress.

FIGURE 8.23 The first paragraph of the discussion section from Campisi et al.'s research paper. The content successfully achieves Strategies 11.1 and 11.2, which entail presenting conclusions and supporting them with an original study's results. The authors summarize results for the magnitude and production rate of a heat shock protein (HSP72) in exercise-trained rats that were exposed to two forms of stress: inescapable tail shock stress (IS) and exercise stress (EXS). Reproduced with permission from the American Physiological Society.

their hypothesis, presented again in the paragraph's first sentence, was supported by the study's results.

Conclusions should be stated with their verbs in the present tense rather than in the past tense. The present tense accurately indicates that conclusions from research are inferences that extend beyond study samples to larger targeted populations. Thus, in the last sentence of their discussion section's first paragraph, Campisi et al. reinforce their conclusion by writing that "physical activity status can modulate the HSP response to stress." Notice that the authors did not write that physical activity status *modulated* the HSP response to stress. In the past tense, the verb *modulate* would problematically restrict the conclusion, indicating that it does not apply beyond Campisi et al.'s study.

Strategy 11.2: Support Your Conclusions by Presenting Your Study's Most Relevant Results and by Developing Any Necessary Warrants

As detailed in the upcoming pages, there are a number of approaches to supporting original conclusions in the discussion section of an IMRAD-structured paper. The primary line of support, however, is developed by applying Strategy 11.2. The strategy entails re-presenting the essential data from a paper's results section—specifically, the data that back the author's conclusions. Much of the content in Figure 8.23, the exemplary paragraph in which Campisi et al. present their conclusion, is devoted to achieving this strategy. After concluding that their hypothesis was confirmed, the authors summarize their study's most relevant supporting results. Notice that Campisi et al.'s hypothesis focuses on the "greater and faster HSP72 responses to stress" in exercise-trained rats. (Recall that HSP72 is a type of heat shock protein, which is a molecule that protects the body's cells against the harmful effects of stress.) In supporting their conclusion, the authors highlight their study findings that involve the *magnitude* and *speed* of the HSP72 responses they observed. Indeed, all of the sentences that summarize the supporting results include adjectives like *greater*, *increased*, and *faster*.

As demonstrated in Figure 8.23, Campisi et al. support their conclusion with a qualitative summary of their key findings. In other words, they do not present any actual data values from their results section. An effective way to strengthen an argument for a conclusion is to present quantitative study data. For example, Campisi et al. might have summarized the specific percentage values for the increased magnitude and rate of HSP72 production that they observed in various cells and tissues.

Sometimes the logical connections between an author's conclusion and supporting data are not immediately obvious to readers. In this case, the argument's warrants should be developed. As defined in Chapter 3, warrants are definitions, assumptions, and explanations that connect claims (or conclusions) with their supporting evidence and reasoning. For an example of an argument that would be strengthened by an explicit warrant, recall this chapter's earlier reference to a hypothetical study on the effects of a carbohydrate-rich diet on endurance performance. Let's say that the subjects were elite endurance cyclists, the caliber of athletes who compete in events like the Tour de France. An experimental group consumed a diet in which 80% of their total calorie intake consisted of carbohydrates. The study's main dependent variable was maximal oxygen consumption, or VO_2max. From Chapter 2, you may recall that VO_2max is a laboratory-based measure of the peak rate at which the body can use oxygen to produce energy. The results of our hypothetical study indicated that subjects on the high-carbohydrate diet experienced a statistically significant 5% increase in VO_2max. In the discussion section of their research paper, the authors used this value to support their conclusion that "a carbohydrate-rich diet enhances endurance performance in elite cyclists."

The preceding argument calls for a warrant. The reason is that the logical connection between VO_2max and endurance performance is questionable. It's obvious that they don't give the first-place prize in the Tour de France to the cyclist with the highest VO_2max determined through laboratory testing. Instead, the prize goes to the cyclist who covers the 2,200-mile race in the shortest amount of time. To develop the necessary warrant for their argument, the authors could present evidence from previous research supporting a high statistical correlation between VO_2max and performance success in actual endurance competitions. Another approach is to explain the physiological reasons for the association between a high VO_2max and peak endurance performance.

Strategy 11.3: Acknowledge and Respond to Viable Alternative Conclusions

It is not unusual for the same set of study data to elicit sharply contrasting interpretations and conclusions from different readers. To avoid weak one-sided arguments in this situation, you must acknowledge the alternatives and convince readers that your data interpretations and conclusions are the strongest. This strategy usually requires identifying and explaining the logical fallacies in the evidence and reasoning that support alternative conclusions to yours. Recall that Chapter 5 presented routines for diagnosing various logical fallacies in scientific arguments. These routines apply to achieving the present strategy.

Strategy 11.4: Support Your Conclusions by Arguing for Your Study's Methodological Strengths

Earlier, in presenting rhetorical goals for methods sections of IMRAD-structured papers, I mentioned that authors often delay arguments for their studies' methodological strengths until the discussion section. In doing so, authors build additional lines of support for their conclusions, complementing the support generated through applying Strategies 11.2 and 11.3. The objective of the present strategy is to convince readers that your conclusions are strong because you obtained their supporting data by using appropriate, valid, precise, and reliable study methods. The strategy encompasses three strategies that we have already covered for making arguments about a study's methodological strengths: Strategies 8.1, 8.2, and 8.3, which were presented beginning on page 408.

Strategy 11.5: Highlight Unanticipated Study Outcomes That Are Especially Interesting and Important

Consider a hypothetical study to determine the effects of regular exercise on academic performance in college students. The researchers originally had no intention to investigate possible influences of the subjects' initial fitness levels. In the course of the experiment, however, an unanticipated pattern surfaced in the results: The subjects whose academic performance improved the most tended to be the least physically fit at baseline. Given that the study was not designed to determine the influence of initial fitness level and that the results were unexpected, should the researchers address them in their discussion section? In this situation, unanticipated results should indeed be discussed if (a) they are at least somewhat relevant to the research issue at hand, (b) they add important knowledge to the research field, and (c) they are likely to be of considerable interest to the audience, the scientific community, and perhaps society at large. Consider these three guidelines, as well as space restrictions, to determine whether and how extensively to discuss unanticipated outcomes of your own research.

Rhetorical Goal 12: Relate Your Study's Outcomes to Those from Previous Studies on Your Research Issue

As explained in the presentation of Rhetorical Goal 11 and its strategies, the core argument in the discussion section of an IMRAD-structured paper is composed of the author's overall conclusions and the supporting results from the reported study. But consider that, on its own, experienced readers will not accept this essential argument as the final resolution to the paper's motivating research issue. Instead, they will reserve judgments until learning how the author's conclusions and the study's supporting results relate to those from previous studies on the issue. Experienced readers know that the outcomes of any one study are usually relatively small pieces of a much larger scientific puzzle. Discussion sections thus call for authors to apply synthesis, the puzzle-building critical thinking skill that was demonstrated back in Chapter 3. Rhetorical Goal 12 is for synthesizing research outcomes.

Figure 8.24 presents a good example of how to apply the goal and its strategies. The passage comes from a published research paper authored by Richard J. Bloomer and coworkers at the University of Memphis (Bloomer et al., 2007). They conducted a study to determine the influence of two separate treatments—a prior bout of resistance exercise and antioxidant vitamin supplementation—on biochemical markers of oxidative stress and muscle injury in subjects who performed a strenuous weightlifting task. Oxidative stress is a potentially harmful consequence of the heightened metabolic processes that occur during intense exercise. Bloomer et al.'s results indicated no significant effects of either treatment on the two dependent variables of interest. These outcomes are similar to those reported in several previously published research papers. However, a number of other published studies have revealed that prior exercise and antioxidant vitamin supplementation significantly affect biomarkers of oxidative stress and muscle injury in exercising subjects. Take a moment, before reading about the strategies for accomplishing Rhetorical Goal 12, to look over Bloomer et al.'s synthesis of the research outcomes.

(1) Oxidative stress biomarkers in the present study were relatively unaffected by exercise or treatment. (2) While several studies have reported increased oxidative stress following pure eccentric and mixed eccentric/concentric resistance exercise as reviewed previously [3], the majority have used untrained individuals as research participants. (3) We have recently reported a minor increase [25], as well as no change [24] in both protein carbonyls and malondialdehyde using highly resistance trained subjects. (4) In agreement with our recent findings, Ramel et al. [26] reported a lower response in lipid peroxidation products (e.g. conjugated dienes) in trained versus untrained subjects following resistance exercise. (5) Additionally, McAnulty et al. [27] reported no change in F_2-isoprostanes or plasma antioxidant potential in strength trained subjects following a two-hour resistance training workout. (6) Taken together, it appears that resistance trained subjects experience minimal change in blood oxidative stress biomarkers (at least as measured by lipid and protein oxidation) following strenuous resistance exercise, during the 48 hour recover period. (7) This may be due to heightened antioxidant defense mechanisms as a result of chronic resistance exercise exposure as reviewed previously [3]…

FIGURE 8.24 An effective synthesis of study outcomes, demonstrating Rhetorical Goal 12. From a published research paper authored by Bloomer et al. (2007). Reproduced with permission.

Strategy 12.1: Synthesize the Outcomes of Previous Studies That Support Your Results and Conclusions

When your study results and conclusions agree with those of related studies in the published literature, synthesis is simply a matter of comparing the matching outcomes. This strategy affords the opportunity to reinforce arguments for your conclusions by supporting them with evidence and reasoning from previous studies on your research issue. Several key sentences in Figure 8.24 illustrate the approach. In the first sentence, the authors appropriately set up their synthesis by summarizing their study's main finding of nonsignificant effects of prior exercise and antioxidant supplementation on biomarkers of oxidative stress in exercising subjects. In sentence 3, Bloomer et al. refer to two of their own previous studies, citations 24 and 25, in which they found little or no effect of prior resistance exercise on two biomarkers (protein carbonyls and malondialdehyde). Sentences 4 and 5 also briefly review studies that support Bloomer et al.'s findings.

Notice that Bloomer et al.'s synthesis is based on a qualitative comparison of study results. When space allows, a quantitative approach to synthesis is ideal. Suppose that, in a discussion section of an original research report, you back your main conclusion with data showing a 30% increase in some key variable. If you are claiming that a previously published study supports your conclusion, your argument will be strengthened if you present the actual percentage increase that the researchers observed. Of course, the increase should be close enough to 30% to convince readers to accept your claim. The point here is that *details* about study outcomes enhance synthesis.

Strategy 12.2: Synthesize the Outcomes of Previous Studies That Contrast Your Results and Conclusions

When your study's outcomes disagree with those of related studies in the literature, it's essential to present the contrasting conclusions and data. In response, experienced readers will expect you to explain the inconsistencies. As discussed in Chapter 3, there are two main approaches to synthesizing studies with outcomes that contrast yours. The first is to argue that the contrasting conclusions are artifacts of their authors' misinterpretations of data. In this case, you must have scientifically and statistically sound reasons for interpreting a study's results differently than the author did.

The second approach, which is more common, is to synthesize contrasting research outcomes by explaining key differences in the methods across the studies. As demonstrated in Figure 8.24, this is the approach that Bloomer et al. took to explain why, in contrast to their findings, some previous studies have revealed significant effects of strenuous prior exercise on biomarkers of oxidative stress. The authors' explanation begins in sentence 2, which identifies an important methodological difference across the studies: Some researchers used *untrained* subjects whereas others used *trained* subjects such as experienced weightlifters. Bloomer et al.'s point is that in previous studies on untrained subjects, prior strenuous exercise was associated with increased oxidative stress. In contrast, studies on trained subjects, including Bloomer et al.'s study, have tended to reveal minimal or no effects of prior strenuous exercise. In sentences 6 and 7, Bloomer et al. explain that resistance trained individuals may be protected against oxidative stress because they have developed "antioxidant defense mechanisms as a result of chronic resistance exercise exposure." The explanation, based on a viable methodological difference across the related studies, nicely synthesizes the contrasting outcomes.

Strategy 12.3: Argue for the Greater Validity of Your Results and Conclusions (When Such an Argument Is Warranted)

This strategy applies when you have sound reasons to argue that your conclusions are actually stronger than contrasting conclusions from related studies. Maybe your methods for screening subjects and assigning them to groups were superior. Or perhaps previous researchers in your field failed to account for extraneous variables that your study controlled effectively. In this case, you should state your claim—that your conclusion is more valid—and then support it by comparing specific strengths of your methods to problematic counterparts in previous studies. Strategy 11.4, as presented on page 423, can be easily adapted to serve this purpose.

Rhetorical Goal 13: Discuss the Mechanisms That Underlie Your Study's Main Results, and Argue for the Most Plausible Underlying Mechanisms (When Such an Argument Is Warranted)

"What's the mechanism?" Engaged readers naturally and appropriately raise this question in response to intriguing findings that are presented in the results sections of IMRAD-structured papers. It's an inquiry into the factors that caused a study's results to turn out as they did. The ideal place to answer questions about underlying mechanisms, applying the rhetorical goal at hand, is the discussion section.

Figure 8.25 presents an excellent model of content that targets the goal. The figure reproduces the third paragraph of the discussion section of a research paper on the effects of exercise and hormone replacement therapy (HRT) on carotid arterial compliance (Moreau et al., 2003). You may recall reading the discussion section's first two paragraphs in Chapter 2, where I presented parts of the paper to demonstrate the activity of deriving rhetorical goals from well-written published articles. The authors, Dr. Kerrie Moreau and colleagues, found that aerobic exercise and HRT were associated with increased compliance, or elasticity, of the carotid arteries in their postmenopausal subjects. This is a favorable outcome because compliant arteries reduce the risk of high blood pressure and stroke. What are the underlying mechanisms by which exercise and HRT increase the compliance of the carotid arteries? Moreau et al. answer this question as follows.

(1) The mechanisms by which HRT and regular aerobic exercise increase carotid arterial compliance have not been established. (2) Arterial compliance can be altered over a short time period by changes in the contractile state of the vascular smooth muscle cells [20,21]. (3) In this context, both HRT and habitual exercise increase nitric oxide bioavailability in the vascular endothelium of conduit arteries [22,23], which would, in turn, reduce expression and release of the key endothelium-derived constricting factor, ET-1 [24,25]. (4) An increase in the NO:ET-1 ratio would act to increase the tonic state of relaxation of vascular smooth muscle cells in the large elastic arteries, thereby increasing arterial compliance. (5) Additionally, HRT either via direct central nervous system actions or via peripheral vasodilation and alterations in baroreflex stimulation, could lower peripheral sympathetic nerve activity. (6) This would reduce tonic α-adrenergic receptor mediated contraction of vascular smooth muscle cells and increase carotid arterial compliance [26–28]. (7) Structural determinates of large artery compliance also may be involved. (8) Both estrogen and regular exercise have

FIGURE 8.25 A demonstration of the rhetorical goal to discuss the mechanisms underlying study results. From a published research paper authored by Moreau et al. (2003). Reproduced with permission from Oxford University Press.

been shown to influence the arterial wall properties by increasing elastin content and inhibiting collagen synthesis [29,30]. **(9)** Although these mechanisms could explain our cross-sectional group differences in carotid arterial compliance, such structural changes likely require a period of years. **(10)** Thus, it is unlikely that the improvements with our 3-month exercise intervention involved these mechanisms, but a reduction in the cross-linking of collagen molecules may have contributed. **(11)** It also is possible that HRT and/or regular exercise may have enhanced the breakdown of advanced glycation end-products in the arterial wall, thereby increasing arterial compliance [31]. **(12)** Future studies should examine the possible mechanisms associated with HRT and habitual exercise-induced improvements in arterial compliance in large elastic arteries within the cardiothoracic region.

FIGURE 8.25 *continued.*

Strategy 13.1: Address All Viable Explanatory Mechanisms

Quite often the results of life science studies can be explained by more than one underlying mechanism. It's essential to address all of the candidates that are viable—that is, the explanatory mechanisms that can be supported with valid scientific knowledge, evidence, and reasoning. By applying Strategy 13.1, authors demonstrate their comprehensive knowledge of the science that defines their research issues. In addition, they establish strong foundations to build convincing arguments for the best causal explanations of study results.

In Figure 8.25, Moreau et al. coherently address four mechanisms that might have accounted for the observed increase in carotid arterial compliance among the postmenopausal subjects who exercised and took HRT. As the authors point out in sentence 2, the short-term mechanisms influence "the contractile state of the vascular smooth muscle cells." Factors that relax smooth muscle contractions naturally reduce arterial stiffness and increase compliance. The first plausible mechanism posited by Moreau et al., introduced in sentence 3, is the increased production and presence of nitric oxide, a molecule that relaxes smooth muscle cells in the linings of blood vessels. The second mechanism (sentences 5 and 6) involves the influence of HRT on the peripheral sympathetic nervous system. The third and fourth mechanisms (sentences 7 through 11) promote structural changes to the arteries, resulting from the direct effects of exercise as well as HRT that contains the hormone estrogen.

Strategy 13.2: In Sufficient Depth, Explain *How* Plausible Mechanisms May Have Influenced Your Study's Results

Explanations of the underlying mechanisms of study results are prompted by readers' *how-does-it-work* questions. Skillful authors address these questions iteratively, narrowing their answers to more and more refined mechanistic levels until reaching the limits of knowledge in their research fields. Consider, for example, Moreau et al.'s point that HRT can increase carotid arterial compliance by lowering peripheral sympathetic nervous system activity (see sentences 5 and 6 in Figure 8.25). In response, experienced readers will ask the following questions (among others).

- How does HRT lower peripheral sympathetic nervous system activity?
- How would a reduction of peripheral sympathetic nervous system activity lead to increased carotid arterial compliance?

Moreau et al. answer the first question in sentence 5, where they explain that HRT can directly influence the central nervous system or the peripheral vasculature, and that it can alter a vascular reflex called the baroreflex. In sentence 6, the authors answer the second question by explaining how the mechanisms summarized in their previous sentence would ultimately increase carotid arterial compliance. Even if you have not studied vascular physiology, the authors' basic strategy in these sentences is obvious: Explain *how* plausible mechanisms may have influenced their study's results.

In published research papers such as Moreau et al.'s, space limitations can unfortunately prohibit authors from developing extensive explanations of the mechanisms that underlie their study results. It's a bit ironic that student projects often afford greater opportunities for taking on the advanced challenges of Strategy 13.2. Especially for students who are undertaking thesis and dissertation projects based on original studies, it's certainly worthwhile to develop deep explanations of underlying mechanisms. The content reflects the writer's advanced scientific knowledge and concern for helping readers deeply understand research outcomes.

Strategy 13.3: Argue for the Mechanisms That *Best* Explain Your Study's Results

Exceeding the challenges of the previous strategy, this one calls for constructing original arguments about underlying mechanisms. The strategy comes into play when the candidate mechanisms that you are proposing have been debated in the scientific literature without sufficient resolution. In addition, the strategy applies when you have strong evidence and reasoning to back the claim that a selected mechanism, or a set of related mechanisms, is more viable than alternatives. Data-driven support might come from your own study, if it was designed to address the roles of specific mechanisms. Or you might rely on data from related studies in the published literature. Your argument should also be supported with well-established knowledge and theories about mechanistic processes. For well-rounded and ultimately convincing arguments, you will need to acknowledge and refute viable alternatives to the mechanisms that you are proposing.

By necessity, some arguments for underlying mechanisms demand concept-driven support. This is the case for Moreau et al., as demonstrated in Figure 8.25. The paragraph's first and last sentences indicate that the mechanisms by which exercise and HRT increase arterial compliance had not been definitively determined before Moreau et al.'s paper was published. Nonetheless, the authors developed a conceptual argument for selected mechanisms that likely accounted for their results. In sentences 7 through 11, Moreau et al. speculate on the most probable causes of increased carotid arterial compliance in their subjects who participated in a 3-month exercise program. The authors rule out changes in the structural properties of artery walls, reasoning that these changes would likely take longer than the study's 3 months. (In sentence 9, it's worth noting that the authors do not cite a source to support their contention that structural changes in artery walls likely take place over years rather than months. Neither do the authors elaborate on the long-term adaptations. Apparently, the peer reviewers of Moreau et al.'s article accepted the authors' contention as common knowledge. Keep in mind, however, that convincing arguments concerning underlying mechanisms generally should be supported with references to sound studies and sufficiently detailed explanations.)

By applying established scientific knowledge and the process of elimination, Moreau et al. narrow in on more probable explanatory mechanisms in sentences 10 and 11. Notice, however, Moreau et al.'s use of conditional language, including phrases such as *could lower*, *may be involved*, and *is possible that*. Under the circumstances—specifically that no previous studies had been conducted to determine the underlying mechanisms for their results—Moreau et al.'s speculative arguments are perfectly appropriate. Experienced readers would not automatically tag them as weak arguments. As demonstrated, the key to developing strong speculative arguments about underlying mechanisms is to support logical hypotheses with established conceptual knowledge and sound scientific reasoning.

Rhetorical Goal 14: Acknowledge Significant Methodological Shortcomings to Your Study, and Explain How They May Have Influenced Its Outcomes

In Chapter 3, where we covered the skill of critically evaluating research methods, I emphasized the point that methodological shortcomings come with the territory of conducting research. It's fair to say that every life science study has at least an imperfection or two in its design and execution. Some procedural limitations are unavoidable due to economic, ethical, and technological factors. So an essential rhetorical goal for discussion sections is to acknowledge and explain methodological shortcomings. The content generated to accomplish this goal boosts the credibility of authors and strengthens arguments for their conclusions by establishing the boundaries within which they are valid. This might seem paradoxical: How can admitting your study's faults have such positive consequences? There really is no paradox if you consider that Rhetorical Goal 14 serves a higher goal in science, which is to advance knowledge through an ongoing process of refining research methods. By discussing methodological limitations to their studies, authors help peer researchers overcome and avoid the barriers in future studies.

For students writing about original research, the rhetorical goal at hand is an especially useful tool for generating advanced content. Science professors greatly value the critical thinking skills that are required to successfully explain how problematic research methods actually influence study outcomes. In addition, explanations of methodological problems are fundamental to developing creative and convincing arguments for their solutions.

Strategy 14.1: Acknowledge Only Those Methodological Shortcomings That May Have Influenced Your Study's Outcomes Significantly

Imagine that, for an exercise physiology course, you and your classmates conducted a study to determine whether maximal oxygen consumption (VO_2max), measured during a high-intensity treadmill-running test, differs in males versus females. We'll say that the results indicated significantly higher values in males. In planning the discussion section of an assigned lab report, you might ponder any number of possible methodological limitations to your study. Suppose, for example, that you did not account for what the subjects ate for breakfast, for whether they consumed caffeine, for how many hours they slept the previous night, for their level of motivation, and for the quality of their running shoes. Could any or all of these factors be significant, or are they simply inconsequential glitches?

There is no reason to discuss a study method as a *limitation* if it was unlikely to have led to invalid results and conclusions. To discover whether a candidate limitation is indeed significant in this sense, run it through two tests. First, reflect on whether there is any sound scientific knowledge to support possible confounding effects. Suppose, for example, that no convincing evidence supports any association between VO_2max and running shoe quality. In this case, there is no reason to bring up the lack of control for subjects' running shoes as a methodological limitation.

The second test involves determining whether the potentially confounding method was adequately controlled. Suppose that researchers have confirmed that caffeine consumption increases VO_2max. In the discussion section of your hypothetical lab report, you should discuss the lack of control for caffeine intake if you suspect or know that (a) some of your subjects consumed it prior to the exercise test and (b) the amount consumed might have differed across the subject groups. The latter condition indicates that the study did not adequately control for caffeine intake. Let's say that you can safely surmise that, on the morning of the exercise test, the males consumed more caffeine than the females. To discuss this potentially unfair advantage, you would explain that the higher VO_2max values observed in the males might not have been due to any sex-specific advantage. Instead, it might have been a consequence of the greater amount of caffeine they consumed.

The take-home message is to avoid discussing methodological shortcomings unless you have sound scientific reasons to think that they could have compromised your study's results and conclusions.

Strategy 14.2: Explain *How* Significant Methodological Shortcomings Might Have Influenced Your Study's Outcomes

Some authors take a laundry-list approach to acknowledging methodological limitations, noting one after the next without any explanation. This is a major problem if readers cannot figure out how the listed flaws could have actually influenced a study's outcomes. The acknowledgement of some methodological shortcomings may thus require elaboration. Consider our hypothetical lab report on differences in VO$_2$max between males and females. Taking a laundry-list approach to Rhetorical Goal 14, you would simply write the following sentence about potential confounding effects of caffeine consumption: *One limitation to the present study was that caffeine consumption was not adequately controlled.* But what if your audience lacks the background knowledge to understand how the proposed shortcoming could have caused confounding effects? Or what if your audience is a professor who is reading your lab report primarily to judge *your* scientific knowledge and reasoning skills? To develop the appropriate content, you would need to explain the underlying mechanisms by which caffeine increases VO$_2$max. This explanation should be supported with evidence from research. In addition, you would need to clarify the point that "caffeine consumption was not adequately controlled." This would require explaining and justifying the possibility that the males consumed more caffeine than the females.

You might ask, why go to such lengths to explain flaws in your own study? Of course, it would be much easier to briefly list the possible shortcomings and move on. But consider how your readers are likely to respond to a list without elaboration. Uncertain about the influences of the listed limitations on your results and conclusions, readers might jump to flawed interpretations that lead them to dismiss your arguments completely.

Strategy 14.3: Speculate on How Your Study Might Have Turned Out If Its Methodological Shortcomings Had Been Avoided

This strategy complements the preceding one by giving readers another view on the implications of methodological limitations. This view is revealed through speculation about what might have happened if your study had been based on more ideal methods. The pattern of reasoning works this way:

If _____ [some methodological limitation] had been eliminated or effectively controlled in my study, then the results might have shown _____ [some speculative results that differ from the study's actual results], and I might have concluded [some speculative conclusion that differs from the one you reached].

Strategy 14.4: Convince Readers That, Despite Your Study's Methodological Shortcomings, Its Outcomes Are Still Valid

In response to your acknowledgment and explanation of study limitations, readers will justly question the validity of your results and conclusions. Anticipating this criticism, or receiving it directly from peer reviewers, how should you respond? If you truly believe that the implicated methods did not significantly compromise your study's outcomes, you should argue accordingly. This is a common challenge for scientists writing research papers for publication. If they cannot convincingly counter reviewers' concerns about potentially flawed methods, their articles will be rejected. The situation is often different for students, particularly when the methods used in class experiments are less than ideal for practical reasons. In this case, if you have scientifically sound reasons to believe that selected flaws were severe enough to yield invalid results, you should not hesitate to explain why.

Rhetorical Goal 15: Discuss the Practical Implications and Applications of Your Study's Results

As presented earlier in the chapter, a key rhetorical goal for the introduction section to all scientific papers is to argue for the importance of their motivating research issues. The content associated with this introductory goal reflects authors' ideas about the meaningfulness of their studies *before* conducting them. The question of a study's importance comes up again after the researchers have interpreted their results and reached their conclusions. Addressed in the discussion section, the question is, "What are the practical implications and applications of your study's results?" In other words, the question is "So what?" Sound answers come through applying Rhetorical Goal 15 and its strategies.

Strategy 15.1: Discuss the Practical Value of Your Study's Outcomes to Society and Nature

Too often, discussion sections unfortunately overlook the implications and applications of their study outcomes to society and nature. To avoid this problem, you must contemplate the real-world value of your study's results.

- Who will benefit from the knowledge derived through your research?
- What specific advantages might be gained?
- What are the practical applications of your discoveries for the general public and for practitioners who rely on knowledge in your research field?
- If your study focused on problems in nature, what solutions do your results provide?

All implications and applications that are relevant to the concerns of society and nature are worth raising, whether they have world-shattering potential or not. Experienced readers appreciate useful insights large or small.

For an excellent demonstration of Strategy 15.1, we'll turn to the second paragraph of Campisi et al.'s discussion section. Recall that in the section's first paragraph the authors summarized their main results, which indicated that exercise-trained rats experienced large and rapid increases in HSP72 when they were exposed to stressful conditions. We should most certainly ask, *so what*? Beyond the factual knowledge that physical activity causes HSP72 induction in rats, is there any practical value to Campisi et al.'s discovery? The authors' well-developed answer is reproduced in Figure 8.26, on the following page.

One implication of the current data is that the greater increase in HSP72 could protect cells in both the brain and periphery from stress-induced damage. It has been previously reported that physical activity can prevent many of the deleterious consequences of stress (2, 9, 11, 12, 15, 23, 35). A greater and faster induction of HSP72 after stress, therefore, could be an important cellular mechanism for the stress-buffering effects of freewheel running. Indeed, there is evidence in the literature to suggest that elevated HSPs can promote cell survival after stressor exposure in brain, peripheral, and immune tissues. Fink et al. (14) were among the first to demonstrate that HSP72 alone protects neurons in the brain after various neuronal insults, and others have subsequently demonstrated the HSP72 upregulation in brain tissues protects those tissues from subsequent lethal damage (30, 48). High levels of HSP72 in peripheral tissues have also been reported to prevent 1) pancreatic cell injury after a supramaximally stimulating dose of cholecystokinin (3), 2) tissue damage in the liver after heat stress (19), 3) myocardial injury after ischemia/reperfusion (31), and 4) endotoxin-driven apoptosis in rat cells (6). Furthermore, underexpression of HSP72 increases susceptibility to hypoxia and reoxygenation injury (38). The immune system benefits from HSP induction as well. For example, HSP72 has been found to confer protection to human peripheral blood monocytes from bacteria-induced apoptosis (18), protect tumor cells from TNF-mediated monocyte cytotoxicity (21), protect monocytes from their own cytotoxicity (22), and provide thermotolerance for murine T cells (7). Thus it is reasonable to suggest that an increased induction of HSPs after stress would be a beneficial and adaptive response of the physically active organism.

FIGURE 8.26 From Campisi et al.'s research paper, a demonstration of the rhetorical goal to discuss the practical implications and applications of study results. Reproduced with permission from the American Physiological Society.

Campisi et al.'s main message, as directly expressed in the first and last sentences of the model text is that exercise-induced increases in HSP72 might protect various cells and physiological systems throughout the body against deleterious effects of stress. Note that, by intention and design, the researchers did *not* conduct their study to determine the long-term protective effects of exercise-induced HSP72 production against negative outcomes such as cell death and disease. So Campisi et al.'s argument for the positive implications of their results relies on an effective synthesis of previous studies in which protective effects have been documented. The model content in Figure 8.26 reflects the critical and creative thinking that underlie success in accomplishing Rhetorical Goal 15.

Strategy 15.2: Discuss the Practical Value of Your Study's Outcomes to the Scientific Community

To develop ideas for this strategy, you must think about how scientists who specialize on your research issue might apply your study's findings to advance knowledge in the field. In addition, the strategy may involve discussing how your study's outcomes might be applied to resolve related issues in peripheral fields. If your brainstorming inspires extensive ideas for future research, Strategy 15.2 will blossom into the following rhetorical goal.

Rhetorical Goal 16: Propose Future Studies on Your Research Issue

Positioned at or close to the end of discussion sections, proposals for future studies are prompted by several situations. For example, let's say that your study turned out perfectly as planned, yielding crystal clear results and definitive conclusions. Along the way, however, interesting new questions and hypotheses came to your mind serendipitously, inspiring ideas for new studies. On the other end of the spectrum, suppose that you ran into unanticipated methodological flaws that significantly compromised efforts to resolve your study's motivating issue. It's another situation for suggesting future research—specifically, research that overcomes the original study's flawed methods. Or maybe your study was just one planned step toward fulfilling a more comprehensive research agenda. In this case, you would write about the next logical steps.

Considering the important roles that Rhetorical Goal 16 serves, you might expect lengthy proposals for new studies in the discussion sections of published research papers. Quite often, however, scientists say little more than "Further research is necessary." To be honest, I have to say that when this perfunctory sentence ends a published research paper without any elaboration, I am reminded of how school children routinely complete their papers by writing "The End." But to be fair, we should recognize that many career scientists have a very good reason for curbing their suggestions for future research in published research papers. They are saving their most innovative ideas for new grant applications. In hopes of obtaining funding to carry out groundbreaking studies, scientists don't want to be scooped by their peers. This is usually not a concern, however, for students. In student papers, extensive proposals for new studies are worthwhile because they afford opportunities to demonstrate highly valued creative thinking and problem-solving skills.

Strategy 16.1: Explain the Problems That Underlie Your Suggestions for Future Research

Rhetorical Goal 16 essentially entails proposing and arguing for solutions to research problems. A logical way to introduce these solutions is to explain their underlying problems, which are the factors that inspire suggestions for new studies. Use the following questions to guide your identification and explanation of such problems.

- What interesting new insights, questions, and hypotheses surfaced in the course of conducting your study, warranting future investigation?
- What knowledge gaps still need to be filled in your research field?
- Which debates in your research field will remain unsettled until they are resolved through future studies?

- What are the next logical steps to take in your research field?
- In your study, as well as related studies in the published literature, which methodological limitations must be overcome through new studies?

Well-developed answers to these questions establish the conceptual rationale for future research. In addition, they are essential for building convincing arguments for new studies.

Strategy 16.2: Present Your Hypotheses for Proposed New Studies

This strategy overlaps with ones covered earlier for presenting hypotheses and their rationale in introduction sections (see Strategies 3.1–3.4). For instance, in the introduction section to an IMRAD-structured paper or a research proposal, recall that a hypothesis should be accompanied by supporting scientific knowledge, evidence, and reasoning (Strategy 3.3, page 392). In the discussion section of IMRAD-structured papers, the same approach applies to presenting hypotheses for future studies. In addition, as is true in introduction sections, hypotheses presented for future research in discussion sections should not be *slam dunks*. In other words, to convince readers that your proposed new studies are truly worth conducting, you will need to demonstrate that while your hypotheses are sound, they are not foregone conclusions (Strategy 3.4, page 392).

Strategy 16.3: Describe and Justify the Key Methods for Your Proposed Future Studies

This strategy is for presenting details about the procedures and analyses for conducting your proposed new studies. Given the numerous rhetorical goals and strategies to accomplish in discussion sections, you won't have space to describe and justify every methodological detail. So the key is to highlight those that are especially relevant to solving the major motivating problems. Let's say that you are suggesting a new study primarily to address the problem that previous investigations on your research issue have not included subjects who represent an important segment of society. After explaining the problem (Strategy 16.1), you should describe the specific characteristics of the representative subjects whom you propose to study. If your readers are likely to question *how* your described methods afford solutions to key motivating problems, you will need to develop appropriate justifications. This pattern of describing and justifying study procedures is essentially the same one that we discussed earlier for generating content in the methods sections of research proposals (see Rhetorical Goals 7 and 8, which are presented beginning on page 399).

Strategy 16.4: Speculate on Potential Beneficial Outcomes of Your Proposed Research

At its best, Rhetorical Goal 16 is a tool for constructing arguments to convince readers that new studies truly need to be conducted to resolve debatable issues, test innovative hypotheses, and fill vital knowledge gaps. To complement the elements of argument in the previous three strategies, this one is for convincing readers about the potential for positive outcomes of the future research that you are suggesting. It's a matter of speculating on how the knowledge to be gained from your proposed new study (or studies) will benefit nature, society, and the scientific community.

RHETORICAL GOALS FOR THE BODY OF REVIEW PAPERS

As introduced in Chapter 1, review papers comprehensively summarize and synthesize research-derived knowledge on focused topics and issues. Recall that there are several different types of review papers:

- Informative reviews mainly summarize research findings.
- Conceptual reviews mainly synthesize research findings to support original concepts and theories.
- Critical reviews mainly address the validity of scientific knowledge by evaluating strengths and weaknesses in research methods and arguments in the published literature.
- Position papers mainly advance original arguments to resolve debatable and controversial research issues.

In each definition the word *mainly* is telling because most well-written review papers blend elements of all four types of documents. For instance, to develop strong arguments in position papers, authors must provide straightforward summaries of research (as in an informative review), establish and support original concepts (as in a conceptual review), and critique the relevant research literature (as in a critical review).

Most published review papers are structured by three major sections: an introduction, a body, and a conclusion. In all of the document types, the introduction section is based on a common set of rhetorical goals, which we covered earlier. The goals involve presenting research issues (Rhetorical Goal 1), convincing readers that research issues are important (Rhetorical Goal 2), introducing the novel and unique features of research and writing projects (Rhetorical Goal 4), and presenting the specific purposes of papers (Rhetorical Goal 5). In argumentative review papers, specifically conceptual reviews and position papers, introduction sections also present overall claims (Rhetorical Goal 6). Conventional rhetorical goals and strategies for the body section of review papers are outlined in Figure 8.27 (page 438) and explained as follows.

Rhetorical Goal 17: Provide essential background knowledge about the studies, critical evaluations, and arguments that are central to your review paper. (439)

Strategies

17.1: Define and explain essential terms and concepts underlying your review paper's topic or research issue. (439)

17.2: Describe the primary methods used to conduct studies on your topic or research issue. (439)

17.3: Highlight the major historical events that have shaped your topic or research issue. (440)

Rhetorical Goal 18: Summarize the published studies on your topic or research issue. (440)

Strategies

18.1: Briefly introduce the research issues that inspired the published studies you are reviewing. (442)

18.2: Summarize the key methods of the published studies you are reviewing. (442)

18.3: Summarize the most relevant results of the published studies you are reviewing. (442)

18.4: Summarize the main conclusions of the published studies you are reviewing. (442)

Rhetorical Goal 19: Synthesize the published studies on your topic or research issue. (443)

Rhetorical Goal 20: Explain and argue for the mechanisms underlying the results of the published studies you are reviewing. (444)

Rhetorical Goal 21: Convince readers to accept your original arguments. (444)

Strategies

21.1: Back your claims with data-driven lines of support. (445)

21.2: Argue for the methodological strengths of studies that support your claims. (446)

21.3: Back your claims with concept-driven lines of support. (447)

21.4: Acknowledge and refute viable counterarguments (to your argument). (447)

21.5: Acknowledge and respond to any significant limitations to your argument. (448)

FIGURE 8.27 Rhetorical goals and strategies for the body section of review papers. The pages on which the goals and strategies are introduced are shown in parentheses.

Rhetorical Goal 17: Provide Essential Background Knowledge about the Studies, Critical Evaluations, and Arguments That Are Central to Your Review Paper

In all types of scientific papers, the introduction section is a logical place to present brief background knowledge about topics and research issues. In review papers, however, more extensive presentations of background knowledge are sometimes necessary. This is because review papers take wide-ranging views on their subjects and they are often intended for diverse audiences. When extensive background knowledge is warranted, the content fits in nicely at the beginning of a review paper's body section. Rhetorical Goal 17 serves this purpose.

Strategy 17.1: Define and Explain Essential Terms and Concepts Underlying Your Review Paper's Topic or Research Issue

This strategy applies when the background information presented in a review paper's introduction section is not sufficient to meet readers' knowledge needs. Take the case of a critical review paper that a student named Libby is writing. Her research issue concerns the effects of long-distance running on the risk of developing osteoarthritis, the joint disease characterized by deterioration of protective cartilage tissue. Whether running causes osteoarthritis is debatable, as evidenced by contradictory results from published studies on this issue. Libby is targeting her paper to primary care physicians. She reasons that her readers will know a lot about osteoarthritis, but they might not be up to date on the biomechanical and physiological mechanisms by which long-distance running could potentially influence the risk of developing the joint disease. Throughout the body of her paper, Libby reviews numerous studies that focused on elucidating these complex mechanisms. To give her readers the necessary background knowledge for understanding the studies' methods and outcomes, Libby begins the body of her paper with a well-developed subsection devoted to achieving Strategy 17.1. The content includes fairly extensive descriptions of how the stresses of running affect joint anatomy and the composition of cartilage tissue. To complement the text, Libby uses graphics—including line drawings and photographs of the leg joints and cartilage—that she obtained with permission from reputable Web sites.

This strategy for presenting background knowledge generates content that reads like a science textbook. In fact, reputable textbooks are usually appropriate sources of background information for the body of review papers.

Strategy 17.2: Describe the Primary Methods Used to Conduct Studies on Your Topic or Research Issue

The core content in the body of most review papers summarizes study results and conclusions. To understand these summaries, the audience must clearly grasp the original studies' methods. Authors of review papers can provide the key methodological details on a study-by-study basis—that is, integrated throughout the body section. Or the methods can be described in a concentrated subsection early in the body. The latter strategy is fitting for describing fundamental methods that are used in many, or all, of the published studies on a focused topic or issue. For example, in numerous studies on long-distance running and osteoarthritis, researchers take x-rays of subjects' leg joints and then measure a common set of dependent variables. They include (a) the width of the joint space, which narrows when cartilage deteriorates in osteoarthritis; (b) the formation of osteophytes, or bone spurs; and (c) a condition called sclerosis, which is the hardening and thickening of cartilage tissue. Following her subsection devoted to Strategy 17.1, Libby included a subsection in which she described the general procedures for collecting and analyzing data for these variables. In doing so, Libby provided key background information to help her readers understand the outcomes of studies that she summarized later in her paper's body.

Strategy 17.3: Highlight the Major Historical Events That Have Shaped Your Topic or Research Issue

The current state of knowledge in all scientific fields has been shaped by noteworthy historical events. Chronicles of these events provide useful background information to help readers better understand the topics and research issues on which review papers are based. This strategy may include summarizing landmark studies, accounting for technological advances in a research field, and profiling the contributions of leading scientists.

Rhetorical Goal 18: Summarize the Published Studies on Your Topic or Research Issue

To serve the informative roles of review papers, authors need a rhetorical goal that generates straightforward summaries of the motives, methods, and outcomes of published studies. This goal is effectively demonstrated in Figure 8.28, which is an excerpt from Libby's critical review paper. Libby organized the main subsections of the paper's body according to the different approaches that researchers have taken to designing and executing studies on long-distance running and osteoarthritis. Among other subsections, one reviewed experimental studies on animals, while another focused on studies in which mechanical models of the joints were subjected to the forces of running. Figure 8.28 presents the first two paragraphs of a subsection that Libby devoted to reviewing observational studies conducted on humans.

Observational Studies on Human Subjects

(1) Due to a lack of long-term experimental studies on human subjects, conclusions about the effects of long-distance running on the risk for osteoarthritis must be drawn from observational studies with cross-sectional and longitudinal designs. (2) In a cross-sectional study,

FIGURE 8.28 An excerpt from a student's critical review paper on long-distance running and osteoarthritis. The content demonstrates Rhetorical Goal 18, which involves summarizing published studies on review paper topics and issues.

Lane et al. [14] aimed to determine whether runners exhibit indicators of osteoarthritis in the knee joints. **(3)** To introduce their study, Lane et al. explained the common view that the excessive forces absorbed by the joints during running might cause the cartilage damage that occurs in osteoarthritis. **(4)** The authors also recognized the opposing views that the effects of running might be negligible or even preventive, because the activity strengthens the bones and muscles, which can protect the joint against excessive forces. **(5)** Lane et al. recruited 41 older runners (n = 18 women and 23 men; mean age = 57.5 years) and 41 non-running controls (n = 18 women and 23 men; mean age = 57.7 years). **(6)** All subjects filled out questionnaires to report their running history, diet, and general health status. **(7)** On average, the runners reported covering 12,547 miles over a period of 9.2 years. **(8)** X-ray exams of the knee joints indicated no significant differences between the runners and non-runners for measures of joint space width, osteophytes, and sclerosis. **(9)** Therefore, Lane et al. concluded that runners are not at risk for clinical osteoarthritis.

(10) In a 15-year longitudinal investigation, Marti et al.[15] studied a group of 27 elite long-distance runners (mean age = 27 years) and 23 non-running controls (mean age = 20 years). **(11)** At baseline, the subjects completed questionnaires to report their running and medical histories. **(12)** Fifteen years later, they completed the same questionnaires and underwent x-ray exams of the hips. **(13)** The runners reported covering an average of 60.63 miles per week over the study period. **(14)** To compare the incidence of osteoarthritis markers in their two groups, Marti et al. used established criteria for grading the severity of joint space narrowing, osteophytes, and sclerosis.[12] **(15)** The researchers conducted non-parametric statistical tests to compare the number of runners versus sedentary control subjects who had moderately high to excessive grades. **(16)** For each outcome, significantly more runners had scores that indicated clinical signs of osteoarthritis. **(17)** The results were as follows: joint space narrowing (4 runners versus 0 controls); osteophytes (4 runners versus 0 controls); and sclerosis (16 runners versus 3 controls). **(18)** Based on these findings, Marti et al. concluded that long-distance runners are at increased risk for developing osteoarthritis of the hips.

FIGURE 8.28 *continued.*

Strategy 18.1: Briefly Introduce the Research Issues That Inspired the Published Studies You Are Reviewing

A logical and effective way to begin summarizing published studies is to introduce their motivating research issues. This includes summarizing the questions, problems, hypotheses, and purposes that underlie the studies. This strategy is demonstrated in sentences 1–4 of the excerpt from Libby's review paper shown in Figure 8.28.

Strategy 18.2: Summarize the Key Methods of the Published Studies You Are Reviewing

The same rationale and strategies that we discussed earlier for describing study methods in IMRAD-structured papers generally apply to review papers. However, because a review paper summarizes numerous studies, descriptions of their methods must be condensed. In Libby's review of the study by Lane et al., for example, the key procedures and analyses are conveyed in only two sentences. In sentences 5 and 6, we learn about the number of subjects studied, their ages, their sex, and the means by which they provided data for the study's main independent variables.

The objective for Strategy 18.2 is to present the most essential details about how reviewed studies were designed and executed. To isolate these details, take the perspective of audience members who have not already read the published research papers that you are summarizing. What specific information will these readers need about the studies' subjects, general design, independent and dependent variables, and procedures for collecting and analyzing data?

Strategy 18.3: Summarize the Most Relevant Results of the Published Studies You Are Reviewing

As described earlier, the results sections of IMRAD-structured papers convey the overall findings, specific data, and outcomes of statistical tests for individual studies. The same strategies for writing about results in IMRAD-structured papers generally apply to review papers (see Strategies 9.1–9.5, presented beginning on page 413). In review papers, however, presentations of results must be more succinct, focusing on the most relevant data that support the conclusions from original studies. In Figure 8.28, good examples of this strategy are Libby's summaries of the results from Lane et al.'s study (sentences 7 and 8) and Marti et al.'s study (sentences 13, 16, and 17). Notice that sentence 17 presents the actual data that Marti et al. used to support their conclusions. When space allows, such quantitative summaries of results are extremely useful for readers. It can also be helpful to include the outcomes of statistical tests for key results.

Strategy 18.4: Summarize the Main Conclusions of the Published Studies You Are Reviewing

When your overall goal is to *inform* readers about the outcomes of published studies, your task is simply to *tell it like it is*. This requires paraphrasing the conclusions that authors presented in the discussion sections of their original research reports. The final sentences of Libby's two paragraphs in Figure 8.28 reflect this straightforward strategy. Elsewhere in her paper, as you will soon see, Libby takes on the greater challenges of synthesis and argument, which involve presenting and supporting her independent interpretations of published study results.

Rhetorical Goal 19: Synthesize the Published Studies on Your Topic or Research Issue

Review papers that do nothing more than summarize study methods, results, and conclusions usually fall far short of their potential. This is especially true for papers addressing research issues that are unresolved and debatable. To go beyond simple summaries, authors of review papers must apply the skills of synthesizing research outcomes. We have already covered a rhetorical goal that can be easily adapted to serve this purpose. As presented earlier for the discussion section of IMRAD-structured papers, it's the goal to relate your results and conclusions to those from previous studies on your research issue (Rhetorical Goal 12, page 424). Recall that the strategies for this goal entail comparing studies with similar, supporting outcomes (Strategy 12.1); presenting and explaining contrasting study outcomes (Strategy 12.2); and arguing for the most valid results and conclusions from the contrasting studies (Strategy 12.3). In applications of these strategies to review papers, the only difference is that authors usually do not have their own study outcomes to include in their syntheses.

An excellent synthesis of research is presented in Figure 8.29, which is an excerpt from a published review paper on exercise and osteoarthritis (Urquhart et al., 2008). Authored by Dr. Donna Urquhart and coworkers at Monash University in Australia, the paper addresses numerous factors that explain why physical activity may pose risks for developing the joint disease in some individuals. One of the factors involves the interaction between obesity, as measured by body mass index (BMI), and physical activity. In the first two sentences of Figure 8.29, Urquhart et al. explain the hypothesis that overweight individuals (whose BMI measures

---

Body mass index

Although obesity is a major risk factor for development of knee osteoarthritis (OA) [21], it is unclear whether overweight individuals who exercise may further increase their risk for joint damage. It is possible that excess mass may, in the presence of activity, impart axial loads that stress joint structures beyond their physiological capabilities and cause accelerated joint degeneration. Several studies of physical activity and OA have investigated high BMI as a potential effect modifier [10,11,14,15]. A 12-year longitudinal study conducted by Hootman and colleagues [10] demonstrated that BMI did not modify the relationship between moderate physical activity and risk for self-reported, physician-diagnosed

FIGURE 8.29 An excellent model of content that synthesizes study outcomes (Rhetorical Goal 19). From a published review paper authored by Urquhart et al. (2008). Reproduced with permission.

knee OA. Similarly, Felson and coworkers [11] also identified no increased risk for radiographic knee OA in middle-aged and elderly persons who had a higher BMI and participated in recreational exercise. However, McAlindon and colleagues [14] found that obese, elderly individuals who participated in heavy physical activity (including lifting objects >5 lb [> 2.27 kg], gardening with heavy tools, brisk cycling and other strenuous sports) were at greater risk for knee OA than those individuals in the lower tertile of BMI. Similarly, increased risk for knee OA was reported in former elite-level athletes with a higher BMI (at age 20 and 30 years) [15,22]. These findings suggest that individuals with a higher body mass may participate in moderate/recreational activity without increased risk for knee OA, but involvement in heavy physical activity increases their risk. Further longitudinal investigation that accurately examines the level of physical activity in obese individuals is required to confirm these preliminary results.

FIGURE 8.29 *continued.*

are high) might be at particularly high risk for developing osteoarthritis if they exercise regularly. However, as the authors' summaries of relevant studies indicate, the evidence for this hypothesis is contradictory. In their synthesis of six separate studies, cited in the model paragraph, Urquhart et al. conclude that the contrasting findings may be attributable to differences in the intensity at which subjects in the various studies performed exercise and daily physical activity.

Rhetorical Goal 20: Explain and Argue for the Mechanisms Underlying the Results of the Published Studies You Are Reviewing

As presented earlier, one of the rhetorical goals for IMRAD-structured papers involves discussing and arguing for the plausible mechanisms that underlie an original study's results. This same goal, in a slightly adapted form, applies to the body of review papers. The obvious difference is that in review papers, discussions and arguments involving underlying mechanisms apply to the results from numerous studies. One approach to the rhetorical goal at hand is to integrate short explanations of plausible mechanisms on a study-by-study basis throughout the body of a review paper. Another approach is to devote entire subsections to more elaborate explanations, syntheses, and arguments. The previously presented strategies for writing about mechanisms in IMRAD-structured papers can be adapted for these applications (see Strategies 13.1, 13.2, and 13.3, which are presented beginning on page 428).

Rhetorical Goal 21: Convince Readers to Accept Your Original Arguments

This is the main rhetorical goal for the body of conceptual reviews, critical reviews, and position papers. All of these documents contain elements of scientific argument, including claims and lines of support for them. To back the claims that are central to argumentative review papers, authors present knowledge, evidence (data), and reasoning from established research. Rhetorical Goal 21 applies to generating original arguments in the body of review papers. The goal relies partly on several strategies that we have discussed for summarizing published studies. For example, to introduce the published studies that you are using to back your claims, you should summarize their motivating issues (Strategy 18.1, page 442). Then, to help readers understand how your supporting data were obtained, you will summarize the supporting studies' methods (Strategy 18.2, page 442). Of course, you will also need to provide straightforward synopses of the studies' results (Strategy 18.3, page 442). The following strategies add vital elements of argument for the most powerful applications of Rhetorical Goal 21.

Strategy 21.1: Back Your Claims with Data-Driven Lines of Support

The most convincing support for claims in argumentative review papers comes with the presentation of relevant and significant data from well-conducted studies. This is the same approach that we discussed for supporting conclusions in the discussion section of IMRAD-structured papers (see the strategies for Rhetorical Goal 11, presented beginning on page 420). Figure 8.30 presents a strong data-driven argument from a published conceptual review paper. The model comes from Dr. Brian Williams and colleagues' article (which I introduced earlier) on physical activity behaviors in children and adolescents who have asthma. The main subsections of the paper's body were organized by lines of support for the authors' overall argument for a multifaceted approach to increasing physical activity levels among these youth. In Figure 8.30, Williams et al. present data to support their claim that children and adolescents with asthma are less physically active than age-matched healthy peers.

Evidence for reduced participation in physical activity among children and young people with asthma

Despite the importance of physical activity to children and young people with asthma, and although some studies have failed to demonstrate a significant difference in physical activity levels between asthmatic and non-asthmatic children [27], there is growing evidence that they are in fact *less* active than their non-asthmatic peers. Lack of participation in physical activity by children and young people with asthma is unsurprising given that physical activity levels are known to be falling among young people in most industrialised nations [28, 29]. These rates fall even

FIGURE 8.30 From Williams et al.'s conceptual review paper, an effective argument supported by data and conceptual reasoning (Rhetorical Goal 21). Reproduced with permission.

further as children reach adolescence [16], particularly in teenage girls. At secondary school age, boys report more physical activity at 11, 13 and 15 than girls and are more likely to meet the guideline of 60 minutes moderate activity per day [30].

Although the majority of evidence points to reduced physical activity among young people with asthma, conceptual and methodological problems, particularly in relation to confounding, have resulted in contradictory evidence [31]. These contradictions arise partly as a result of methodological and disease detection/diagnostic issues. Engagement in significant levels of physical activity can lead to improved detection, particularly exercise induced asthma (EIA). Consequently, activity rates in children and young people with asthma may appear artificially high because already-active children and young people are more likely to be diagnosed as having asthma.

Notwithstanding this methodological problem, studies in a range of countries suggest that children and young people with asthma *are* less active than their healthy peers and that they attribute their limited participation to their asthma. In a comparative study of 112 children with and without asthma Glazebrook et al found that children with asthma reported fewer physical activities than the non-asthma group (median 4 per day vs 6 per day). Furthermore, asthma was the strongest predictor of lower activity scores, followed by younger age [32]. In addition, a study of 137 United States (US) children aged 6–12 years found that those with asthma were less active than their peers in all classifications of activity level used [33]. In Germany, a study of 254 teachers in 46 schools across a range of ages found that only 60% of children with asthma took part in physical education lessons on the same basis as their healthy peers; the remaining 40% did not participate, or only participated sometimes or to a limited extent [34]. An Australian survey also found reduced activity and a belief that asthma was responsible. Parents reported that as a result of asthma 31% did not participate in sport; 21% did not ride a bicycle; 20% did not swim; and 18% did not take part in break time play at school [35]. Young people with asthma also believe that their physical activity is reduced as a result. A survey of 123,000 US young people aged 12–14 found that 52% of those with a diagnosis of asthma reported that their activities were curtailed because of their asthma [4].

FIGURE 8.30 *continued.*

Demonstrating Strategy 21.1, the excerpt's third paragraph contains relevant, effectively synthesized, and convincing research-derived support for the authors' claim that children and adolescents with asthma are less physically active than their healthy peers. The authors' presentation of the supporting studies' quantitative data is an especially strong feature of the example.

Strategy 21.2: Argue for the Methodological Strengths of Studies That Support Your Claims

This strategy is especially useful for review papers on debatable issues in which supporting data exist for contrasting claims. To convince readers that the data supporting your claims are strongest, you will need to justify the methodological strengths of the studies from which the data were derived. This entails arguing for the greater relevance, validity, and precision of the methods compared to those used in studies supporting alternative claims. Strategies 8.2 and 8.3, presented earlier for the methods sections of IMRAD-structured papers and research proposals, can be adapted for this purpose.

Strategy 21.3: Back Your Claims with Concept-Driven Lines of Support

As explained in Chapter 3, concept-driven lines of support for scientific arguments are based on established knowledge and theories rather than on numerical data. Consider the example of a position paper intended to convince readers that long-distance running increases the risk for developing osteoarthritis. In concept-driven lines of support, the author would explain the biomechanical and physiological mechanisms by which running might damage cartilage and thereby promote the disease.

Strategy 21.4: Acknowledge and Refute Viable Counterarguments (to Your Argument)

To build well-rounded and unbiased arguments, experienced scientific authors acknowledge viable alternatives, or counterarguments. This strategy begins with conveying the claims on which counterarguments are based and presenting their supporting evidence and reasoning. The resulting content communicates an important message to proponents of arguments that oppose yours: It's a way of saying *I know where you're coming from*. To take your acknowledgment a helpful step farther, you might even address the strongest points of the counterarguments. By doing so, you will convince your opponents that you are not biased. If staunch supporters of counterarguments question whether you fully comprehend and appreciate their views, they may dismiss your argument from the outset. Of course, after acknowledging viable counterarguments, you face the major challenge of refuting them, which requires arguing for their significant flaws. The most productive approaches to refuting counterarguments include (a) pointing out methodological limitations of studies that support them, (b) explaining common misinterpretations of their supporting data, and (c) revealing logical fallacies in the evidence and reasoning used to back them.

An example of Strategy 21.4 in action is presented as follows in Figure 8.31.

Contrary to my argument, the results of many cross-sectional studies support the conclusion that long-distance running does not increase the risk of developing osteoarthritis.[2, 6, 8, 9, 10, 14, 17] In general, these studies effectively control for confounding factors that are associated with osteoarthritis, such as age, body weight,

FIGURE 8.31 An effective demonstration of Strategy 21.4, which involves supporting original arguments in review papers by acknowledging and refuting counterarguments.

and previous joint injuries. However, the study findings and conclusions are questionable due to problematic methods for recruiting subjects. In a typical cross-sectional study on running and osteoarthritis, researchers recruit volunteers from local running clubs. The subjects are usually middle-age or older individuals who have been running on a regular basis for many years. After filling out questionnaires to report their running history, the subjects undergo x-ray exams of their leg joints. The typical finding is that the severity of osteoarthritis markers in the runners is no greater than in age-matched sedentary controls.

The common result of cross-sectional studies would be convincing if the subjects accurately represented the entire population of runners who have been involved in the activity for many years. However, the cross-sectional design does not account for the possibility that the older runners represent a unique subset of the population who can run long distances without experiencing cartilage deterioration. These runners may have genetic advantages that prevent osteoarthritis. As explained by Johnson and Cohen,[7] this phenomenon is called the "healthy runner effect." Cross-sectional studies may eliminate a large subset of the population of runners who cover long distances at a relatively young age, develop osteoarthritis, and then stop running to avoid pain and further cartilage deterioration. If this subset of the population were accounted for, the results of cross-sectional studies might conceivably indicate a greater severity of osteoarthritis markers in the runners versus the non-runners.

FIGURE 8.31 *continued.*

The passage comes from Libby's review paper, after she presents an original claim that long-distance running increases the risk of developing osteoarthritis. To acknowledge and refute an opposing argument, which is that runners are at no greater risk than normally active people, Libby explains a methodological shortcoming in the counterargument's supporting studies.

Strategy 21.5: Acknowledge and Respond to Any Significant Limitations to Your Argument
Scientific audiences generally do not expect all argumentative review papers to *prove* their claims definitively and to resolve their motivating issues completely. This is

especially true for papers based on highly debatable issues that are unresolved due to a lack of research or methodological challenges. Experienced readers usually expect such papers to contribute to resolving their underlying issues but still have limitations to their arguments. Strategy 21.5 is for admitting these limitations, explaining their influences, and demonstrating that they do not severely undermine the arguments at hand. In its intentions and applications, the present strategy is similar to strategies for discussing limitations in the discussion section of IMRAD-structured papers (see the strategies for Rhetorical Goal 14, beginning on page 430).

RHETORICAL GOALS FOR THE CONCLUSION SECTION OF REVIEW PAPERS

The third and final major section of review papers is devoted to summarizing the main messages of their body sections and to proposing ideas for future research (Figure 8.32). In published review papers, these ending sections commonly begin with one of the following headings: *Conclusions, Summary and Conclusions,*

Rhetorical Goal 22: Briefly reiterate the key information, ideas, and arguments that were central to the body of your review paper.

Strategies

22.1: Refocus readers' attention on your review paper's central topic or research issue.

22.2: Summarize the overall conclusions from the published studies that you have reviewed.

22.3: Summarize your main arguments.

Rhetorical Goal 23: Suggest future directions and new studies concerning your paper's topic or research issue.

FIGURE 8.32 Rhetorical goals and strategies for the conclusion section of review papers.

or *Conclusions and Future Directions.* Conclusion sections tend to be relatively short, typically not exceeding two or three paragraphs.

Rhetorical Goal 22: Briefly Reiterate the Key Information, Ideas, and Arguments That Were Central to the Body of Your Review Paper

A simple question underscores this rhetorical goal: *What are the most important take-home messages that readers should glean from your review paper?* After isolating the main points, use them to generate and organize the content of a summary paragraph or two. No new information, ideas, and arguments should be included.

Conclusions and Future Directions

The overall purpose of this paper was to critically evaluate previous research on the effects of long-distance running on the development of osteoarthritis. Although many studies have been published on this topic, little consensus has been reached regarding the risks of running. As emphasized throughout this paper, inconsistent findings can be attributable to flaws in study design and methodology. In an ideal investigation on this topic, teenage or young adult subjects would be randomly assigned to running and control groups. On a regular basis over many decades, the subjects would undergo x-ray exams to identify changes in their joint anatomy and cartilage structure. Since this experimental approach is both impractical and unethical, researchers must rely on cross-sectional studies and short-term longitudinal studies. In summary, there are several limitations to the approaches taken in these studies:

- In individual studies, the extent of running has not varied much across subjects; therefore, researchers have been unable to determine with certainty whether there is a dose-response relationship between running and osteoarthritis.
- Many observational studies have been limited because subjects have self-reported their participation in running. Self-reported data may not be accurate.
- Cross-sectional studies have failed to account for the "healthy runner" effect, which may bias their results to falsely indicate that running does not increase the risk for developing osteoarthritis.
- Previous studies have failed to document the clinical significance of joint space narrowing, osteophytes, and sclerosis in runners.

Future studies should address these limitations directly. For instance, to determine whether a dose-response relationship exists between running and osteoarthritis, researchers should recruit

FIGURE 8.33 A model conclusion section from a student's review paper on long-distance running and osteoarthritis.

recreational, intermediate-level, and highly competitive runners
who cover a wide range of distances in their exercise and training
programs. The spectrum of running involvement is necessary
to reveal whether increases in the stresses absorbed by the leg
joints, with increased running distances, are correlated with more
severe markers of osteoarthritis. To avoid the problems associated
with self-reported data, future studies should incorporate direct
and objective methods for assessing running involvement, such as
the use of telemetric accelerometers. Ultimately, the most valuable
studies will be designed to relate changes in x-ray markers of osteoarthritis with clinical outcomes such as gait function, muscle
strength, and joint mobility in older individuals who have run long
distances throughout their adult lives.

FIGURE 8.33 *continued.*

This goal is demonstrated in Figure 8.33 (pages 450–451), which is the entire
conclusion section from Libby's critical review paper.

Strategy 22.1: Refocus Readers' Attention on Your Review Paper's Central Topic or Research Issue

This strategy generates content that briefly introduces the summary of key messages from a review paper's body. Libby's approach, as demonstrated in the first
two sentences of her conclusion, is to reorient readers to the central purpose and
problem that motivated her paper.

Strategy 22.2: Summarize the Overall Conclusions from the Published Studies That You Have Reviewed

For each subtopic or specific issue addressed in the body of your paper, identify
the most representative and influential studies that you have reviewed. Then, simply summarize their overall conclusions to generate the essential content for this
strategy.

Strategy 22.3: Summarize Your Main Arguments

This strategy is for briefly reiterating the key elements of argumentative review
papers, including their claims, lines of support, counterarguments and approaches
to refuting them, and acknowledgments of potential limitations. In the first paragraph of Figure 8.33, for example, Libby reiterates her claims about methodological
shortcomings of observational studies on running and osteoarthritis. The bulleted
list effectively highlights the take-home messages of Libby's critical review.

Rhetorical Goal 23: Suggest Future Directions and New Studies on Your Paper's Topic or Research Issue

This rhetorical goal is practically identical to Rhetorical Goal 16, which applies to proposing future research in the discussion section of IMRAD-structured papers. As presented earlier (beginning on page 435), the strategies for Rhetorical Goal 16 can be adapted for achieving the present goal. In the conclusion section of a review paper, this goal entails offering suggestions for new research to overcome specific limitations of previous studies and to fill knowledge gaps that were highlighted in the paper's body. The last paragraph of Libby's conclusion is an excellent example of Rhetorical Goal 23 in action.

SUMMING UP AND STEPPING AHEAD

This chapter began by recalling our analogy of rhetorical goals as power tools for crafting the content of scientific papers. As skilled workers in all fields know very well, generating the highest-quality products demands top-of-the-line tools and considerable experience in wielding them for the greatest advantages. Reflecting on this analogy in the context of this book's focus on *learning* to write successfully in science, I leave you with this last bit of advice: Keep working hard to enhance your power tools for producing superior scientific writing. To do so, you'll need to practice using the rhetorical goals and strategies that work well for you regularly, sharpen the ones that are productive only some of the time, and discard those that never yield good results. And keep in mind that you can always add new power tools to your mental toolbox by adopting the rhetorical goals and strategies that you derive from excellent models in the scientific literature.

Appendix A:
Guidelines for Preparing and Delivering Oral Presentations and Poster Presentations

In addition to its written media—that is, journal articles, books, and Web site literature—science is traditionally communicated through oral presentations and a special form of writing called poster presentations. These are the primary modes of communicating science in professional conferences, seminars, lectures, and thesis and dissertation meetings. Entire books have been written on how to prepare and give scientific presentations. This is quite fitting because successful outcomes in this form of communication depend on truly extensive knowledge and skill. Given this book's primary focus on scientific writing, we don't have space to address the craft of presentations in depth. Nonetheless, the following guidelines provide foundations for success in preparing and delivering effective oral and poster presentations.

PREPARING THE CONTENT, STRUCTURE, AND STYLE OF ORAL PRESENTATIONS

Like a scientific paper, an oral presentation is a form of communication primarily defined by its content, structure, and style. Scientific papers and oral presentations are produced through similar planning, drafting, and revising activities. Recall, for example, the activities that we covered in Chapter 1 for defining a writing project: They included *analyzing your task, selecting a topic and refining a research issue, learning about scientific discourse conventions, analyzing your audiences,* and *searching for scientific literature.* These same activities are essential to defining an oral presentation. It thus makes sense that many of the following guidelines for preparing oral presentations overlap with our approach to the scientific writing process. However, a number of specialized guidelines for preparing an oral presentation account for unique challenges of the medium. These include the direct interaction of speakers and audiences, the requirement that audiences experience oral presentations in a strict linear path, and the greater reliance on graphics, as opposed to text, in oral presentations.

Carefully Attend to Assignment Instructions, Guidelines, and Evaluation Criteria

As a first step to preparing your oral presentation, make sure that you clearly understand your assignment. What topic areas and research issues are appropriate to present on? Exactly how long should you plan to speak? How formal is the presentation supposed to be? What sorts of visual aids are expected or preferred?

Should you prepare computer-based slides, overhead transparencies, or live demonstrations? Will the audience have the opportunity to ask questions during and/or after your talk? On what criteria will your presentation be evaluated? Raise these questions, along with others that you deem as essential for defining your presentation assignment, to whoever gave it and will evaluate it.

Refine the Focus on Your Topic or Research Issue

One of the most common and troublesome flaws in oral presentations is content that covers too much ground—in other words, too many disparate topics and research issues. On the delivery end of such a presentation, audience members feel overwhelmed and frustrated with the speaker. The most successful oral presentations focus narrowly on topics and research issues. Even for relatively long talks that last up to an hour, it's perfectly acceptable and appropriate to address only one refined subject or one focused research question, problem, or hypothesis.

Take a Goal-Directed Approach

As described throughout this book, expert scientific authors develop rhetorical goals for the major sections of their papers and then use their goal-based plans to guide processes of generating, organizing, and revising content. The same approaches are ideal for crafting the content of oral presentations. Indeed, most of the conventional rhetorical goals for scientific papers (as presented in Chapter 8) apply to oral presentations. However, given key differences in the communication dynamics between papers and presentations, some rhetorical goals play special roles in the latter. Consider, for example, that attendees of a seminar or a thesis defense cannot get a quick and detailed overview of the presentation's structure by perusing its contents at will, as they would do by leafing through the pages of a manuscript. (An exception to the preceding point is a presentation in which the speaker gives the audience a handout of his or her slides.) So an especially important rhetorical goal for introducing oral presentations is to give the audience advance information about the content and structure of your talk. This goal reflects the first part of the standard advice to *tell them what you're going to tell them, tell them, and then tell them what you told them*. Along these lines, a key goal for the body of oral presentations is to regularly orient readers to how the content you are presenting at the moment fits into your talk's overall structure.

Consider a few more examples of rhetorical goals that take special precedence in oral presentations:

- Convince your audience that your topic or research issue is important to society, to science, and to *them*. By doing so, you will capture their attention and reward their attendance.
- Provide the essential background knowledge to prepare your audience for any advanced content in your presentation. By doing so, you will avoid losing attendees, figuratively as well as literally.

- Introduce the logistics of your presentation—for example, inform the audience about how much time you will devote to covering major sections of your talk, and clarify whether attendees should feel free to raise questions in the middle of your talk or whether questions should be reserved for the end.
- Respectfully acknowledge and respond to alternatives to your conclusions and arguments, especially the alternatives that specific audience members are likely to raise.
- To conclude your presentation, summarize its main messages, conclusions, and arguments so that attendees leave with tangible and memorable insights.

Take an Audience-Centered Approach

It is by no means trivial to point out that oral presentations put communicators literally face-to-face with their audiences. The primary explanation for many botched presentations is that the speakers simply failed to carry out sufficient and accurate audience analyses. A fairly common flaw is overestimating the audience's knowledge. In this case, the presenter speaks over the heads of attendees, erroneously addressing them as if they were peer specialists on the topic. While the speaker is feeling pride and glibness about his expertise and technical acuity, the audience is hardly impressed. They have either zoned out or are fuming mad at the presenter for wasting their time with what they perceive as gobbledygook. On the other hand, some presenters underestimate their audiences' knowledge and harp on obvious content. It's a very uncomfortable experience for the speaker to watch as attendees file out of the meeting room, begin reading newspapers, or glare back with facial expressions that exclaim, "Why did we need to come to this talk? You're just telling us what we already know!"

Chapter 1 presented audience analysis as a key activity for defining a writing project. Recall that the process involves raising questions about the characteristics of intended readers. Writers apply their characterizations of audiences to gain insights about the appropriate content, structure, and style for scientific papers. The same approach applies to preparing oral presentations. For example, an essential audience-analysis question is, "Who are your primary and secondary attendees?" The answer sheds light on topics to cover, rhetorical goals to accomplish, arguments to emphasize, and counterarguments to refute. It's especially important to characterize your audience by their relationship with you and by their reasons for attending your presentation. Do they view you as an expert or a novice on your topic or research issue? Are they in attendance to gain knowledge for themselves or to evaluate your knowledge and research skills? Are they friendly colleagues who are likely to agree with your conclusions and arguments? Or is your audience composed of strangers, some of whom are likely to oppose your views?

To appreciate the importance of the preceding questions, consider the case of a graduate student presenting his thesis research to faculty members who are leading scientists in the field. The student had every reason to be confident that his presentation would be highly successful. But in preparing and delivering it, the student overlooked critical qualities defining his true relationship with the audience. He mistakenly assumed the role of the singular expert in the room. This attitude was obvious from the student's swagger and somewhat patronizing tone of voice. In response, the faculty felt justified to play hardball. They asked questions and leveled criticisms that they would normally reserve for professional colleagues. Overwhelmed by a sense of intimidation, the student was unable to demonstrate his truly superior knowledge and research skill.

Organize the Content of Your Presentation by Recognizable Themes

As presented in Chapter 4, a number of conventional themes are used to structure the content of scientific papers. Examples are chronological, sequential, most-to-least-important, and part-to-whole themes. Matched appropriately with the content at hand, the same general themes can be used to structure oral presentations. For example, if you are presenting a biological process, the content would be appropriately organized by a sequential theme. Or if you are presenting an argument, your lines of support might be best organized in a most-to-least-important pattern, or vice-versa. When oral presentations are based on clearly recognizable organizing themes, attendees can easily integrate newly presented information and predict the nature of upcoming information. These logical insights prevent the audience from being sidetracked by struggling to figure out a presentation's structure.

Choose an Efficient and Reliable Strategy for Delivering Your Talk

Regarding the logistics of preparing and delivering the oral content of scientific presentations, common options are as follows.

- After scripting and memorizing your entire presentation, give it word for word without external aids such as note cards.
- After scripting your entire presentation on manuscript paper or note cards, read it word for word.
- On manuscript paper or 3 × 5 cards, write brief reminders of what to say; then use the notes to guide your talk.
- On your presentation slides, include bullet points to read verbatim or to cue more elaborate discussion.
- Just wing it! That is, don't plan what to say at all. Just present your slides and say whatever comes to mind.

Each of the preceding strategies has its advantages and disadvantages. For example, if you read your entire presentation from a script, you will be certain to communicate everything you planned, doing so precisely on time. But when read from a script, presentations often strike audiences as artificial and dull. Compared to extemporaneous speech, script reading can elicit a monotonous and detached feeling, especially when the speaker is fixating on the script rather than interacting with the audience.

Individual circumstances determine the most efficient and reliable preparation and delivery strategies. If you lack experience giving oral presentations, and if you tend to get nervous speaking in front of crowds, it's best to prepare some scripted material, at least for the most complex and challenging parts of your talk. With practice and familiarity, you may choose to adopt the logistical strategy of experienced scientific speakers: Rather than memorizing their presentations or reading from scripts, they use cues on their presentation slides to prompt points, ideas, and discussions. Experienced speakers deliver their content precisely as planned because they are intimately familiar with it and because they have practiced delivering it.

Prepare to Meet the Time Requirement for Your Presentation

This guideline reflects one of the major challenges of preparing successful oral presentations. It's critical to meet their assigned time allotments, especially to avoid two dreaded occurrences. One is failing to prepare a sufficient amount of content to fill the scheduled time. The other is preparing too much content and having to end a presentation without telling a complete story or making a well-rounded and convincing argument. Both situations wind up embarrassing presenters and inconveniencing their audiences. Veteran speakers have a sharp sense of how their content scales in terms of presentation time. Through experience, they come to intuitively know approximately how long they will need to explain concepts, summarize studies, and present supporting evidence for their arguments. But for experts and novices alike, the keys to preparing the right amount of material are rehearsal and revision. Through trial and error, successful speakers tailor the content of their presentations to perfectly fit assigned time allotments.

Consider the Setting and Purpose of Your Presentation

Picture the following two presentation scenarios.

(1) An Honors thesis proposal, held on a Friday afternoon in a campus lounge, in which a student presents exploratory ideas for a research project to her advisor and 10 lab mates. The presenter's main purpose is to get feedback on her proposal ideas.

(2) An Honor thesis defense, held at 8:00 on a Monday morning in a university classroom, in which the same student presents her completed research project to three faculty members, two of whom are strangers (to the

student). The presenter's main purpose is to convince her audience that her work has great merit, warranting graduation with an Honors degree.

The contrasting settings and purposes call for preparing very different presentations. The first will be considerably more informal in its structure and tone, encouraging the audience's extensive participation and input throughout the talk. The second presentation will be more highly scripted and precise in its content and organization.

Go Easy on the Humor, If You Go with Humor at All

You might be tempted to spice up your presentations by telling a few jokes or funny stories. Be warned, however, that using humor in scientific presentations is risky business, especially for students giving thesis and dissertation talks. You might consider adding humor in the following cases.

- You are a well-recognized world's leading expert in your field and your audience will be delighted and awed by *anything* you say.
- Your audience is composed of colleagues who justly expect an informal presentation, given its setting and purpose.
- Your humor is brief, appropriate, tasteful, relevant to your talk's subject, and truly funny.

Make Sure to Test-Drive Your Presentation Technology

Take to heart the following spin on Murphy's Law: *Anything that can go wrong will go wrong with presentation technologies.* If you have given or attended a fair number of academic talks, whether presented by students or by world-renowned scholars, you have seen this version of Murphy's Law in operation. In an all-too-frequent occurrence, a presenter's PowerPoint slides do not display properly because he originally prepared them on a computer with a different operating system. In another nightmarish but common enough scenario, the presenter finds out too late that the power supply for his projector is missing from its case. In addition to computer glitches, all sorts of technological difficulties can arise in presentation rooms: lights don't dim properly, air conditioning or heating systems are on the fritz, or the room's acoustics are terrible. Experienced presenters address and solve these sorts of problems long before scheduled talks. They do so by testing their presentation technology ahead of time and, when it's possible to do so, by checking out their presentation rooms and requesting necessary changes.

Prepare Answers to the Questions That Your Audience Is Likely to Ask

Most scientific presentations end with question-and-answer (Q and A) sessions, which can be extremely engaging and useful for everyone involved. Experienced presenters anticipate questions that their audiences are likely to raise, and they prepare well-developed answers ahead of time. This requires taking your audience's perspective; considering what information, ideas, and arguments are most relevant to your audience's concerns and interests; and accounting for the views and responses of your most critical audience members. If you are presenting original research, be especially mindful of methodological limitations to your study and of knowledge gaps in the field. Critical audience members will certainly ask about these matters. If you are planning a thesis or dissertation defense, talk with your advisor about the questions that your committee is likely to raise. Then, practice your answers and seek your advisor's feedback on them.

Practice, Practice, and Then Practice Some More

There are so many excellent reasons for practicing oral presentations at least a few times before giving them. As mentioned, two reasons are to make sure that talks meet required time allotments and to resolve potential technological problems with presentation equipment and rooms. In addition, repeated rehearsal is the antidote to the nervousness and anxiety that often arise when we speak in front of crowds. It's important to practice with purpose. For example, in a trial run for an upcoming presentation, you might focus on the goals of being physically relaxed, hitting time targets for certain parts of your talk, and avoiding digressions. If it's possible to do so, practice in front of an audience—perhaps your advisor, class mates, or lab mates. Solicit their feedback on specific qualities of your presentation. Did you speak loudly enough? Were you making good eye contact? Did you interject any distracting *ums* and *you knows*? Were there any graphics that confused your audience? For especially useful feedback, try videotaping your practice sessions so that you can self-critique your presentation.

PREPARING YOUR PRESENTATION SLIDES

Few scientific presentations can succeed without visual aids, the most common of which are computer-projected slides that present text, static graphics (tables and figures), and dynamic images (such as animations and video). In the majority of scientific presentations, speakers spend most of their time talking about their slides, which audience members spend most of their time closely examining. The rhetorical goals for oral presentations are accomplished through the coupling of words and visual representations. So, as addressed by the following guidelines, the effective design of slides contributes greatly to the success of oral presentations.

Learn the Ins-and-Outs of Your Graphics and Slide Preparation Programs

These days, the best design principles and sensibilities mean little if you lack practical knowledge and skill in using computer programs for developing graphics and presentation slides. Earlier in the book, Table 4.2 presented commonly used software programs for creating original graphics, including tables, data graphs, line drawings, and concept maps. These sorts of graphics can be imported into slide presentation programs such as Microsoft PowerPoint and OpenOffice Impress (http://www.openoffice.org/product/impress.html). If you are not already an expert at using these programs, it's worth taking time to work through the tutorials that come with the software. Through technology departments on your campus, or on the Internet, you might even take a class or two on designing and developing computer-based presentations.

Be aware that in all-purpose graphics and slide presentation programs, many default settings and optional features are not aligned with the best principles for communicating science. One example is the elaborate decoration in some of PowerPoint's design themes. The embellishments serve no direct purpose for conveying scientific content and achieving the rhetorical goals for a presentation. In addition, PowerPoint includes the optional feature to accompany transitions, or changes in slide elements, with various sounds such as chimes, hammers, whooshes, and explosions. At least for scientific presentations, these sorts of sound effects are inappropriate.

Follow the Best Principles for Designing Graphics

The bread-and-butter slides of most scientific presentations communicate information in the form of graphics, or tables and various types of figures. Chapter 4 presented a set of principles to guide the design of effective graphics for scientific papers (pages 188 through 203). The same principles apply to designing graphics for slides in oral presentations. Recall, for example, the Goldilocks principle for designing graphics: *Not too little, not too much, but just the right amount of information.* The *not too much* part of this principle is especially important for presentations because the audience has a limited amount of time to view slide graphics. As a general guideline, each graphic in your presentation should have a fair amount of open space, and its contents should convey no more than four or five major points in support of a unified, overall message.

Use Attractive, Easy to View, and Unostentatious Colors

In all forms of scientific communication, content rules. Essential matters of meaning and logic should come through without obstruction or embellishment. This is why the text of scientific papers is black and white, both figuratively and literally. However, the relatively heavy emphasis on visuals in scientific presentations underscores the importance of choosing and using colors to complement content.

Limit the number of colors in your slide presentation. Slides that include an excessive number of colors are distracting, especially when the colors are used inconsistently from slide to slide. Any given slide will have a main background color and foreground color. One or two additional colors can be used to highlight foreground elements, such as key words and phrases in strings of text. So for designing most slide presentations, a maximum of four colors is sufficient. One exception occurs when the objects or organisms that the presenter is speaking about are multicolored.

Avoid colors that elicit charged emotions. Some colors naturally give rise to subtle emotional responses that can divert the audience's attention from essential information and, in some cases, negatively influence interpretations of slide content. Such emotionally charged colors—including bright reds, oranges, and yellows—should not be used for the background or foreground in scientific presentations.

Use colors that have a calming effect. Unostentatious colors that tend to elicit calming responses include relatively dark blues and greens as well as off-whites. These colors work well as slide backgrounds.

Use high-contrast background and foreground colors. Color *contrast* refers to the difference in qualities such as hue and intensity between two colors. For background and foreground colors, a high contrast is ideal. For example, it's much easier to read black text on a white background than black text on a dark red background. For a slide presentation in a completely darkened room, a dark background with a light foreground works best. When the room is partially lighted, a light background with a dark foreground is optimal. Examples of background-foreground combinations with good contrast are blue-yellow, blue-white, green-white, white-black, and beige-dark blue.

Avoid color combinations that impair perception in people who are color blind. Approximately 5%–9% of males and 0.5% of females are color blind. These individuals have difficulty perceiving information when it is presented in the following color combinations: red-green, red-blue, orange-blue, and yellow-green.

Use Attractive and Highly Legible Fonts

Regarding the design of slide text, there are three main factors to consider: typeface, font style, and font size. Typeface refers to the design features that distinguish sets, or "families," of alphabetic letters, numbers, punctuation marks, and other typewritten symbols. Word processing and slide presentation programs offer numerous choices of typefaces. Here are a few examples, presented as they appear in 12-point font: Arial, Comic Sans MS, Garamond, Lucida Sans Unicode, Times New Roman, and Verdana.

One feature that distinguishes typefaces is whether they have serifs; as defined in Chapter 4, these are the short lines and hook-like markings that are added to the tops and bottoms of letters. The legibility of serif typeface is relatively poor on computer-projected slides because the resolution of digitized letters is low. So a sound conventional guideline is to use a sans-serif typeface for *all* of the text in presentation slides. Commonly used typefaces include Arial, Calibri, Lucida Sans Unicode, Tahoma, and Verdana.

Regarding font size, the main guideline is to make the text large enough so that everyone in your audience can read your slides without straining their eyes. Depending on the typeface, font size may need to be in the range of approximately 30–40 points for titles and approximately 18–28 points for the main text. It's also important to avoid using excessively large fonts because they waste screen space and have an unsophisticated quality.

The general rule for special font styles—such as bold, italic, and underline—is to use them sparingly, if at all. Italicized text is usually discouraged because it is difficult to read on a presentation screen. The same is true of long strings of underlined text as well as words in which all letters are capitalized. Bold font can be used effectively to accentuate key words and phrases. However, the excessive use of bold font can give text a heavy, dense look and feel.

Use Animations and Video to Present Dynamic Scientific Concepts and Processes

Advances in technology have made computer-based presentations fairly routine, replacing the outdated approach of showing transparencies on overhead projectors. A great advantage of projecting computer images is that they can be dynamic, as presented in computer animations and video. It's not always feasible for students to create original dynamic images for their presentations. These days, however, the Internet is an amazing resource for free, high-quality animations and videos that demonstrate dynamic concepts and processes in all life science fields. With permission granted from the authors of these resources, you can download and import them into your presentation. Or you can always create links from your presentation to the Web sites on which the resources are displayed.

Be Consistent with the Color Schemes, Fonts, and Layout of All Your Slides

A lack of consistency in slide design features can undermine the highly desirable sense of conceptual unity that characterizes effective oral presentations. For example, audiences can be distracted by slide-to-slide inconsistencies in background and foreground colors, placements of titles, and typefaces and font sizes.

Limit the Presentation Time of Each Slide to No More Than a Few Minutes—Say 2 or 3 Minutes Tops.

Effective oral presentations have a lively and progressive tempo. Skilled speakers very deliberately avoid spending an inordinately long time on any one slide. This is fitting for two main reasons. First, slides that take more than a few minutes to present are often overloaded with content, which overwhelms the audience. Second, a relatively fast tempo keeps attendees on the edge of their seats in anticipation. Certainly, there are exceptions to the guideline at hand. For example, it might take longer than 2 or 3 minutes to adequately explain a complex biological process that, for the sake of the audience's comprehension, needs to be displayed graphically on a single slide. However, by and large, it's a good practice to distribute content so that no slide can cause a presentation to stall or bog down. One more point about timing: To add a bit of variety, which your audience will surely appreciate, you might intersperse slides that take relatively short versus long times to present.

Limit Your Use of Bullet Points

Among savvy audiences, the #1 complaint about oral presentations involves the overuse of bullet points. Perhaps you have attended presentations in which one slide after the next displayed one bullet point after the next, and the speaker simply read the screen contents, saying little or nothing more. An apt response to the speaker would be, "Next time, just send me an e-mail attachment with your PowerPoint file, and I'll read the bullet points at my own convenience!" It's true that for accomplishing certain rhetorical goals, slides that display bullet points can be effective. An example is a slide that presents three main conclusions from a study. As described, however, the problem is *overusing* bullet points. To avoid the problem, follow these more specific guidelines (presented, of course, as bullet points!).

- Limit the number of bullet-point slides to a maximum of 20%–25% of all your presentation slides.
- On any one slide, limit the number of bullet points to a maximum of 5–7.
- Don't present more than 2 or 3 bullet-point slides in succession.

Prepare an Informative and Engaging Title Slide

On the title slide for your oral presentations, include the following: an engaging title, prepared according to the guidelines presented in Chapter 4 for titling scientific papers (page 172); your name and the name of your educational institution; the date of your presentation; and perhaps an image, such as a line drawing or photograph, that captures the essential subject of your presentation.

Title Each Slide with Its Take-Home Message

When reading scientific papers, we can take our time to figure out the take-home messages of their paragraphs, sections, graphics, and other units of content. This advantage, however, is not afforded by oral presentations. So we count on speakers to directly indicate key points. An effective tactic for reinforcing a presentation's take-home messages is to display them prominently as slide titles. Imagine, for example, that you are giving a talk on the health benefits and risks of consuming caffeine. One of your slides presents data from a study indicating that people who drink caffeinated beverages have a relatively low risk of developing Alzheimer's disease. You might title the slide as follows: *Caffeinated Beverage Consumption Reduces the Risk of Alzheimer's Disease.*

Keep Your Audience Grounded to Your Presentation's Structure and Rhetorical Goals

Given the dynamics of oral presentations on scientific topics, audiences usually must devote a considerable amount of mental energy to keep up with speakers and to understand everything fully. A major challenge is figuring out how the content being presented at any given moment fits into the presentation's overall structure and contributes to accomplishing the speaker's rhetorical goals. To help your audience meet this challenge by reducing its cognitive burden, you can design your slides to continually reinforce your presentation's structure and goals.

The best way to ground an audience to a talk's structure is to present slides with outlines. You might, for example, present a general outline of your entire presentation when you introduce it. Then, along the way, include slides that display more detailed outlines of unified sections of your talk. Upon reaching major structural landmarks, re-present your outlines in a way that clearly indicates the content that you have covered, the content that remains, and the relationships between sections. It's also extremely helpful to include slides that reinforce your rhetorical goals. Suppose, for example, that one of your overall goals is to support an original hypothesis. And let's say that the major sections of your presentation are organized by different lines of support. To begin each new line of support, you might display a slide that re-presents your hypothesis, briefly summarizes the arguments that you have already presented for it, and outlines its upcoming supporting evidence and reasoning.

Distribute Your Slides in Handouts

Slide presentation programs such as PowerPoint and OpenOffice Impress enable users to create handouts that reproduce slides, specifying the number of slides to print on a page. If it's possible to provide a handout for every member of your audience, consider the advantages of doing so. During your talk, audience members will be able to spend extra time examining slides that are most relevant to their concerns. To better understand a slide that you are presenting at any given moment, your audience will be able to refer to essential background information presented in earlier slides. Handouts facilitate efficient note taking. And, of course, they enable audiences to revisit presentations any time in the future.

GUIDELINES FOR GIVING ORAL PRESENTATIONS

The quality of oral presentations depends on much more than what speakers have to say. Successful outcomes also depend on how speakers perform, or deliver their presentations. As addressed in the following guidelines, skillful presenters communicate through movement and gesture, by controlling characteristics of voice, and by interacting with their audiences.

Observe and Adopt the Performance Characteristics of Expert Presenters

Certain qualities of performance distinguish the presentations of excellent teachers and speakers. Through their facial expressions, postures, and movements, the experts convey comfort, confidence, enthusiasm, proficiency, and a genuine concern for whether they are meeting their audiences' needs. For example, in responding to a student's question about challenging lecture material, a skillful teacher might stop in her tracks for a moment to ponder the inquiry deeply. Then, with a bounce in her step, the teacher will move toward the questioner and establish eye contact before beginning to answer. These sorts of nonverbal actions elevate levels of attention and motivation among audience members, which ultimately enhances their comprehension and learning experience. The advice at hand is to be mindful of the performance characteristics of excellent speakers. Take note of their mannerisms and means of engaging their audiences. In your own presentations, try out different methods that you have observed in experts' presentations and, into your regular repertoire, incorporate the methods that work especially well for you.

Avoid Digressions: Stay on Target and on Time

In the midst of presentations that are going well, speakers sometimes gain a sense of confidence that leads them to indulge in extemporaneous digressions from their plans. They become like tightrope walkers who decide to go without a net. It's usually a big mistake to entertain digressions while giving a scientific talk. In addition to breaking the coherence of presentations, digressions waste time and can prevent speakers from covering everything they planned. It's challenging to avoid digressions because they can arise and take over without our immediate awareness. If you are inclined to entertain digressions, you may need to practice attending to and dismissing them deliberately. In practice presentations, cue your audience to interrupt you anytime you begin to go off course. Another tactic is to set time goals for completing segments of your presentation, such as the introduction or the half-way point. By concentrating on reaching these landmarks on time, and checking your progress every so often, you can prevent digressions from taking root.

Communicate through Movement

Skillful speakers enrich their messages with movement. An arm gesture can accentuate a vital idea. A slow stroll toward the audience alerts everyone to focus their attention on an important upcoming conclusion. A shrug of the shoulders highlights a particularly perplexing problem, and a knowing smile prefaces the delivery of an especially elegant solution. To appreciate the importance of communicating through movement, contrast the preceding examples with an image of a speaker who stands motionless behind a podium throughout the duration of his/her talk. Even if the content is quite interesting, it's easy for the audience to be lulled by the speaker's immobility.

Note, however, that excessive movement can be problematic, particularly when it is repetitive. Audiences can get sidetracked, and even somewhat irritated, by a presenter's nervous rocking and pacing movements. An especially distracting movement is tremor of the hand holding a laser pointer. While the speaker intends for the audience to focus on specific details on the presentation screen, the tremor captures most of their attention. Very often, speakers make these excessive movements unconsciously. So, to correct the problem, they must first learn about it from someone else, like a helpful colleague observing a practice presentation. Videotaped practice presentations are great for making speakers aware of any problematic movements.

Communicate through Eye Contact

Another essential form of nonverbal communication for successful oral presentations is eye contact. If a speaker spends an inordinate amount of time looking at her notes or the presentation screen, the audience gets the message that their presence and understanding are relatively inconsequential. Or if a speaker has a glazed-over look in the audience's general direction, the attitude of aloofness and a lack of enthusiasm are conveyed. To ensure that you are communicating through good eye contact, think of your audience as individuals rather than as a collective entity. Just as you would naturally establish eye contact with a friend in a one-on-one conversation, aim to connect with each member of your audience in this way. Of course, this might not be possible to do if you are speaking to a very large group. In this situation, scan your audience—from left-to-right and back-to-front—establishing eye contact with a representative number of attendees in each segment of the room.

Communicate by Varying the Characteristics of Your Speech

This guideline is similar to the one for communicating with movement. It's a matter of enriching a presentation's content, as well as the audience's experience, through using nonverbal cues—in this case, variations in qualities of vocalization. In general, make sure to vary the pitch of your voice—in other words, avoid a monotone presentation, which squelches any sense of enthusiasm and lulls audiences. To accentuate specific conclusions, you might slow your rate of speech, raise the volume of your voice, and enunciate key words. To encourage your audience to reflect on especially important ideas, interject relatively long pauses. To convey subordinate information, pick up the pace of your speech ever so slightly and tone down your volume.

Take Your Time to Find the Best Words and Phrasing

If you have not scripted or memorized your talk, you may face on-the-spot dilemmas about how, ideally, to phrase certain ideas. Novice presenters often choose too hastily, thinking that their audiences will respond negatively to pauses and reflections. In contrast, experienced speakers know that audiences appreciate occasional pauses, during which they engage in their own reflections about the presentation material. So when ideas do not roll off your tongue easily, it's okay to take a deep breath, focus your attention inwardly, and silently try out a phrase or two before addressing your audience again.

Carefully Guide Your Audience through Complicated Slides

This guideline addresses an especially common problem that greatly frustrates audiences of scientific talks. It's a case in which speakers present complicated slides, such as extremely busy data graphs, without explaining them sufficiently. In response to one disorienting slide after the next, the audience gains the impression of the speaker saying, "I'm not even going to make the effort to help you understand—you figure it out for yourselves." In offering a sufficient explanation of a complicated slide, the presenter does the following:

- Announces the slide's overall purposes and general take-home messages.
- Orients viewers to the slide's key navigational landmarks, such as the variables and measurement units on graph axes.
- States the slide's specific take-home messages.
- For each specific message, directly points to the supporting information, guiding viewers through essential trends and patterns in step-by-step fashion.
- Summarizes how the slide serves its overall purposes and supports its take-home messages.

Attend to and, If Necessary, Respond to Your Audience's Feedback

During an oral presentation, when you are concentrating intently on what to say and how to say it, the skill of attending to what's going on in your audience is extremely challenging. It is, however, a skill truly worth practicing and developing. Adjustments made in response to *real-time* audience analysis are often essential for successful presentation outcomes. Suppose, for example, that you become aware of many perplexed faces in your audience. In response, you might stop to reflect on the source of the problem and then elaborate its solution. Or you might speak directly to obviously confused audience members, in order to figure out how to best clarify the matter. If, in another scan of your audience, you note nodding heads in agreement with certain ideas and arguments, you might reinforce them later in the talk. If folks look bored and are on the verge of falling asleep, you might pick up the pace or skip to more exciting segments of your talk.

In Q and A Sessions, Repeat Each Question So That the Entire Audience Can Hear

Applying this practical bit of advice, experienced speakers ward off an especially frustrating experience for audiences: It's being clueless in trying to understand the presenter's answer to a question because they did not hear the question clearly in the first place. By taking the time to repeat a question so that everyone can hear, the presenter engages the entire audience in the ensuing answer and in any further discussion that it inspires.

Request Clarification of Ambiguous Questions

Don't go headlong into answering a question that you don't understand thoroughly. Instead, take a moment to reflect on the question. If it's still not clear in your mind, politely let the asker know. You might also explain your (mis)understanding to help the asker clarify the question.

If You Don't Know, by All Means Say So!

Novice presenters sometimes make the major mistake of bluffing responses to questions that they lack the essential knowledge to answer. In contrast, expert presenters respond to such questions unapologetically, simply telling the audience that they do not know the answer. Of course, this can be a problem in a Q and A session that is expressly intended to test the presenter's knowledge, as in the case of a thesis or dissertation defense. Even so, the presenter's credibility will be damaged considerably more by bluffing answers than by admitting the truth. In seminar settings, experienced speakers often return questions that they cannot answer to the audience at large, expecting that an attendee or two might know the answers. In another expert approach, presenters promise to seek out the answers to questions that have stumped them. Of course, this obliges the presenter to obtain the questioner's contact information, such as an e-mail address, and then to follow through on his or her promise.

PREPARING POSTER PRESENTATIONS

As their name indicates, poster presentations communicate science in text and graphics on paper posters. Common options for the dimensions of scientific posters are 36 × 48 inches, 36 × 56 inches, 42 × 72 inches, and 48 × 96 inches. Most professional meetings, such as the annual conferences of scientific organizations, feature poster presentations along with oral presentations. Poster presentations are also common vehicles for students to communicate their research at university functions as well as professional meetings. In a typical presentation session, posters are attached to easels or to the walls of hallways or auditoriums. Attendees consult conference programs to identify the poster sessions that interest them. Then, at the scheduled times, attendees quite literally visit with presenters, who stand by their posters to discuss their research and answer questions about it. The occasion offers excellent opportunities for sharing knowledge, networking, and developing ideas for new collaborative projects.

In the old days, scientists and science students created their posters by using word processing programs and then printing out the contents on 8 ½ × 11 inch or 8 ½ × 14 inch sheets of paper, which they glued to sheets of poster board. For a poster communicating an original study, which is the most common type of scientific poster presentation, one sheet was typically devoted to introducing the research issue; another sheet or two summarized the methods and results, respectively; and one more sheet presented the main conclusions or discussion points. The old-fashioned production approach still adequately serves the essential communication goals of poster presentations. However, these days, most scientists create their posters using slide presentation software (such as PowerPoint and OpenOffice Impress) or page design and publishing software (such as Adobe InDesign and QuarkXPress). In addition, special printers produce all of a poster's contents on one large sheet of paper. Students may have access to these printers through their science departments or through general printing services on campus.

Poster presentations are a special form of scientific writing, so all of the planning, drafting, and revising activities that we covered in Chapters 1 through 7 apply to preparing posters. In addition, their content should accomplish the key rhetorical goals for scientific writing, as presented in Chapter 8. Many of the preceding guidelines for preparing oral presentations also apply to poster presentations. For example, the ideal content for a poster is generated through goal-directed and audience-centered approaches to writing. And a principled approach to designing graphics is essential for the best outcomes. A few specific guidelines for preparing and delivering poster presentations come next.

Design Your Poster According to Expert Models

Especially for students who have little or no experience with this form of scientific communication, the best way to learn how to produce effective poster presentations is to observe models and to adopt their strongest features. You might find scientific posters displayed on the walls of hallways in science buildings on your campus. Or your science professors may have sample posters in their laboratories. In addition, excellent examples are presented on numerous educational Web sites. To locate these sites, do an Internet search with the phrase "excellent examples of scientific poster presentations." Make sure to check out the superior Web resources provided by Colin Purrington, a biology professor at Swarthmore College; do an Internet search with the phrase "Colin Purrington advice on designing scientific posters."

To gain helpful information from model posters, raise the following questions:

- What features describe their overall layout?
- How much text is devoted to each major section of the IMRAD format?
- What techniques do their authors use to communicate ideas succinctly (for example, note the uses of lists and bullet points)?
- How many graphic elements (tables and figures) are included, and how are they positioned in relation to explanatory text?
- Which color schemes are most attractive, easy to view, and unostentatious?
- Which typefaces do the authors use for different text elements, such as titles, headings, and the main text body?

Design Your Poster with a Logical and Organized Layout

Successful posters are easy to follow as a consequence of their logical and organized overall layouts. Consider, for example, the layout of contents in the poster template in Figure A-1, which is available to download from Colin Purrington's Web site (http://www.swarthmore.edu/NatSci/cpurrin1/posteradvice.htm). The body of the poster is logically organized by the major sections of an IMRAD-structured research paper. The sections are grouped in easy-to-follow columns, which are separated by ample empty space. While you can certainly get creative with the layout of your posters, the template in Figure A-1 would be perfectly suitable for many presentations of original studies.

Design Your Poster with Attractive and Legible Text

As is true for the text in slide presentations, poster text must be large enough for audiences to view without straining their eyes. The titles and body text of posters should be legible from approximately 12–15 feet and 5–8 feet, respectively. Following publication conventions, poster titles, section headings, and graphic elements (figure captions, legends, axis labels, and so on) may be displayed in a sans-serif typeface; good options are Helvetica, Arial, or Lucida Sans Unicode.

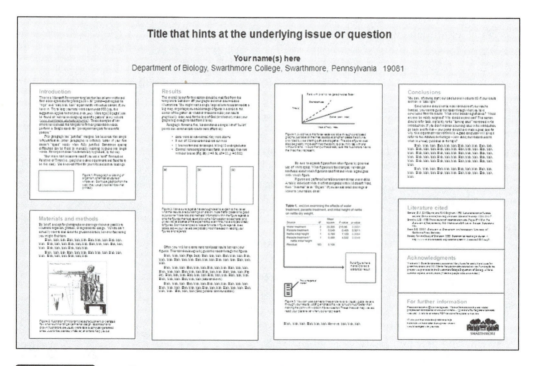

FIGURE A-1 An effective layout for a scientific poster. Reproduced with permission.

The body text should be displayed in a serif typeface such as Palatino or Times New Roman. Avoid underlining, all caps, and any embellished font styles such as shadowing and outlining words. To emphasize important words and phrases, use italics and bold font conservatively.

Limit the Amount of Text on Your Poster

One of the main complaints about poorly designed scientific posters is that they have too much text, which usually means too few graphics. Audiences feel inconvenienced by having to spend more than 10 minutes or so reading line after line of poster text. For good reasons, many poster viewers are looking for quick take-home messages, which they can then discuss with the presenters to gain more elaborate insights. Here's a bit of Colin Purrington's advice for curbing the amount of text in poster presentations (http://www.swarthmore.edu/NatSci/cpurrin1/posteradvice.htm):

- Limit the entire text of your poster to fewer than 800 words.
- For selected purposes, such as summarizing study results, use lists and bullet points rather than long paragraphs.
- Limit the width of text boxes to approximately 40 characters, which corresponds to approximately 11 words per line.
- Do not exceed 10 sentences in a block of text.
- Single space lines of text.
- When it's feasible to do so, use graphics in place of text (e.g., instead of detailing your study's sequential methods in lines of text, present a flow chart).

Use Appropriate Poster Colors

The same guidelines for choosing and using colors in slide presentations apply to poster presentations. For example, a high contrast between background and foreground colors is advised; the best combination for posters is a light background and dark foreground. Limit the number of colors you use (two to four colors should generally be the maximum), and avoid the emotion-eliciting colors—the bright reds, oranges, and yellows.

DELIVERING POSTER PRESENTATIONS

Even at major scientific meetings, poster presentations tend to be much less formal than oral presentations. In fact, many scientific poster sessions have something of a party-like atmosphere, complete with wine and cheese servings. Attendees seek to learn about new research in their fields, but they also view poster sessions as opportunities to catch up with old friends and meet new ones. Nonetheless, as reflected in the following guidelines, there is a conventional etiquette to delivering poster presentations.

Dress for the Occasion

While poster sessions tend to be informal, attendees generally expect presenters to stand out as the best dressed folks in the crowd. Especially for student presenters, it's a good idea to wear fairly formal attire, consistent with how you might dress for a job interview in an academic setting.

Reach Out to Your Audience

Make sure to greet people as they approach your poster. Once they have spent a few minutes studying it, offer a personal tour. You might reach out by asking the following sorts of questions.

- Would you like further explanation?
- Do you have any questions about my research?
- So what do you think?

General interpersonal skills, such as establishing eye contact and engaging people in dialogue (rather than lecturing them), make all the difference for successful poster presentations.

Be Prepared to Answer Questions about Your Research

The guidelines presented earlier for responding to questions in oral presentations generally apply to poster presentations. For example, you should be prepared to answer questions that attendees will likely ask about your poster and your research project. And if you don't know the answer to a question, don't fake it. You might offer to seek the answer in the upcoming week and send it in an e-mail message to the asker.

Appendix B:
Glossary of Sentence Grammar Terms

Adjectives clarify and describe nouns and pronouns. Most adjectives answer one of these questions: *How many? Which kind or type? Which one?*

> **Few** studies have been conducted with **valid** methods for evaluating the prevalence of **clinical** depression in people with insomnia.

Adverbs clarify and describe verbs, adjectives, other adverbs, and entire sentences. Most adverbs answer one of these questions: *How? Why? When? Where? To what extent or degree?*

> The number of marital complaints increases **significantly** when a spouse snores **loudly**.

Clauses are sentence parts or whole sentences that contain a subject and predicate. There are two major types of clauses: **independent clauses** and **dependent clauses**. Independent clauses convey complete thoughts and can therefore stand alone as sentences.

> Weiss et al. found that **sleep deprivation diminishes the ability to learn**.

In the preceding sentence, the word group in bold font is an independent clause. It has a subject (*sleep deprivation*) and a predicate (*diminishes the ability to learn*), and it expresses a complete thought.

Dependent clauses, which are also called subordinate clauses, do not express a complete thought, even though they contain a subject–predicate unit.

> **Although normal sleep promotes good mental health**, sleep deprivation can sometimes elevate mood.

Indicated by bold font, the dependent clause has a subject (*normal sleep*) and a predicate (*promotes good mental health*), but it cannot stand alone in meaning. The reason involves the word *although*, which is a subordinating conjunction. It sets up the requirement for an independent clause (*sleep deprivation can sometimes elevate mood*) to complete the meaning of the dependent clause.

Conjunctions join parts of sentences and indicate their relationships in meaning. **Coordinating conjunctions** (*and, but, or, nor, for, so,* and *yet*) join independent clauses. Here's an example using the coordinating conjunction *but*:

> Five study participants were dismissed due to negative side effects of the drug treatment, **but** all of the remaining participants completed the study.

Subordinating conjunctions introduce dependent, or subordinate, clauses. There are many subordinating conjunctions, including these: *after, although, because, before, even though, if, once, since, unless, when, where, whereas, whether,* and *while*. (See the entry for **Clauses** to learn about the role of subordinating conjunctions in dependent clauses.)

Conjunctive adverbs connect independent clauses and serve as transitions. They include *accordingly, also, consequently, furthermore, however, moreover, therefore,* and *thus*.

> Salazar [14] found that subjects who took the medicinal herb Valerian experienced improvements in sleep quality; **however**, the clinical significance of this research is questionable because the subjects did not suffer from insomnia.

Dependent clauses (see **Clauses**).
Direct objects reflect the target action of verbs.

> Many people experience **insomnia** during psychologically stressful times in their lives.

The direct object, *insomnia*, reflects the target action of the verb *experience*.
Independent clauses (see **Clauses**).
Nouns name people, places, things, qualities, or actions.

> To ensure normal **growth**, **children** need to sleep at least 9 **hours** per **night**.

Phrases are word groups that cannot stand alone as a sentence because they lack a subject and a predicate. The different types of phrases include **noun phrases**, **verb phrases**, and **prepositional phrases**. A noun phrase contains a noun (or pronoun) and all of the words that modify it: *certain sleep medications, short-term insomnia, individuals with sleep disorders*. A verb phrase contains a verb and its auxiliaries, or helping words: *had been sleeping, cannot remember, was found to be*. A prepositional phrase contains a preposition and its object. The preposition begins the phrase, and its object is usually a noun or noun phrase: *of the study, in the results section, with sleep disorders*.

Predicates assert the action or condition of subjects. A predicate contains a verb and its complements, or the words that complete its meaning. In the following example, the subject is italicized and the predicate is in bold font.

> *The optimal duration of nightly sleep* **differs across individuals**.

Prepositions come before nouns, noun phrases, and pronouns, forming prepositional phrases that modify other words or phrases. The long list of prepositions includes *about, above, after, among, around, as, at, before, below, between, by, during, for, from, in, of, on, out, than, to, under, up, with,* and *without.*

> Smith et al. investigated the effects of moderate sleep loss on brain wave activity **during** motor task performance.

Pronouns refer to, or substitute for, nouns. The long list of pronouns includes these words: *it, this, which, they, their, those, these, I, we, who, whom, whose, she, he, any, both, few, many, several,* and *neither.*

> Although scientists know the effects of the drug Modafinil, **they** are uncertain about how **it** promotes wakefulness.

Subjects are what sentences or clauses are about. A **simple subject** is the single noun or pronoun that tells what its sentence or clause is about.

> The **combination** of bright light and caffeine significantly improved mental alertness.

A **complete subject** consists of the simple subject and all of the words that modify it.

> **The combination of bright light and caffeine** significantly improved mental alertness.

Verbs express the condition or action of their subjects. In the following sentence, the complete subjects are italicized and the verbs are in bold text.

> *The researchers* **discovered** that *the time to sleep onset* significantly **decreased** in patients who received massage therapy.

References

Alesci, S., Martinez, P.E., Kelkar, S., Ilias, I., Ronsaville, D.S., Listwak, S.J., et al. (2005). Major depression is associated with significant diurnal elevations in plasma interleukin-6 levels, a shift of its circadian rhythm, and loss of physiological complexity in its secretion: clinical implications. *The Journal of Clinical Endocrinology & Metabolism, 90*(5), 2522–2530.

American Psychological Association (2001). *Publication manual of the American Psychological Association (5th edition).* Washington, DC: APA.

Bereiter, C., & Scardamalia, M. (1987). *The psychology of written communication.* Hillsdale, NJ: Lawrence Erlbaum Associates.

Black, C.D., O'Connor, P.J., & McCully, K.K. (2005). Increased daily physical activity and fatigue symptoms in chronic fatigue syndrome. *Dynamic Medicine, 4*:3. doi:10.1186/1476-5918-4-3

Bloomer, R.J., Falvo, M.J., Schilling, B.K., & Smith, WA. (2007). Prior exercise and antioxidant supplementation: effect on oxidative stress and muscle injury. *Journal of the International Society of Sports Nutrition, 4*:9. doi:10.1186/1550-2783-4-9

Campisi, J., Leem, T.H., Greenwood, B.N., Hansen, M.K., Moraska, A., Higgins, K., et al. (2003). Habitual physical activity facilitates stress-induced HSP72 induction in brain, peripheral, and immune tissues. *American Journal of Physiology: Regulatory, Integrative, and Comparative Physiology, 284*, R520–R530.

Council of Science Editors (2006). *The CSE manual for authors, editors, and publishers (7th edition).* Reston, VA: Council of Science Editors.

Flower, L.S., & Hayes, J.R. (1981). A cognitive process theory of writing. *College Composition and Communication, 32*, 365–387.

Flower, L.S., Hayes, J.R., Carey, L., Schriver, K., & Stratman, J. (1986). Detection, diagnosis, and the strategies of revision. *College Composition and Communication, 37*, 16–55.

Gopen, G.D., & Swan, J.A. (1990). The science of scientific writing. *American Scientist, 78*, 550–558.

Lips, K.R., Brem, F., Brenes, R., Reeve, J.D., Alford, R.A., Voyles, J., et al. (2006). Emerging infectious disease and the loss of biodiversity in a Neotropical amphibian community. *Proceedings of the National Academy of Science of the United States of America, 103*, 3165–3170.

Logan, A.C. (2004). Omega-3 fatty acids and major depression: a primer for the mental health professional. *Lipids in Health and Disease, 3*, 25–32.

Martin, R.A. (2001). Humor, laughter, and physical health: methodological issues and research findings. *Psychological Bulletin, 127*, 504–519.

Moreau, K.L., Donato, A.J., Seals, D.R., DeSouza, C.A., & Tanaka, H. (2003). Regular exercise, hormone replacement therapy and the age-related decline in carotid arterial compliance in healthy women. *Cardiovascular Research, 57,* 861–868.

Nassar E., Mulligan C., Taylor, L., Kerksick C., Galbreath M., Greenwood M., et al. (2007). Effects of a single dose of N-Acetyl-5-methoxytryptamine (Melatonin) and resistance exercise on the growth hormone/IGF-1 axis in young males and females. *Journal of the International Society of Sports Nutrition, 4*:14. doi:10.1186/1550-2783-4-14

Reilly, S.M., & Blob, R.W. (2003). Motor control of locomotor hindlimb posture in the American alligator (Alligator mississippiensis). *The Journal of Experimental Biology, 206,* 4327–4340.

Roberts, T.J. & Scales, J.A. (2002). Mechanical power output during running accelerations in wild turkeys. *The Journal of Experimental Biology, 205,* 1485–1494.

Tanaka, H., Seals, D.R., Monahan, K.D., Clevenger, C.M., DeSouza, C.A., & Dinenno, F.A. (2002). Regular aerobic exercise and the age-related increase in carotid artery intima-media thickness in healthy men. *Journal of Applied Physiology, 92,* 1458–1464.

Urquhart, D.M., Soufan, C., Teichtahl, A.J., Wluka, A.E., Hanna, F., & Cicuttini, F.M. (2008). Factors that may mediate the relationship between physical activity and the risk for developing knee osteoarthritis. *Arthritis Research & Therapy, 10*:203. doi:10.1186/ar2343

Vredenburg, V.T. (2004). Reversing introduced species effects: Experimental removal of introduced fish leads to rapid recovery of a declining frog. *Proceedings of the National Academy of Sciences of the United States of America, 101,* 7646–7650.

Williams, B., Powell, A., Hoskins, G., & Neville, R. (2008). Exploring and explaining low participation in physical activity among children and young people with asthma: a review. *BMC Family Practice, 9*:40. doi:10.1186/1471-2296-9-40

Wong, M. & Licinio, J. (2001). Research and treatment approaches to depression. *Nature Reviews Neuroscience, 2,* 343–351.

Index